Marine Lipids 2017

Special Issue Editors

Rosário Domingues
Ricardo Calado
Pedro Domingues

MDPI • Basel • Beijing • Wuhan • Barcelona • Belgrade

MDPI

Special Issue Editors

Rosário Domingues
University of Aveiro
Portugal

Ricardo Calado
University of Aveiro
Portugal

Pedro Domingues
University of Aveiro
Portugal

Editorial Office
MDPI AG
St. Alban-Anlage 66
Basel, Switzerland

This edition is a reprint of the Special Issue published online in the open access journal *Marine Drugs* (ISSN 1660-3397) in 2017 (available at: http://www.mdpi.com/journal/marinedrugs/special_issues/marine_lipids_2017).

For citation purposes, cite each article independently as indicated on the article page online and as indicated below:

Lastname, F.M.; Lastname, F.M. Article title. *Journal Name* **Year**, *Article number, page range.*

First Edition 2018

ISBN 978-3-03842-799-5 (Pbk)
ISBN 978-3-03842-800-8 (PDF)

Table of Contents

About the Special Issue Editors

Rosário Domingues, Ph.D., is presently Associated Professor with Habilllitation in the Department of Chemistry & CESAM, University of Aveiro (UA), Aveiro, Portugal. She is Vice Head of the Department of Chemistry and Director of the Doctoral Program of Biochemistry at UA. She graduated in Pharmaceutical Sciences, University of Coimbra (1990), received her Ph.D. degree in Chemistry (1998), and Habilitation in Biochemistry (2014) at the University of Aveiro. Since 2016, she has held the contract of Associated Professor with Habilitation in the Mass Spectrometry Centre, Department of Chemistry, University of Aveiro (UA). From 1991 to 1998, she was Assistant at UA, and Auxiliar Professor between 1998–2016. She has over 25 years of research experience in the field of mass spectrometry, and is a well-established researcher in the field of lipidomics, glycomics and changes in biomolecules associated with oxidative stress monitored by mass spectrometry. She is one of the co-leaders of the Lipidomics Laboratory of the Mass Spectrometry Centre of the UA. Her major research interests are focused on Lipidomics, namely, Marine Lipidomics, Oxidative Lipidomics and Lipidomics in Health and Disease. She is/has already been PI and /or team member of more than 20 projects funded by national and European programs. At present, she is the coordinator at UA of two European projects: the project H2020-MSCA-ITN-2015 Innovative Training Networks (ITNH2020-MSCA-ITN-2015) "MASS spectrometry TRaining in Protein Lipoxidation ANalysis for Inflammation; MASSTRPLAN (Grant number 675132) and the project GENIALG - Genetic diversity exploitation for Innovative macro-algal biorefinery (ANR-15-MRSE-0015), Innovation actions BG-01-2016: Large-scale algae biomass integrated biorefineries. To date, she has authored one book, six book chapters and more than 220 articles published in international journals with peer review.

Ricardo Calado, Ph.D., is presently Principal Researcher in the Department of Biology, University of Aveiro (UA), Aveiro, Portugal. He is Vice Head of the Department of Biology & CESAM and Coordinates the Marine Technology Platform of UA. He graduated in Animal Biology, University of Lisbon (1999) and received his Ph.D. degree in Animal Biotechnology (2005) by the same university. He is an independent researcher at UA since 2008 and develops his research in the fields of marine aquaculture and biotechnology, with an emphasis on the role of marine lipids. He is one of the co-leaders of the Lipidomics Laboratory of the Mass Spectrometry Centre of the UA. One of his major research interests is to add value to marine lipids for food, feed and biotechnological uses (including pharma and cosmetics). He is/has already been PI and/or team member of research projects funded by national and international programs accounting for >5 million euros. To date, he has authored 6 books, 9 book chapters and more than 150 articles published in international journals with peer review.

Pedro Domingues is presently Assistant Professor with Habilitation at Aveiro Mass Spectrometry Center at University of Aveiro, Portugal. He received his BA in Pharmaceutical Sciences at the University of Coimbra in 1990 and his PhD in Chemistry- Mass Spectrometry at Aveiro University in 2000. He has been working in mass spectrometry applications since 1996. His research is concerned with applications of mass spectrometry for the study of oxidative stress. His main work has been developed in the detection and characterization of free radicals by mass spectrometry, the study of oxidative stress in lipids and phospholipids and amino acids. Currently, his main interests are the study and characterization of oxidized lipids, peptides and proteins using mass spectrometric approaches. He is one of the co-leaders of the Lipidomics Laboratory of the Mass Spectrometry Centre of the UA. He is author or co-author of more than 150 publications. He was the Director of the Master in Biomolecular Methods of the University of Aveiro (2007–2009). Currently, he is the Director of the Master in Biochemistry of the University of Aveiro, a position he has held since 2010, member of the board of the Portuguese Biochemistry Society and member of the Pedagogical Board of the University of Aveiro.

Preface to "Marine Lipids 2017"

In recent years, there is an increasing interest in marine organisms as valuable sources of lipids with high nutritional value, such as n-3 fatty acids (e.g., 20:5 and 22:6), and lipids with bioactive properties (e.g., polar lipids such as glycolipids and phospholipids). Lipids from several marine sources are considered high added value bioactive molecules with health benefits, promoting wellbeing and being paramount for the prevention of cardiovascular and other chronic diseases. A total of 10 scientific articles were published in this Special Issue of "Marine Lipids 2017", with both review and research works addressing a number of different marine organisms (e.g., fungi, macroalgae, shrimp, krill, sea urchins…). Lipids bioactivity was screened, with emphasis on anti-inflammatory, anti-proliferative, hypolipidemic and anti-microbial activities, along with their potential use in food, nutraceutics, feed, pharma and cosmetic industries. This Special Issue presents the reader with the latest advances in analytical approaches to study marine lipids, including emerging chromatography and mass spectrometry techniques that allow an unprecedented insight into marine lipidomes. Their enhanced resolution and high throughput performance allows researchers to identify and quantify lipid species present in marine organisms under a wider and more integrative framework, allowing the identification of lipid signatures and their potential valorisation for a number of increasing biotechnological applications.

Rosário Domingues, Ricardo Calado and Pedro Domingues

Special Issue Editors

marine drugs

MDPI

Article

Valorization of Lipids from *Gracilaria* sp. through Lipidomics and Decoding of Antiproliferative and Anti-Inflammatory Activity

Elisabete da Costa [1], Tânia Melo [1], Ana S. P. Moreira [1], Carina Bernardo [2], Luisa Helguero [2], Isabel Ferreira [3], Maria Teresa Cruz [3], Andreia M. Rego [4], Pedro Domingues [1], Ricardo Calado [5], Maria H. Abreu [4] and Maria Rosário Domingues [1,*]

[1] Centro de Espectrometria de Massa, Departamento de Química & QOPNA, Universidade de Aveiro, Campus Universitário de Santiago, 3810-193 Aveiro, Portugal; elisabetecosta@ua.pt (E.d.C.); taniamelo@ua.pt (T.M.); ana.moreira@ua.pt (A.S.P.M.); p.domingues@ua.pt (P.D.)
[2] Instituto de Biomedicina (IBIMED), Departamento de Ciências Médicas, Universidade de Aveiro, 3810-193 Aveiro, Portugal; carinabernardo@ua.pt (C.B.); luisa.helguero@ua.pt (L.H.)
[3] Centro de Neurociências e Biologia Celular (CNC), Universidade de Coimbra, 3004-517 Coimbra & Faculdade de Farmácia, Universidade de Coimbra, 3000-548 Coimbra, Portugal; isabelcvf@gmail.com (I.F.); trosete@ff.uc.pt (M.T.C.)
[4] ALGAplus-Produção e Comercialização de Algas e seus Derivados, Lda., 3830-196 Ílhavo, Portugal; amrego@algaplus.pt (A.M.R.); htabreu@algaplus.pt (M.H.A.)
[5] Departamento de Biologia & CESAM, Universidade de Aveiro, Campus Universitário de Santiago, 3810-193 Aveiro, Portugal; rjcalado@ua.pt
* Correspondence: mrd@ua.pt

Academic Editor: Peer B. Jacobson
Received: 11 November 2016; Accepted: 13 February 2017; Published: 2 March 2017

Abstract: The lipidome of the red seaweed *Gracilaria* sp., cultivated on land-based integrated multitrophic aquaculture (IMTA) system, was assessed for the first time using hydrophilic interaction liquid chromatography-mass spectrometry and tandem mass spectrometry (HILIC–MS and MS/MS). One hundred and forty-seven molecular species were identified in the lipidome of the *Gracilaria* genus and distributed between the glycolipids classes monogalactosyl diacylglyceride (MGDG), digalactosyl diacylglyceride (DGDG), sulfoquinovosyl monoacylglyceride (SQMG), sulfoquinovosyl diacylglyceride (SQDG), the phospholipids phosphatidylcholine (PC), lyso-PC, phosphatidylglycerol (PG), lyso-PG, phosphatidylinositol (PI), phosphatidylethanolamine (PE), phosphatic acid (PA), inositolphosphoceramide (IPC), and betaine lipids monoacylglyceryl- and diacylglyceryl-*N,N,N*-trimethyl homoserine (MGTS and DGTS). Antiproliferative and anti-inflammatory effects promoted by lipid extract of *Gracilaria* sp. were evaluated by monitoring cell viability in human cancer lines and by using murine macrophages, respectively. The lipid extract decreased cell viability of human T-47D breast cancer cells and of 5637 human bladder cancer cells (estimated half-maximal inhibitory concentration (IC_{50}) of 12.2 μg/mL and 12.9 μg/mL, respectively) and inhibited the production of nitric oxide (NO) evoked by the Toll-like receptor 4 agonist lipopolysaccharide (LPS) on the macrophage cell line RAW 264.7 (35% inhibition at a concentration of 100 μg/mL). These findings contribute to increase the ranking in the value-chain of *Gracilaria* sp. biomass cultivated under controlled conditions on IMTA systems.

Keywords: glycolipids; phospholipids; betaine lipids; seaweeds; bioactivity; mass spectrometry; hydrophilic interaction liquid chromatography–electrospray ionization–mass spectrometry HILIC–ESI–MS

1. Introduction

Red seaweeds within the genus *Gracilaria* are one of the world's most cultivated and valuable marine macrophytes. This group of seaweeds is well adapted to cultivation on land-based integrated multitrophic aquaculture (IMTA) systems, allowing its sustainable production under controlled and replicable conditions that provide a secure supply of high-grade seaweed biomass for demanding markets (e.g., food, pharmaceuticals) [1–3]. *Gracilaria* sp. is a source of multiple products, among which lipids, namely polyunsaturated fatty acids (PUFAs) such as arachidonic (20:4(*n*-6), AA) and eicosapentaenoic (20:5(*n*-6), EPA) acids, are emerging as valuable components [4]. Fatty acids (FAs) are mainly esterified to polar lipids such as glycolipids, phospholipids, and betaine lipids. Polar lipids are nowadays recognized as an important reservoir of fatty acids with nutritional value, e.g., *n*-3 FAs [5,6], and they are also considered high-value novel lipids with beneficial health effects such as antitumoral [7,8], antiviral [8,9], antifungal [10], antibacterial [11], and anti-inflammatory [11,12], with potential applications in the nutraceutical and pharmaceutical industries [7,8,11]. However, in spite of their recognized potential, they are still scarcely studied [12–15]. Some studies reported that polar lipids isolated from seaweeds can promote growth-inhibiting effects on human hepatocellular carcinoma cell lines (HepG2) [16] and thus can act as inhibitors of DNA polymerases with capability to inhibit tumor cell proliferation [17]. Moreover, they have been associated with anti-inflammatory properties through the inhibition of pro-inflammatory cytokines interleukin IL-6 and IL-8 production [15] and/or by the inhibition of nitric oxide (NO) production [18–21]. Lipid-based agents are therefore emerging molecules in therapeutics aimed to regulate inflammatory pathways or even impair downstream tumorigenic processes [22–24].

To fully explore the bioactive properties of seaweed lipids and thus contribute to seaweed valorization, it is fundamental to characterize their structure and understand how it modulates bioactivity [13,14]. Nowadays, the detailed structural characterization of lipids can be accomplished by using mass spectrometry (MS) coupled with liquid chromatography (LC). This lipidomic approach has the advantage of providing a detailed analysis of the lipid profile and affording the identification and quantification of more than 200 lipid molecular species in one single LC–MS run [25,26]. This detailed information on the specificity of molecular species and corresponding classes of polar lipids cannot be achieved using traditional approaches, typically based on the previous separation of polar lipid classes by thin-layer chromatography (TLC) and silica gel on column chromatography, followed by off-line gas chromatography-mass spectrometry (GC–MS) analysis of FAs [27–33]. MS-based technologies allowed researchers to explore the full lipidomic signature of distinct matrices [34–36]. To date, they allowed for the identification of the full lipidome signature of cultivated seaweeds *Ulva lactuca* Linnaeus, 1753 [37], *Chondrus crispus* Stackhouse, 1797 [26], and *Codium tomentosum* Stackhouse, 1797 [25]. These novel approaches based on specific identification and quantification at the molecular level using high-throughput analysis are promising tools for bioprospection [3,38,39].

The main goal of the present study was to identify and characterize the polar lipid profile of *Gracilaria* sp. cultivated under controlled conditions on a land-based integrated multitrophic aquaculture (IMTA) system, using hydrophilic interaction liquid chromatography-electrospray ionization-mass spectrometry (HILIC–ESI–MS). The lipid extract of this red seaweed was also screened for its growth inhibitory effects in human breast and bladder cancer cell lines, as well as anti-inflammatory effects by inhibiting the production of NO.

2. Results and Discussion

The lipid extract of *Gracilaria* sp. obtained by chloroform:methanol extraction accounted for about 3000 ± 600 mg/kg dry mass (relative standard deviation (RSD) < 20%). The lipid extract was mainly composed of glycolipids (1980 ± 148 mg/kg of biomass) and phospholipids (165 ± 53 mg/kg of biomass), and the remaining lipid extract corresponded to betaine lipids and others (Table 1).

Table 1. Composition of lipid extract of *Gracilaria* sp. (mean and SD of triplicate).

Composition	Mean	SD
Lipids (mg/kg biomass)	3000	600
Glycolipids (mg/kg biomass)	1980	148
Phospholipids (mg/kg biomass)	165	52.7
Betaines and others [1]	855	-

[1] Betaines and others were determined by the difference of lipid content and the sum of content of glycolipids and phospholipids.

2.1. Polar Lipidome

The profile of *Gracilaria* polar lipidome was determined by HILIC–ESI–MS and allowed for the identification of molecular species of glycolipids, phospholipids, and betaine lipids. Overall, the lipidome of *Gracilaria* sp. comprised 147 molecular species (Figure 1).

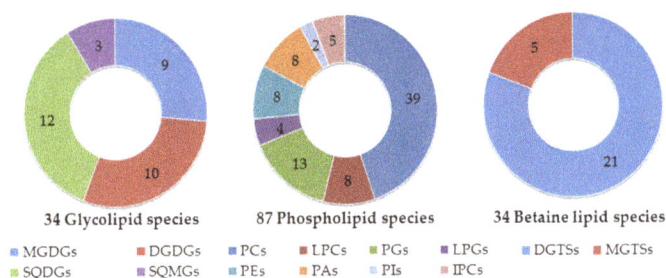

Figure 1. Number of molecular species identified by HILIC–ESI–MS, distributed by the classes of glycolipids: monogalactosyl diacylglyceride (MGDG), digalactosyl diacylglyceride (DGDG), sulfoquinovosyl monoacylglyceride (SQMG), sulfoquinovosyl diacylglyceride (SQDG), phospholipids: phosphatidylcholine (PC) and lyso-PC (LPC), phosphatidylglycerol (PG) and lyso-PG (LPG), phosphatidylinositol (PI), phosphatic acid (PA), phosphatidylethanolamine (PE), inositolphosphoceramide (IPC), and betaine lipids: monoacylglyceryl- and diacylglyceryl-*N,N,N*-trimethyl homoserine (MGTS and DGTS).

The glycolipids of the classes monogalactosyl diacylglyceride (MGDG) and digalactosyl diacylglyceride (DGDG) were identified in the LC–MS spectra in positive mode as [M + NH₄]⁺ ions [25]. Detailed structure of MGDG and DGDG molecular species was accomplished by LC–MS/MS analysis of [M + NH₄]⁺ ions and analysis of ESI–MS/MS of the [M + Na]⁺ ions after solid phase extraction (SPE) fractionation of lipid extract (fraction 3 rich in glycolipids). Overall, 34 molecular species were identified, as described in Table 2. Galactolipids contained nine MGDG molecular species and 10 DGDG molecular species (Table 2, Figure S1a,b). The most abundant MGDG molecular species were found at *m/z* 774.3 and 796.3 [M + NH₄]⁺, corresponding to MGDG (18:1/16:0) and to MGDG (20:4/16:0), with a minor contribution from MGDG (18:2/18:2), respectively. Other MGDG molecular species identified contained in their composition 14-, 16-, and 18-carbon saturated fatty acids (SFAs) and monounsaturated fatty acids (MUFAs) and 18:2, 20:4, and 20:5 polyunsaturated fatty acyl (PUFAs) moieties (Table 2). Regarding DGDGs, the most abundant molecular species were identified as [M + NH₄]⁺ ions at *m/z* 936.3, corresponding to DGDG (18:1/16:0), followed by DGDG (20:4/16:0) with minor contribution of DGDG (18:2/18:2) at *m/z* 958.2. Moreover, other DGDG molecular species were identified containing 14-, 16-, 18-, and 20-carbon fatty acids (FAs) such as 20:4 and 20:5 PUFAs.

Concerning sulfolipids, 12 sulfoquinovosyl diacylglycerides (SQDGs) and three sulfoquinovosyl monoacylglycerides (SQMGs) were identified as negative [M − H]⁻ ions. The most abundant species were attributed to SQDG (14:0/16:0), SQDG (16:0/16:0) and SQDG (16:0/20:4), observed as [M − H]⁻ ions at *m/z* 765.5, 793.5, and 841.6, respectively. The fatty acyl signature of SQDGs included 14-, 16-, 18-carbon SFAs and MUFAs and 18- and 20-carbon PUFAs (Table 2, Figure S1c). Three SQMGs were identified as SQMG (14:0), SQMG (16:0), and SQMG (16:1). SQMGs were never before reported in the lipidome of seaweeds from the genus *Gracilaria*. Glycolipids have already been identified for members of the Rhodophyta (red seaweeds), namely in the genus *Gracilaria* [21,30,40–42]. However, the majority of published works only identified a few species of glycolipids, either by using offline TLC–MS [30,40,42,43] or selected solvent extraction and MS analysis [15,20,29]. More recently, a detailed profile of *Chondrus crispus* was reported using LC–MS and MS/MS [26].

Table 2. Identification of MGDG and DGDG molecular species observed by HILIC–ESI–MS, as $[M + NH_4]^+$ ions and SQDG and SQMG molecular species observed as $[M − H]^-$ ions [2].

$[M + NH_4]^+$	Lipid Species	Fatty Acyl Chains
m/z	(C:N)	
Monogalactosyl diacylglyceride (MGDG)		
746.3	MGDG (32:1)	16:1/16:0 and 14:0/18:1
748.3	MGDG (32:0)	16:0/16:0 and 14:0/18:0
774.3	**MGDG (34:1)**	18:1/16:0
776.3	MGDG (34:0)	18:0/16:0
794.3	MGDG (36:5)	20:5/16:0
796.3	**MGDG (36:4)**	20:4/16:0 and 18:2/18:2
Digalactosyl diacylglyceride (DGDG)		
908.3	DGDG (32:1)	16:1/16:0 and 14:0/18:1
910.3	DGDG (32:0)	16:0/16:0 and 14:0/18:0
934.3	DGDG (34:2)	18:2/16:0 and 18:1/16:1
936.3	**DGDG (34:1)**	18:1/16:0
956.3	DGDG (36:5)	20:5/16:0
958.3	**DGDG (36:4)**	20:4/16:0 and 18:2/18:2
$[M − H]^-$	Lipid Species	Fatty Acyl Chains
Sulfoquinovosyl diacylglyceride (SQDG)		
763.6	SQDG (30:1)	14:0/16:1
765.6	**SQDG (30:0)**	14:0/16:0
791.6	SQDG (32:1)	16:1/16:0 and 14:0/18:2
793.6	**SQDG (32:0)**	16:0/16:0 and 14:0/18:0
813.6	SQDG (34:4)	18:4/16:0
817.6	SQDG (34:2)	18:2/16:0
819.6	SQDG (34:1)	18:1/16:0
839.6	SQDG (36:5)	20:5/16:0
841.6	**SQDG (36:4)**	20:4/16:0
857.6	SQDG (36:4-OH)	20:4-OH/16:0
Sulfoquinovosyl monoacylglyceride (SQMG)		
527.4	SQMG (14:0)	
553.4	SQMG (16:1)	
555.4	SQMG (16:0)	

[2] The assignment of the fatty acyl composition of molecular species was made according to the interpretation of the corresponding MS/MS spectra. Bold *m/z* values correspond to the most abundant species detected in the LC–MS spectrum; C means the number of carbon atoms; N represents double bonds in the fatty acyl chains; MGDG: monogalactosyl diacylglyceride; DGDG: digalactosyl diacylglyceride; SQMG: sulfoquinovosyl monoacylglyceride; SQDG: sulfoquinovosyl diacylglyceride; and HILIC–ESI–MS: hydrophilic interaction liquid chromatography–electrospray ionization–mass spectrometry.

Gracilaria sp. lipidome included 87 molecular species of phospholipids (PLs) within eight classes, namely phosphatidylglycerol (PG) and lyso-PG (LPG), phosphatidylcholine (PC) and lyso-PC (LPC), phosphatidylethanolamine (PE), phosphatidylinositol (PI), inositolphosphoceramide (IPC) and phosphatidic acid (PA). PC is a main component of extraplastidial membranes, while PG is found in chloroplastic membranes [5].

The PL classes PG, lyso-PG, PA, PI, and IPC were identified as negative $[M − H]^-$ ions, while PC and lyso-PC were identified as negative $[M + CH_3COO]^-$ ions. PC, LPC and PE were also identified as positive $[M + H]^+$ ions. The identity of all molecular species identified (Table 3) was confirmed by LC–MS/MS, as described in the literature [25,26]. About 39 PCs were identified by LC–MS (Figure S2a). The most abundant ions were observed at *m/z* 760.6 and at *m/z* 782.6, respectively attributed to PC (16:0/18:1) and to PC (16:0/20:4) with a minor contribution of PC (18:2/18:2). Other PC molecular species were identified and contained 14- to 22-carbon fatty acids. Lyso-PC consisted of eight molecular species (Table 3, Figure S2b) and the most abundant was LPC (20:4), observed at *m/z* 544.4. All molecular species identified are described in Table 3. Thirteen PGs and four lyso-PGs species were identified by LC–MS as $[M - H]^-$ ions (Table 3, Figure S2c,d). The most abundant ion was observed at *m/z* 769.4, mainly corresponding to PG (16:0/20:4), with a minor contribution from PG (18:2/18:2). The prominent lyso-PG at *m/z* 483.3 was LPG (16:0). PI species were observed as $[M - H]^-$ ions at *m/z* 833.5 and 835.5 and attributed to PI (16:1/18:1) and PI (16:0/18:1), respectively. Eight PAs were identified (Table 3, Figure S2e), with the most abundant species identified as PA (20:4/20:4) at *m/z* 743.3, while the other PA molecular species were esterified to 16:0, 18:1, 18:2, 18:3, 20:3, 20:4, and 20:5 FAs. PEs contained eight molecular species, identified as $[M + H]^+$ (Table 3, Figure S2f). The most abundant ion was observed at *m/z* 716.4 and identified as PE (16:1/16:1) and PE (16:0/18:2). The phospholipids from *Gracilaria* sp. hold PCs and LPCs, PGs, LPGs, PIs, PEs, and PAs, already reported for the lipidome of other Rhodophyta [20,26,31,32]. Fatty acids esterified in the PLs included saturated and unsaturated 16-, 18-, and 20-carbon FAs, and PCs and PAs were the only PLs classes that included 20:3(*n*-6) FA. The 20:3(*n*-6) FA is usually a minor component of the whole pool of FAs in red seaweeds [31,32] but is an important intermediate compound in the biosynthesis of 20:4(*n*-6) FA.

Table 3. Identification of phospholipid molecular species observed by HILIC–ESI–MS, as $[M + H]^+$ ions for PC, LPC, and PE and as $[M - H]^-$ ions for PG, LPG, PI, PA, and IPC [2].

$[M + H]^+$	Lipid Species	Fatty Acyls Chain
m/z	(C:N)	
Phosphatidylcholine (PC)		
732.6	PC (32:1)	16:0/16:1 and 14:0/18:1
734.6	PC (32:0)	16:0/16:0 and 14:0/18:0
754.6	PC (34:4)	14:0/20:4 and 16:2/18:2
756.6	PC (34:3)	16:0/18:3 and 14:0/20:3
758.6	PC (34:2)	16:0/18:2 and 16:2/18:1
760.6	**PC (34:1)**	16:0/18:1
762.6	PC (34:0)	16:0/18:0
780.6	PC (36:5)	16:0/20:5 and 18:2/18:3
782.6	**PC (36:4)**	16:0/20:4 and 18:2/18:2
784.6	PC (36:3)	16:0/20:3 and 18:1/18:2
786.6	PC (36:2)	18:0/18:2 and 18:1/18:1
788.6	PC (36:1)	18:0/18:1
798.5	PC (37:3)	16:0/21:3 and 18:1/19:2
804.5	PC (38:7)	18:3/20:4 and 18:2/20:5
806.5	PC (38:6)	18:2/20:4 and 18:1/20:5
808.5	PC (38:5)	18:1/20:4 and 18:2/20:3
810.5	PC (38:4)	18:1/20:3 and 16:0/22:4
812.5	PC (38:3)	18:0/20:3 and 18:1/20:2
814.5	PC (38:2)	16:0/22:2 and 18:1/20:1
818.5	PC (38:0)	18:0/20:0 and 16:0/22:0
840.4	PC (40:3)	18:1/22:2
844.4	PC (40:1)	18:1/22:0
Lyso-phosphatidylcholine (LPC)		
494.4	LPC (16:1)	
496.4	LPC (16:0)	
518.4	LPC (18:3)	
520.4	LPC (18:2)	
522.4	LPC (18:1)	
524.4	LPC (18:0)	
542.4	LPC (20:5)	
544.4	**LPC (20:4)**	

Table 3. *Cont.*

[M + H]⁺	Lipid Species	Fatty Acyls Chain
m/z	(C:N)	
Phosphatidyletanolamine (PE)		
716.4	PE (34:2)	16:1/18:1 and 16:0/18:2
718.3	PE (34:1)	16:1/18:0 and 16:0/18:1
740.4	PE (34:0)	16:0/18:0
742.4	PE (36:3)	18:1/18:2
744.4	PE (36:2)	18:1/18:1
746.3	PE (36:1)	18:0/18:1
[M − H]⁻	Lipid Species	Fatty Acyl Chains
Phosphatidylglycerol (PG)		
717.4	PG (32:2)	16:1/16:1 and 16:0/16:2
719.4	PG (32:1)	16:0/16:1
721.4	PG (32:0)	16:0/16:0
741.4	PG (34:4)	16:0/18:4
743.5	PG (34:3)	16:0/18:3
745.5	PG (34:2)	16:1/18:1
747.5	PG (34:1)	16:0/18:1 and 16:1/18:0
767.5	PG (36:5)	16:0/20:5
769.4	**PG (36:4)**	16:0/20:4 and 18:2/18:2
773.5	PG (36:2)	18:1/18:1
Lyso-phosphatidylglycerol (LPG)		
481.3	LPG (16:1)	
483.3	**LPG (16:0)**	
509.3	LPG (18:1)	
531.3	LPG (20:4)	
Phosphatidylinositol (PI)		
833.5	PI (34:2)	16:1/18:1
835.5	PI (34:1)	16:0/18:1
Phosphatidic acid (PA)		
693.4	PA (36:5)	16:0/20:5
695.4	PA (36:4)	16:0/20:4
717.4	PA (38:7)	18:3/20:4
719.4	PA (38:6)	18:2/20:4
721.4	PA (38:5)	18:1/20:4
741.3	PA (40:9)	20:4/20:5
743.3	**PA (40:8)**	20:4/20:4
745.3	PA (40:7)	20:3/20:4
Inositolphosphoceramide (IPC)		
810.5	**IPC (*t*35:0)**	*t*18:0/17:0
908.6	IPC (*d*42:0)	*d*18:0/24:0
920.6	**IPC (*t*42:2)**	*t*18:1/24:1
922.6	IPC (*t*42:1)	*t*18:0/24:1
924.6	IPC (*t*42:0)	*t*18:0/24:0

[2] The assignment of the fatty acyl composition of molecular species was made according to the interpretation of the corresponding MS/MS spectra. Bold *m/z* values correspond to the most abundant species detected in the LC–MS spectrum; C means the number of carbon atoms; N represents double bonds in the fatty acyl chains; PC: phosphatidylcholine; LPC: lyso-PC; PG: phosphatidylglycerol; LPG: lyso-PG; PI: phosphatidylinositol; PA: phosphatic acid; PE: phosphatidylethanolamine; and IPC: inositolphosphoceramide.

Five molecular species were assigned as IPCs and the most abundant ones were IPC (*t*18:0/17:0), observed at *m/z* 810.5, and IPC (*t*18:1/24:1), observed at *m/z* 920.6 (Figure S3a). LC–MS/MS spectrum of the [M − H]⁻ ions of IPCs, as exemplified for IPC at *m/z* 920.6 in Figure S3b, showed the typical fragmentation pathways of IPCs such as the losses of 162 Da and 180 Da, due to the fragmentation pathways that lead to the elimination of inositol, the product ion at *m/z* 538.3 resulting from the loss of fatty acyl chains, and the product ion at *m/z* 259.0 that corresponded to an inositol monophosphate anion. IPCs were identified in lipid extract of *Gracilaria* sp. and are considered an important biomarker of Rhodophyta taxonomy, in accordance with what was already reported for *Chondrus crispus* using a lipidomic approach [26]. IPCs are required to maintain membrane properties such as viscosity and electrical charge and participate in the control of enzymatic activity or act as membrane anchors for some proteins [44].

Twenty-one DGTS and five MGTS molecular species were identified by LC–MS and MS/MS as [M + H]+ ions (Table 4, Figure S4a,b, respectively). The most abundant DGTS species were found at *m/z* 710.7, corresponding to DGTS (16:0/16:1), with a minor contribution from DGTS (14:0/18:1) species, followed by DGTS (18:1/18:1) observed at *m/z* 764.8. Overall, DGTSs combine distinctive molecular species bearing different combinations of FAs, ranging between 14- and 20-carbon FAs, as reported on Table 4. MGTSs comprised MGTS (14:0), MGTS (16:1), MGTS (16:0), MGTS (18:2), and MGTS (18:1) species, identified at *m/z* 446.5, 472.5, 474.5, 498.6, and 500.6, respectively.

Table 4. Identification of DGTS and MGTS molecular species observed by HILIC–ESI–MS as [M + H]+ ions [2].

[M + H]+	Lipid Species	Fatty Acyls Chain
m/z	(C:N)	
Diacylglyceryl trimethyl homoserine (DGTS)		
656.7	DGTS (28:0)	14:0/14:0
682.7	DGTS (30:1)	14:0/16:1
684.8	DGTS (30:0)	14:0/16:0
708.7	DGTS (32:2)	16:1/16:1 and 14:0/18:2
710.7	**DGTS (32:1)**	16:0/16:1 and 14:0/18:1
712.7	DGTS (32:0)	16:0/16:0 and 14:0/18:0
732.7	DGTS (34:4)	16:2/18:2 and 14:0/20:4
734.7	DGTS (34:3)	16:1/18:2
736.7	DGTS (34:2)	16:0/18:2 and 16:1/18:1
738.7	DGTS (34:1)	16:0/18:1 and 16:1/18:0
740.7	DGTS (34:0)	16:0/18:0 and 14:0/20:0
760.6	DGTS (36:4)	16:0/20:4
764.8	**DGTS (36:2)**	18:1/18:1
766.8	DGTS (36:1)	18:0/18:1
Monoacylglyceryl trimethyl homoserine (MGTS)		
446.5	MGTS (14:0)	
472.5	MGTS (16:1)	
474.5	MGTS (16:0)	
498.6	MGTS (18:2)	
500.6	MGTS (18:1)	

[2] The assignment of the fatty acyl composition of molecular species was made according to the interpretation of the corresponding MS/MS spectra. Bold *m/z* values correspond to the most abundant species detected in the LC–MS spectrum; C means the number of carbon atoms; N represents double bonds in the fatty acyl chains; MGTS: monoacylglyceryl-N,N,N-trimethyl homoserine; and DGTS: diacylglyceryl-N,N,N-trimethyl homoserine.

Betaine lipids are components of extraplastidial membranes [45]. They are naturally occurring lipids not found in higher plants, but are widely distributed in algae [45,46]. Betaines are a class of acyl glycerolipids that have a quaternary amine alcohol ether-linked to a diacylglycerol moiety, lacking in phosphorous. Interestingly, DGTSs are described herein for the first time in the lipidome of genus *Gracilaria* and MGTS species were not reported before in the lipidome of red seaweeds. This may be due to the lack of sensitivity of most reported analytical tools based on TLC and GC–MS approaches, since the co-elution of betaines and PC by TLC approaches could have prevented their discrimination [47]. Only recently, through the use of MS-based tools, betaine lipids were identified in the lipidome of red seaweed *Chondrus crispus* [26] and green seaweed *Codium tomentosum* [25].

2.2. Fatty Acid Profile

The fatty acid profile of the lipid extract was characterized by GC–MS analysis of fatty acid methyl esters (FAMEs). The profile of fatty acids included 14:0, 16:0, 18:0, 18:1(*n*-9), 18:2(*n*-6), 20:4(*n*-6), and 20:5(*n*-3), among which 16:0 (48.5% $\pm$ 1.1%), 18:1(*n*-9) (14.4% $\pm$ 0.38%), and 20:4(*n*-6) (13.6% $\pm$ 0.46%)

were the most abundant (Figure 2). Overall, SFAs accounted for 57.5% of the total FAs identified, followed by MUFAs (18.3%) and PUFAs (18.4%). The fatty acids identified by GC–MS were also reported to be esterified to polar lipids of *Gracilaria* sp. by LC–MS and MS/MS analysis.

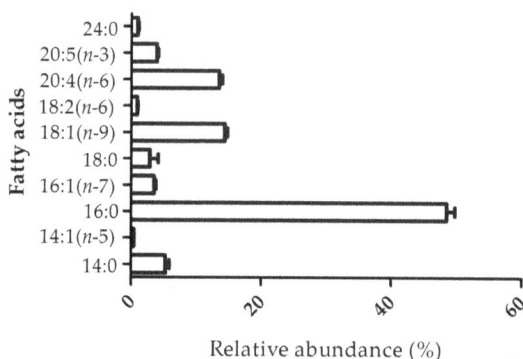

Figure 2. Fatty acid profile of lipids from *Gracilaria* sp. determined by GC–MS analysis of fatty acid methyl esters (FAMEs). Mean ± SD (%) of triplicate, traces < 0.1% not shown.

The *n*-6/*n*-3 ratio determined for our *Gracilaria* sp. sample was 3.6. The World Health Organization (WHO) recommends an optimal balance intake of *n*-6 PUFAs and *n*-3 PUFAs to prevent chronic diseases and that this balance should be maintained with an adequate daily dosage of *n*-6 PUFAs (5%–8% of daily energy intake) and *n*-3 PUFAs (1%–2% of daily energy intake) [48]. With this recommendation in mind, it is possible to estimate that a suitable *n*-6/*n*-3 ratio is less than 5. Also, some authors reported that a ratio of *n*-6/*n*-3 less than 4 is adequate in the prevention of several diseases such as cardiovascular [49], autoimmune [50], and inflammatory diseases [50,51], and cancer [49,50]. These findings support the use of *Gracilaria* sp. for human consumption.

2.3. Bioactivity of Lipid Extract of Gracilaria sp.

The bioactivity of *Gracilaria* sp. lipid extract was assessed, specifically its antiproliferative effect in two human cancer cell lines (breast cancer—T-47D and bladder cancer—5637) and its anti-inflammatory effect in a mouse leukemic monocyte macrophage cell line (RAW 264.7) stimulated with LPS.

2.3.1. Activity of Lipid Extract on Human Cancer Cell Viability

The growth inhibitory effect induced by the lipid extract on cancer cells is shown in Figure 3. A lipid extract of *Gracilaria* sp. reduced cell viability in both cell lines in a dose-dependent manner at concentration range of 10 to 20 µg/mL ($p < 0.001$), with a calculated half-maximal inhibitory concentration (IC_{50}) of 12.2 µg/mL and 12.9 µg/mL for T-47D (Figure 3A) and 5637 (Figure 3B) cancer cells, respectively.

The anti-tumor effects of polar lipids were previously reported as affecting angiogenesis and solid tumor growth via inhibition of replicative DNA polymerase activities [22,52]. Extracts rich in glycolipids isolated from distinct seaweeds inhibited the growth of a human hepatocellular carcinoma cell line (HepG2) (IC_{50} of 126 µg/mL) [16] and were found to induce apoptosis of human colon carcinoma Caco-2 cells when associated with sodium butyrate [53]. Otherwise, SQDG isolated from *Gigartina tenella* Harvey, 1860, accepted as *Chondracanthus tenellus* (Harvey) Hommersand, 1993, inhibited DNA polymerase α, DNA polymerase β, and HIV-reverse transcriptase type 1 or downregulated *Tie2* gene expression in tumors [17,54]. It has been hypothesized that the biological properties of glycolipids such as SQDG are closely related to the sugar moiety and the presence of PUFA chains.

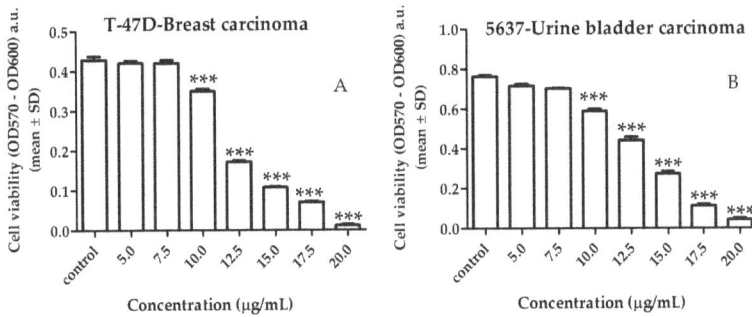

Figure 3. Effect of lipid extracts of *Gracilaria* sp. on T-47D breast (**A**) and 5637 bladder (**B**) cancer cell lines, after 96 h incubation. Results are shown as mean ± SD of three independent determinations (*** $p < 0.001$, compared to control). OD: optical density; a.u.: arbitrary units.

2.3.2. Activity of the Lipid Extract on Nitric Oxide Production

The anti-inflammatory activity of the lipid extract of *Gracilaria* sp. was assessed based on its ability to inhibit nitric oxide (NO) production in RAW 264.7 macrophages stimulated with LPS. For a range of concentrations between 25 and 100 μg/mL, the lipid extract did not compromise the cellular viability of macrophages (Figure 4A). The extract showed a dose-dependent NO inhibition of 35% attained at the concentration of 100 μg/mL (Figure 4B). Therefore, the concentration exhibiting anti-inflammatory activity also presented a safety profile to macrophages (Figure 4A). Meanwhile, at lower concentrations (≤ 50 μg/mL), the extract had no significant inhibitory effect.

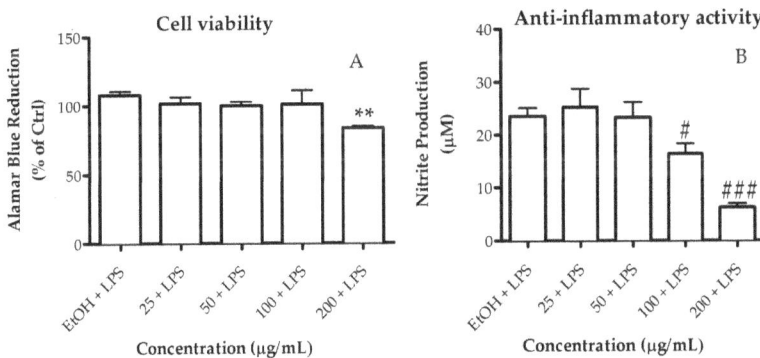

Figure 4. Cell viability and anti-inflammatory activity of *Gracilaria* sp. lipid extract. (**A**) Assessment of metabolically active cells was performed using a resazurin bioassay. Results are expressed as a percentage of resazurin reduction relative to the control (Ctrl); (**B**) Anti-inflammatory activity was measured as inhibition of NO production, quantified by the Griess assay. Nitrite concentration was determined from a sodium nitrite standard curve and the results are expressed as concentration (μM) of nitrite in a culture medium. Each value represents the mean ± SD from at least three independent experiments (** $p < 0.01$ compared to Ctrl; # $p < 0.05$, ### $p < 0.001$ compared to ethanol (EtOH, vehicle) plus lipopolysaccharide (LPS)).

Previous works have reported that polar lipid may be beneficial for inflammatory diseases [11,20,55,56]. Accordingly, polar lipids isolated from red algae have demonstrated strong anti-inflammatory activity, even higher when compared with pure 20:5(*n*-3) FA isolated from the same extracts [43], suggesting that the polar lipid itself may contribute to the anti-inflammatory activity. In the cases of *Chondrus crispus* and *Palmaria palmata* (Linnaeus) Weber & Mohr, 1805, the polar lipids such as glycolipids and phospholipids showed NO inhibitory activity through downregulation of inducible nitric oxide synthase (iNOS) [20,21,57]. Moreover, extracts rich in glycolipids bearing high proportions of PUFA, isolated from the red seaweeds *Palmaria palmata*, *Porphyra dioica* J. Brodie & L. M. Irvine, 1997, and *Chondrus crispus*, downregulated LPS-induced pro-inflammatory responses in human macrophages through the inhibition of IL-6 and IL-8 production, thus inferring their potential anti-inflammatory activity [15]. Therefore, as for other red seaweeds, the lipid extract from *Gracilaria* sp. proved to have effective anti-inflammatory activity.

Lipid extract from *Gracilaria* sp. showed antiproliferative and anti-inflammatory activity. However, it was not possible to determine exactly which lipid components are responsible for these bioactivities. Even in the literature, the majority of studies have also addressed biological activities of lipid extracts rather than pure lipid molecules, which hampers the determination of a relationship between structure and bioactivity. This is due to the fact that the isolation of a pure lipid molecule is very difficult and even pure lipid standards are not available for several lipid classes. Some authors have put some effort into this issue, scarcely addressed for the lipid extracts from seaweed, and isolated enriched extracts in some classes of lipids to further test their bioactivity. Ohta et al. [17] reported that SQDG (20:5/16:0), isolated from red seaweed *Gigartina tenella*, was a potent inhibitor of eukaryotic DNA polymerases [17]. Tsai et al. also reported that enriched extract with SQDG isolated from red seaweeds, with high levels of PUFAs such as 20:4(*n*-6) FA and 20:5(*n*-3) FA, inhibited the growth of human hepatocellular carcinoma cell line (HepG2), rather than enriched extracts with MGDG or DGDG [16]. This research group has also showed that the sulfolipids isolated from seaweed exhibited higher inhibitory effect than sulfolipids isolated from spinach, previously reported as inhibitors of DNA polymerases and of the proliferation of human cervix carcinoma (HeLa) [22]. The aforementioned SQDG-enriched extracts displayed strong inhibitory effects and contained SQDG (20:5/16:0) [17] or contained SQDG assembling PUFAs [16], which are also found in the extract of *Gracilaria* sp. analyzed within this work. Thus, SQDG (18:2/16:0), SQDG (18:4/16:0), SQDG (20:4/16:0), and SQDG (20:5/16:0), identified in the extract of *Gracilaria* sp., can contribute to the observed antiproliferative effects.

In what concern anti-inflammatory activities, Banskota et al. reported that the extracts rich in MGDG and DGDG isolated from red seaweed *Chondrus crispus* inhibited NO production through downregulation of iNOS [21]. The enriched extract contained MGDG (20:5/20:5), MGDG (20:5/20:4), MGDG (18:4/16:0), MGDG (20:4/16:0), and MGDG (20:5/16:0) and the respective DGDG analogues. Interestingly, the majority of these molecular species were also found in the extract of *Gracilaria* sp. analyzed in the present work. Moreover, the same group of researchers isolated MGDG, DGDG, SQDG, PC, and PG molecular species from the lipid extract of *Palmaria palmata* and all the polar lipids showed NO inhibitory activity [20]. The isolated polar lipids identified were MGDG (20:5/20:5), MGDG (20:5/16:0), DGDG (20:5/20:5), DGDG (20:5/14:0), DGDG (20:5/16:0), SQDG (20:5/14:0), PG (20:5/16:0), PG (20:5/16:1) and PC (20:5/20:5). All the molecular species contained 20:5(*n*-3) FA, and showed higher activity than the free FA 20:5(*n*-3), suggesting that the entire polar lipid structure (e.g., sulfolipid, phospholipid, or galactolipids) is essential for the extension of NO inhibition. Aside from the PC, the reported glycolipids and PG were also found in the lipidome of *Gracilaria* sp. Thus, the presence of these glycolipids and PGs in the lipid extract of *Gracilaria* sp. can contribute to the observed anti-inflammatory properties.

The presence of several polar lipids with recognized bioactive polar lipids in *Gracilaria* sp. can be related to the bioactivity observed in this work. However, more studies are needed to understand the structural/bioactivity relation of seaweed polar lipids, which deserve to be explored.

3. Experimental Section

3.1. Biomass

Dried samples (25 °C, up to 12% moisture content) of *Gracilaria* sp. (*G. vermiculophylla* or *G. gracilis*, pending confirmation by DNA barcode analysis) (harvested in August 2014) were provided by ALGAplus Ltd. (production site located at Ria de Aveiro, mainland Portugal, 40°36'43" N, 8°40'43" W). The biomass is continuously produced by clonal propagation (asexual reproduction strategy) and thus has lower variability than would be expected from wild harvested biomass.

3.2. Reagents

HPLC grade chloroform and methanol were purchased from Fisher Scientific Ltd. (Loughborough, UK). All other reagents were purchased from major commercial sources. Milli-Q water (Synergy, Millipore Corporation, Billerica, MA, USA), RPMI 1640 media from PAA (Pasching, Austria), Phenol-red-free RPMI 1640 medium, penicillin–streptomycin, TrypLE express, fetal bovine serum (FBS), and Presto Blue from Gibco Technologies (Invitrogen Life Sciences, Paisley, UK) were used.

3.3. Lipid Extraction Procedure

A mixture of chloroform/methanol (1:2, v/v) was added to 250 mg of dry weight seaweed. The mixture was transferred to a glass tube with a Teflon-lined screw cap and, after the addition of 3.75 mL of solvent mixture, it was homogenized by vortexing for 2 min and then incubated in ice on an orbital shaker for 2 h 30 min. The mixture was centrifuged at 2000 rpm for 10 min and the organic phase collected. The biomass residue was re-extracted twice with 1.5 mL of solvent mixture and 2.3 mL of water was added to the total collected organic phase to induce phase separation. Following this procedure, samples were centrifuged for 10 min at 1500 rpm, and the organic (lower) phase was collected in a new tube. Three biological replicates were performed, with extractions and analyses taking place on different days. Lipid extracts were dried under a stream of nitrogen gas and the lipid content was estimated as (%) of dry weight. Lipid extracts were stored at −20 °C prior to analysis by LC–MS.

3.4. Quantification of Glycolipids and Phospholipids

Glycolipid quantification was achieved by calculating the hexose content (% glucose) through the orcinol colorimetric method (CyberLipids, [58]). The amount of sugar was read from a calibration curve prepared by performing the reaction on known amounts of glucose (up to 40 μg, from an aqueous solution containing 5 mg/mL of sugar). Phospholipids were quantified by a molybdovanadate method for the simultaneous assay of orthophosphate and some organic phosphates, as described by Bartlett and Lewis, and routinely performed in the authors' laboratory [26,59,60]. Absorbance of standards and samples was measured on a microplate UV-Vis spectrophotometer (Multiskan GO, Thermo Scientific, Hudson, NH, USA).

3.5. Fractionation of Lipid Extract

Isolation of polar lipids from pigments was performed using a modification of Pacetti's method [61]. A sample of lipid extract (1 mg) was dissolved in 300 μL of chloroform and transferred to a Supelclean™ LC–Si SPE Tube (bed wt. 500 mg, volume 3 mL cartridges; SUPELCO, Sigma–Aldrich, St. Louis, MO, USA), followed by sequential elution with 4 mL of chloroform, 3 mL of ether diethyl ether:acetic acid (98:2), 5 mL of acetone:methanol (9:1 v/v), and 4 mL of methanol. Fractions 1 and 2, corresponding to neutral lipids and pigments, were discarded. Fractions 3 and 4, rich in glycolipids and in phospholipids plus betaines, respectively, were recovered, separated, dried under nitrogen, and stored at −20 °C prior to analysis by ESI–MS.

3.6. Hydrophilic Interaction Liquid Chromatography–Electrospray Ionization–Mass Spectrometry (HILIC–ESI–MS)

Lipid extracts were analyzed by hydrophilic interaction liquid chromatography (HILIC) on a Waters Alliance 2690 HPLC system (Waters Corp., Milford, MA, USA) coupled to a Finnigan LXQ electrospray linear ion trap mass spectrometer (Thermo Fisher, San Jose, CA, USA). Mobile phase A consisted of 25% water, 50% acetonitrile, and 25% methanol, with 1 mM ammonium acetate, and mobile phase B consisted of 60% acetonitrile and 40% methanol with 1 mM ammonium acetate. Lipid extracts (12.5 μg) were diluted in mobile phase B (100 μL) and 10 μL of the reaction mixture were introduced into an Ascentis Si HPLC Pore column (15 cm × 1.0 mm, 3 μm; Sigma–Aldrich, St. Louis, MO, USA). The solvent gradient, flow rate through column and conditions used for acquisition of full scan LC–MS spectra and LC–MS/MS spectra in both positive and negative ion modes were the same as previously described [25,26]. The identification of molecular species of polar lipids was based on the assignment of the molecular ions observed in LC–MS spectra. Only ions observed in the LC–MS spectra with a relative abundance >2% were considered for identification. All analyses were performed in analytical triplicate.

3.7. Electrospray–Mass Spectrometry (ESI–MS) Conditions

Fractions 3 and 4 recovered from lipid extract were analyzed by ESI–MS on a Q-Tof 2 quadrupole time of flight mass spectrometer (Micromass, Manchester, UK) operating in positive mode. Each sample, diluted in 195 μL of methanol, was introduced through direct infusion with the following electrospray conditions: flow rate of 10 mL/min, voltage applied to the needle at 3 kV, a cone voltage at 30 V, source temperature of 80 °C, and solvation temperature of 150 °C [62]. The resolution was set to about 9000 FWHM (full width at half maximum). Tandem mass spectra (MS/MS) were acquired by collision induced dissociation (CID), using argon as the collision gas (pressure measured as the setting in the collision cell 3.0×10^5 Torr). The collision energy was between 30 and 60 eV. Both MS and MS/MS spectra were recorded for 1 min. Data acquisition was carried out with a MassLynx 4.0 data system.

3.8. Fatty Acid Analysis by Gas Chromatography-Mass Spectrometry (GC–MS)

Fatty acid methyl esters (FAMEs) were prepared from lipid extracts using a methanolic solution of potassium hydroxide (2.0 M) according to the methodology previously described [26]. Volumes of 2.0 μL of the hexane solution containing FAMEs were analyzed by gas chromatography-mass spectrometry (GC–MS) on an Agilent Technologies 6890 N Network (Santa Clara, CA, USA) equipped with a DB-FFAP column with the following specifications: 60 m long, 0.25 mm internal diameter, and 0.25 μm film thickness (J & W Scientific, Folsom, CA, USA). The GC equipment was connected to an Agilent 5973 Network Mass Selective Detector operating with an electron impact mode at 70 eV and scanning the range *m/z* 40–500 in a 1 s cycle in a full scan mode acquisition. The oven temperature was programmed from an initial temperature of 80 °C, a linear increase to 155 °C at 15 °C/min, followed by linear increase at 8 °C/min to 210 °C, then at 30 °C/min to 250 °C, standing at 250 °C for 18 min. The injector and detector temperatures were 220 and 280 °C, respectively. Helium was used as the carrier gas at a flow rate of 0.5 mL/min. The identification of each FA was performed by mass spectrum comparison with those in the Wiley 275 library and confirmed by its interpretation and comparison with the literature. The relative amounts of FAs were calculated by the percent area method with proper normalization, considering the sum of all areas of the identified FAs.

3.9. Cell Viability Assay on T-47D and 5637 Tumor Cell Lines

The antiproliferative activity of lipid extracts was examined by the effect of *Gracilaria* sp. lipid extracts on the T-47D human breast cancer and urinary bladder cancer cell lines' metabolism using the Prestoblue colorimetric assay (Invitrogen Life Sciences, Paisley, UK). Tumor cells were cultivated in Dulbecco's Modified Eagle Medium (DMEM-F12, Invitrogen Life Technologies, Paisley, UK) with 10% fetal bovine serum (FBS; Gold, PAA) and 5 mg/L 1% penicillin/steptomicin (Invitrogen) in a humidified incubator at 37 °C under an atmosphere of 5% CO_2. Cell were plated on 96-well plates and allowed to attach for 24 h, 100 μL of cell suspension ($1–2 \times 10^4$ cell/mL in complete medium) were used. Following this step, 200 μL of the treatment solution in a range of 25–100 μg/mL were applied to the culture. The lipid extract was dissolved in DMSO and diluted to a final concentration of 0.1% DMSO in a phenol-red free RPMI 1640 medium supplemented with 2% charcoal treated FBS (DCC), 1% glutamate, and 1% PEST. The same concentration of DMSO was used in untreated controls [63]. The treatment medium was changed 48 h later, and was removed from each cell after 48 h for viability assay using PrestoBlue Absorbance measured at 570 nm and 600 nm at 1, 2, 3, 4, and 5 h on a plate reader, which gave a linear absorbance range. Experiments were carried out in quadruplicate and three independent experiments were carried out for each cell line.

3.10. Anti-Inflammatory Activity on Nitrite Production in RAW 264.7 Cells

Test solutions of *Gracilaria* sp. lipid extracts (25 mg/mL) were prepared in ethanol and stored at –20 °C until used. Serial dilutions of tested solutions with culture medium were prepared and sterilized by filtration immediately before in vitro assays. Ethanol concentrations ranged from 0.1% to 0.8% (v/v).

RAW 264.7, a mouse leukemic monocyte macrophage cell line from American Type Culture Collection (ATCC TIB-71), was supplied by Otília Vieira (Centro de Neurociências e Biologia Celular, Universidade de Coimbra, Coimbra, Portugal) and cultured in Dulbecco's Modified Eagle Medium (Invitrogen Life Technologies, Paisley, UK) supplemented with 10% non-inactivated fetal bovine serum, 100 U/mL penicillin, and 100 μg/mL streptomycin at 37 °C in a humidified atmosphere of 95% air and 5% CO_2. During the experiments, cells were monitored through microscope observation to detect any morphological change. Assessment of metabolically active cells was performed using a resazurin bioassay [64]. Briefly, cell duplicates were plated at a density of 0.1×10^6/well, in a 96-well plate and allowed to stabilize overnight. Following this period, cells were either maintained in a culture medium (control) or pre-incubated with various concentrations of *Gracilaria* sp. lipid extracts or its vehicle for 1 h, and later activated with 50 ng/mL LPS for 24 h. After the treatments, resazurin solution (50 μM in culture medium) was added to each well and incubated at 37 °C for 1 h, in a humidified atmosphere of 95% air and 5% CO_2. As viable cells are able to reduce resazurin (a non-fluorescent blue dye) into resorufin (pink and fluorescent), their number correlates with the magnitude of dye reduction. Quantification of resofurin was performed on a Biotek Synergy HT (BioTek Instruments, Winooski, VT, USA) plate reader at 570 nm, with a reference wavelength of 620 nm. The production of nitric oxide was measured by the accumulation of nitrite in the culture supernatants, using a colorimetric reaction with the Griess reagent [65]. Briefly, 170 μL of culture supernatants were diluted with equal volumes of the Griess reagent [0.1% (w/v) N-(1-naphthyl)-ethylenediamine dihydrochloride and 1% (w/v) sulphanilamide containing 5% (w/v) H_3PO_4] and maintained for 30 min in the dark. The absorbance at 550 nm was measured on a Biotek Synergy HT plate reader. Culture medium was used as a blank and nitrite concentration (μM) was determined from a regression analysis using serial dilutions of sodium nitrite as standard. Experiments were carried out at least three times.

3.11. Statistical Analysis

Antiproliferative and anti-inflammatory bioassays were measured in quadruplicate and in three different and independent experiments. Results were expressed as mean ± SD. One-way analysis of variance (ANOVA) followed by Dunnett's multiple comparison tests was used to compare the treatment group to a single control group, after checking for assumptions. Statistical differences were calculated and represented with the following symbols of significance level ** $p < 0.01$, *** $p < 0.001$, # $p < 0.05$, ### $p < 0.001$. Statistical analysis was performed using GraphPad Prism 5.0 for Windows (GraphPad Software, San Diego, CA, USA).

4. Conclusions

The comprehensive elucidation of the *Gracilaria* sp. lipidome has been successfully accomplished for the first time. Liquid chromatography–mass spectrometry–based approach afforded the identification of 147 molecular species of polar lipids, distributed between the glycolipids, phospholipids, and betaine lipids classes. It was possible to identify novel sulfolipids (SQMG) and betaine lipids, among which DGTS were identified for the first time on the genus *Gracilaria* and MGTS within the Rhodophyta. Lipid extracts (~80% polar lipids) from *Gracilaria* sp. cultivated on land-based IMTA were screened for bioactivity and collectively shown to be a natural source of bioactive lipids with antiproliferative and anti-inflammatory activities. The presence of these bioactive polar lipids in *Gracilaria* sp. promotes its consumption as a functional food for the prevention of various diseases. Seaweeds' land-based culture using IMTA is a sustainable solution towards the production of large volumes of biomass displaying replicable bioactive properties. The higher degree of production conditions control enabled by land-based IMTA, versus open-water large-scale culture, allows for the production of higher value products with better positioning in value-chains supplying high-end markets.

Supplementary Materials: LC–MS spectra information are available online at www.mdpi.com/1660-3397/15/3/62/s1.

Acknowledgments: The authors are grateful to ALGAplus—Produção e Comércio de algas e seus derivados, Lda. for supplying the seaweed samples. Thanks are due to the Fundação para a Ciência e a Tecnologia (FCT, Portugal), European Union, QREN, POPH, FEDER, and COMPETE for funding QOPNA research unit (UID/QUI/00062/2013), CESAM (UID/AMB/50017/2013) and the strategic project UID/NEU/04539/2013. We also thank RNEM (REDE/1504/REM/2005) for the Portuguese Mass Spectrometry Network. Elisabete da Costa (SFRH/BD/52499/2014), Tânia Melo (SFRH/BD/84691/2012), and Isabel Ferreira (SFRH/BD/110717/2015) are grateful to FCT for their grants. Luisa Helguero's contribution to this work was financed by national funds through FCT within the project UID/BIM/04501/2013 granted to Institute for Biomedicine. This work is a contribution of the Marine Lipidomics Laboratory.

Author Contributions: Elisabete da Costa conceived and designed the experiments, prepared samples, performed extraction protocols, acquisition and data analyses by GC–MS and HILIC–MS/MS, antiproliferative bioassays, and wrote the paper; Tânia Melo supervised acquisition of data by HILIC–MS/MS and the analyses of data; Ana S. P. Moreira supervised the acquisition of data by GC–MS and the analyses of data and participated on antiproliferative bioassays. Carina Bernardo cultivated human breast cancer and urinary bladder cancer cell lines and supervised the antiproliferative bioassay and the analysis of data; Luisa Helguero coordinated the antiproliferative bioassay experiments, analyzed the data, and co-wrote the paper; Isabel Ferreira conceived and designed the anti-inflammatory bioassay experiment, supervised the analysis of data, and co-wrote the paper; Maria Teresa Cruz coordinated the anti-inflammatory bioassay experiments, analyzed the data, and co-wrote the paper; Andreia M. Rego cultivated and prepared the seaweed samples; Pedro Domingues optimized the LC–MS conditions used; Ricardo Calado supervised the statistical analyses and co-wrote the paper; Maria H. Abreu coordinated the experimental design, provided algal samples, and co-wrote the paper; and Maria Rosário Domingues coordinated all the experiments and data analyses and co-wrote the paper.

Conflicts of Interest: The authors declare no conflict of interest.

References

1. Abreu, M.H.; Pereira, R.; Sassi, J.-F. Chapter 12: Marine algae and the global food industry. In *Marine Algae: Biodiversity, Taxonomy, Environmental Assessment, and Biotechnology*; Pereira, L., Magalhaes, J., Eds.; CRC Press: Boca Raton, FL, USA, 2014.
2. Abreu, M.H.; Pereira, R.; Yarish, C.; Buschmann, A.H.; Sousa-Pinto, I. IMTA with *Gracilaria vermiculophylla*: Productivity and nutrient removal performance of the seaweed in a land-based pilot scale system. *Aquaculture* **2011**, *312*, 77–87. [CrossRef]
3. Leal, M.C.; Rocha, R.J.M.; Rosa, R.; Calado, R. Aquaculture of marine non-food organisms: What, why and how? *Rev. Aquac.* **2016**, 1–24. [CrossRef]
4. Francavilla, M.; Franchi, M.; Monteleone, M.; Caroppo, C. The red seaweed *Gracilaria gracilis* as a multi products source. *Mar. Drugs* **2013**, *11*, 3754–3776. [CrossRef] [PubMed]
5. Guschina, I.A.; Harwood, J.L. Lipids and lipid metabolism in eukaryotic algae. *Prog. Lipid Res.* **2006**, *45*, 160–186. [CrossRef] [PubMed]

6.   Van Ginneken, V.J.T.; Helsper, J.P.F.G.; de Visser, W.; van Keulen, H.; Brandenburg, W.A. Polyunsaturated fatty acids in various macroalgal species from North Atlantic and tropical seas. *Lipids Health Dis.* **2011**, *10*, 104. [CrossRef] [PubMed]

7.   Stengel, D.B.; Connan, S.; Popper, Z.A. Algal chemodiversity and bioactivity: Sources of natural variability and implications for commercial application. *Biotechnol. Adv.* **2011**, *29*, 483–501. [CrossRef] [PubMed]

8.   Mohamed, S.; Hashim, S.N.; Rahman, H.A. Seaweeds: A sustainable functional food for complementary and alternative therapy. *Trends Food Sci. Technol.* **2012**, *23*, 83–96. [CrossRef]

9.   Plouguerné, E.; De Souza, L.M.; Sassaki, G.L.; Cavalcanti, J.F.; Romanos, M.T.V.; da Gama, B.A.P.; Pereira, R.C.; Barreto-Bergter, E.; Villela Romanos, M.T.; da Gama, B.A.P.; et al. Antiviral sulfoquinovosyldiacylglycerols (SQDGs) from the Brazilian brown seaweed *Sargassum vulgare*. *Mar. Drugs* **2013**, *11*, 4628–4640.

10.   Mattos, B.B.; Romanos, M.T.V.; de Souza, L.M.; Sassaki, G.; Barreto-Bergter, E. Glycolipids from macroalgae: Potential biomolecules for marine biotechnology? *Rev. Bras. Farmacogn.* **2011**, *21*, 244–247. [CrossRef]

11.   Plouguerné, E.; da Gama, B.A.P.; Pereira, R.C.; Barreto-Bergter, E. Glycolipids from seaweeds and their potential biotechnological applications. *Front. Cell. Infect. Microbiol.* **2014**, *4*, 1–3. [CrossRef] [PubMed]

12.   Küllenberg, D.; Taylor, L.A.; Schneider, M.; Massing, U. Health effects of dietary phospholipids. *Lipids Health Dis.* **2012**, *11*, 1–16. [CrossRef] [PubMed]

13.   Da Costa, E.; Silva, J.; Mendonça, S.; Abreu, M.; Domingues, M. Lipidomic approaches towards deciphering glycolipids from microalgae as a reservoir of bioactive lipids. *Mar. Drugs* **2016**, *14*, 101. [CrossRef] [PubMed]

14.   Maciel, E.; Leal, M.C.; Lillebø, A.I.; Domingues, P.; Domingues, M.R.; Calado, R. Bioprospecting of marine macrophytes using MS-based lipidomics as a new approach. *Mar. Drugs* **2016**, *14*, 49. [CrossRef] [PubMed]

15.   Robertson, R.C.; Guihéneuf, F.; Bahar, B.; Schmid, M.; Stengel, D.B.; Fitzgerald, G.F.; Ross, R.P.; Stanton, C. The Anti-Inflammatory effect of algae-derived lipid extracts on lipopolysaccharide (LPS)-stimulated human THP-1 macrophages. *Mar. Drugs* **2015**, *13*, 5402–5424. [CrossRef] [PubMed]

16.   Tsai, C.; Pan, B.S. Identification of sulfoglycolipid bioactivities and characteristic fatty acids of marine macroalgae. *JAFC* **2012**, *60*, 8404–8410. [CrossRef] [PubMed]

17.   Ohta, K.; Mizushina, Y.; Hirata, N.; Takemura, M.; Sugawara, F.; Matsukage, A.; Yoshida, S.; Sakaguchi, K. Sulfoquinovosyldiacylglycerol, KM043, a new potent inhibitor of eukaryotic DNA polymerases and HIV-reverse transcriptase type 1 from a marine red alga, *Gigartina tenella*. *Chem. Pharm. Bull.* **1998**, *46*, 684–686. [CrossRef] [PubMed]

18.   Bergé, J.P.; Debiton, E.; Dumay, J.; Durand, P.; Barthomeuf, C. In vitro anti-inflammatory and anti-proliferative activity of sulfolipids from the red alga *Porphyridium cruentum*. *J. Agric. Food Chem.* **2002**, *50*, 6227–6232. [CrossRef] [PubMed]

19.   Lopes, G.; Daletos, G.; Proksch, P.; Andrade, P.B.; Valentão, P. Anti-inflammatory potential of monogalactosyl diacylglycerols and a monoacylglycerol from the edible brown seaweed *Fucus spiralis* Linnaeus. *Mar. Drugs* **2014**, *12*, 1406–1418. [CrossRef] [PubMed]

20.   Banskota, A.H.; Stefanova, R.; Sperker, S.; Lall, S.P.; Craigie, J.S.; Hafting, J.T.; Critchley, A.T. Polar lipids from the marine macroalgae *Palmaria palmata* inhibit lipopolysaccharide-induced nitric oxide production in RAW264.7 macrophage cells. *Phytochemistry* **2014**, *101*, 101–108. [CrossRef] [PubMed]

21.   Banskota, A.H.; Stefanova, R.; Sperker, S.; Lall, S.; Craigie, J.S.; Hafting, J.T. Lipids isolated from the cultivated red alga *Chondrus crispus* inhibit nitric oxide production. *J. Appl. Phycol.* **2014**, *26*, 1565–1571. [CrossRef]

22.   Maeda, N.; Kokai, Y.; Ohtani, S.; Sahara, H.; Hada, T.; Ishimaru, C.; Kuriyama, I.; Yonezawa, Y.; Iijima, H.; Yoshida, H.; et al. Anti-tumor effects of the glycolipids fraction from spinach which inhibited DNA polymerase activity. *Nutr. Cancer* **2007**, *57*, 216–223. [CrossRef] [PubMed]

23.   Mayer, A.M.S.; Rodríguez, A.D.; Berlinck, R.G.S.; Fusetani, N. Marine pharmacology in 2007–2008: Marine compounds with antibacterial, anticoagulant, antifungal, anti-inflammatory, antimalarial, antiprotozoal, antituberculosis, and antiviral activities; affecting the immune and nervous system, and other miscellaneous mec. *Comp. Biochem. Physiol. C Toxicol. Pharmacol.* **2011**, *153*, 191–222. [PubMed]

24.   Pereira, H.; Barreira, L.L.; Figueiredo, F.; Custódio, L.L.; Vizetto-Duarte, C.; Polo, C.; Rešek, E.; Engelen, A.; Varela, J.J. Polyunsaturated fatty acids of marine macroalgae: Potential for nutritional and pharmaceutical applications. *Mar. Drugs* **2012**, *10*, 1920–1935. [CrossRef] [PubMed]

25.   Da Costa, E.; Melo, T.; Moreira, A.S.P.; Alves, E.; Domingues, P.; Calado, R.; Abreu, M.H.M.H.; Domingues, M.R. Decoding bioactive polar lipid profile of the macroalgae *Codium tomentosum* from a sustainable IMTA system using a lipidomic approach. *Algal Res.* **2015**, *12*, 388–397. [CrossRef]

26. Melo, T.; Alves, E.; Azevedo, V.V.; Martins, A.S.; Neves, B.; Domingues, P.; Calado, R.; Abreu, M.H.; Domingues, M.R. Lipidomics as a new approach for the bioprospecting of marine macroalgae-Unraveling the polar lipid and fatty acid composition of *Chondrus crispus*. *Algal Res.* **2015**, *8*, 181–191. [CrossRef]

27. Dembitsky, V.M.; Řezanková, H.; Řezanka, T.; Hanuš, L.O. Variability of the fatty acids of the marine green algae belonging to the genus *Codium*. *Biochem. Syst. Ecol.* **2003**, *31*, 1125–1145. [CrossRef]

28. Khotimchenko, S.Y.V.; Vaskovsky, Y.E.; Titlyanova, T.V.; Vaskovsky, V.E.; Titlyanova, T.V. Fatty acids of marine algae from the Pacific Coast of North California. *Bot. Mar.* **2002**, *45*, 17–22. [CrossRef]

29. Ragonese, C.; Tedone, L.; Beccaria, M.; Torre, G.; Cichello, F.; Cacciola, F.; Dugo, P.; Mondello, L. Characterisation of lipid fraction of marine macroalgae by means of chromatography techniques coupled to mass spectrometry. *Food Chem.* **2014**, *145*, 932–940. [CrossRef] [PubMed]

30. Khotimchenko, S.V. Distribution of glyceroglycolipids in marine algae and grasses. *Chem. Nat. Compd.* **2002**, *38*, 186–191. [CrossRef]

31. Khotimchenko, S.V. Lipids from marine alga *Gracilaria verrucosa*. *Chem. Nat. Compd.* **2005**, *41*, 230–232. [CrossRef]

32. Sanina, N.M.; Goncharova, S.N.; Kostetsky, E.Y. Fatty acid composition of individual polar lipid classes from marine macrophytes. *Phytochemistry* **2004**, *65*, 721–730. [CrossRef] [PubMed]

33. Kendel, M.; Couzinet-Barnathan, G.; Mossion, A.; Viau, M.; Fleurence, J.; Barnathan, G.; Wielgosz-Collin, G. Seasonal composition of lipids, fatty acids, and sterols in the edible red alga *Grateloupia turuturu*. *J. Appl. Phycol.* **2013**, *25*, 425–432. [CrossRef]

34. He, H.; Rodgers, R.P.; Marshall, A.G.; Hsu, C.S. Algae polar lipids characterized by online liquid chromatography coupled with hybrid linear quadrupole ion trap/fourier transform ion cyclotron resonance mass spectrometry. *Energy Fuels* **2011**, *25*, 4770–4775. [CrossRef]

35. Naumann, I.; Darsow, K.H.; Walter, C.; Lange, H.A.; Buchholz, R. Identification of sulfoglycolipids from the alga *Porphyridium purpureum* by matrix-assisted laser desorption/ionisation quadrupole ion trap time-of-flight mass spectrometry. *Rappid Commun. Mass Spectrom.* **2007**, *21*, 3185–3192. [CrossRef] [PubMed]

36. Okazaki, Y.; Kamide, Y.; Hirai, M.Y.; Saito, K. Plant lipidomicss based on hydrophilic interaction chromatography coupled to ion trap time-of flight mass spectrometry. *Metabolomics.* **2011**, *9*, 121–131. [CrossRef] [PubMed]

37. Kumari, P.; Kumar, M.; Reddy, C.R.K.; Jha, B. Nitrate and phosphate regimes induced lipidomic and biochemical changes in the intertidal macroalga *Ulva lactuca* (Ulvophyceae, Chlorophyta). *Plant Cell Physiol.* **2014**, *55*, 52–63. [CrossRef] [PubMed]

38. Popendorf, K.J.; Fredricks, H.F.; Van Mooy, B.A.S. Molecular ion-independent quantification of polar glycerolipid classes in marine plankton using triple quadrupole MS. *Lipids* **2013**, *48*, 185–195. [CrossRef] [PubMed]

39. Leal, M.C.; Munro, M.H.G.; Blunt, J.W.; Puga, J.; Jesus, B.; Calado, R.; Rosa, R.; Madeira, C. Biogeography and biodiscovery hotspots of macroalgal marine natural products. *Nat. Prod. Rep.* **2013**, *30*, 1380–1390. [CrossRef] [PubMed]

40. Pettitt, T.R.; Harwood, J.L. Alterations in lipid metabolism caused by illumination of the marine red algae *Chondrus crispus* and *Polysiphonia lanosa*. *Phytochemistry* **1989**, *28*, 3295–3300. [CrossRef]

41. Trevor, R.; Pettitt, A.; Jones, L.; Harwood, J.L. Lipids of the marine red algae, Chondrus crispus and *Polysiphonia lanosa*. *Phytochemistry* **1989**, *28*, 399–405. [CrossRef]

42. Harwood, J.L. Lipid metabolism in the red marine algae *Chondrus Crispus* and *Polysinphonza Lanosa* as modified by temperature. *Phytochemistry* **1989**, *28*, 1–6. [CrossRef]

43. De Souza, L.M.; Sassaki, G.L.; Romanos, M.T.V.; Barreto-Bergter, E. Structural characterization and anti-HSV-1 and HSV-2 activity of glycolipids from the marine algae *Osmundaria obtusiloba* isolated from Southeastern Brazilian coast. *Mar. Drugs* **2012**, *10*, 918–931. [CrossRef] [PubMed]

44. Khotimchenko, S.V.; Vaskovsky, V.E. An inositol-containing sphingolipid from the red alga *Gracilaria verrucosa*. *Russ. J. Bioorg. Chem.* **2004**, *30*, 168–171. [CrossRef]

45. Sato, N. Betaine Lipids. *Bot. Mag.* **1992**, *1*, 185–197. [CrossRef]

46. Dembitsky, V.M. Betaine ether-linked glycerolipids. *Prog. Lipid Res.* **1996**, *35*, 1–51. [CrossRef]

47. Khotimchenko, S.V. Variations in lipid composition among different developmental stages of *Gracilaria verrucosa* (Rhodophyta). *Bot. Mar.* **2006**, *49*, 34–38. [CrossRef]

48. *Diet, Nutrition and the Prevention of Chronic Diseases*; WHO Technical Report Series 916; World Health Organization: Geneva, Switzerland, 2003; Available online: http://www.who.int/dietphysicalactivity/publications/trs916/en/ (accessed on 10 January 2017).

49. Simopoulos, A. The importance of the omega-6/omega-3 fatty acid ratio in cardiovascular disease and other chronic diseases. *Exp. Biol. Med.* **2008**, *233*, 674–688. [CrossRef] [PubMed]

50. Simopoulos, A. The importance of the ratio of omega-6/omega-3 essential fatty acids. *Biomed. Pharmacother.* **2002**, *56*, 365–379. [CrossRef]

51. Husted, K.S.; Bouzinova, E.V. The importance of *n*-6/*n*-3 fatty acids ratio in the major depressive disorder. *Medicina* **2016**, *52*, 139–147. [CrossRef] [PubMed]

52. Murray, M.; Hraiki, A.; Bebawy, M.; Pazderka, C.; Rawling, T. Anti-tumor activities of lipids and lipid analogues and their development as potential anticancer drugs. *Pharmacol. Ther.* **2015**, *150*, 109–128. [CrossRef] [PubMed]

53. Hossain, Z.; Kurihara, H.; Hosokawa, M.; Takahashi, K. Growth inhibition and induction of differentiation and apoptosis mediated by sodium butyrate in Caco-2 cells with algal glycolipids. *In Vitro Cell. Dev. Biol. Anim.* **2005**, *41*, 154–159. [CrossRef] [PubMed]

54. Zhang, J.; Li, C.; Yu, G.; Guan, H. Total synthesis and structure-activity relationship of glycoglycerolipids from marine organisms. *Mar. Drugs* **2014**, *12*, 3634–3659. [CrossRef] [PubMed]

55. Burri, L.; Hoem, N.; Banni, S.; Berge, K. Marine omega-3 phospholipids: Metabolism and biological activities. *Int. J. Mol. Sci.* **2012**, *13*, 15401–15419. [CrossRef] [PubMed]

56. D'Arrigo, P.; Servi, S. Synthesis of lysophospholipids. *Molecules* **2010**, *15*, 1354–1377. [CrossRef] [PubMed]

57. Banskota, A.H.; Gallant, P.; Stefanova, R.; Melanson, R.; O'Leary, S.J.B. Monogalactosyldiacylglycerols, potent nitric oxide inhibitors from the marine microalga *Tetraselmis chui*. *Nat. Prod. Res.* **2012**, *27*, 37–41.

58. Koch, A.K.; Kappeli, O.; Fiechter, A.; Reiser, J. Hydrocarbon assimilation and biosurfactant production in *Pseudomonas aeruginosa mutants*. *J. Bacteriol.* **1991**, *173*, 4212–4219. [CrossRef] [PubMed]

59. Dória, M.L.; Cotrim, Z.; Macedo, B.; Simões, C.; Domingues, P.; Helguero, L.; Domingues, M.R. Lipidomic approach to identify patterns in phospholipid profiles and define class differences in mammary epithelial and breast cancer cells. *Breast Cancer Res. Treat.* **2012**, *133*, 635–648. [CrossRef] [PubMed]

60. Bartlett, E.M.; Lewis, D.H. Spectrophotometric determination of phosphate esters in the presence and absence of orthophosphate. *Anal. Biochem.* **1970**, *36*, 159–167. [CrossRef]

61. Pacetti, D.; Boselli, E.; Lucci, P.; Frega, N.G. Simultaneous analysis of glycolipids and phospholids molecular species in avocado (*Persea americana Mill*) fruit. *J. Chromatogr. A* **2007**, *1150*, 241–251. [CrossRef] [PubMed]

62. Melo, T.; Silva, E.M.P.; Simões, C.; Domingues, P.; Domingues, M.R.M. Photooxidation of glycated and non-glycated phosphatidylethanolamines monitored by mass spectrometry. *J. Mass Spectrom.* **2013**, *48*, 68–78. [CrossRef] [PubMed]

63. Dória, M.L.; Cotrim, C.Z.; Simões, C.; Macedo, B.; Domingues, P.; Domingues, M.R.; Helguero, L.A. Lipidomic analysis of phospholipids from human mammary epithelial and breast cancer cell lines. *J. Cell. Physiol.* **2013**, *228*, 457–468. [CrossRef] [PubMed]

64. O'Brien, J.; Wilson, I.; Orton, T.; Pognan, F. Investigation of the Alamar Blue (resazurin) fluorescent dye for the assessment of mammalian cell cytotoxicity. *Eur. J. Biochem.* **2000**, *267*, 5421–5426. [CrossRef] [PubMed]

65. Green, L.C.; Wagner, D.A.; Glogowski, J.; Skipper, P.L.; Wishnok, J.S.; Tannenbaum, S.R. Analysis of nitrate, nitrite, and ^{15}N nitrate in biological fluids. *Anal. Biochem.* **1982**, *126*, 131–138. [CrossRef]

marine drugs

MDPI

Article

Krill Oil-In-Water Emulsion Protects against Lipopolysaccharide-Induced Proinflammatory Activation of Macrophages In Vitro

Gabriel A. Bonaterra [1,*], David Driscoll [2,3], Hans Schwarzbach [1] and Ralf Kinscherf [1]

[1]   Department of Medical Cell Biology, Philipps-University Marburg, Robert-Koch-Straße 8, 35032 Marburg, Germany; hans.schwarzbach@uni-marburg.de (H.S.); ralf.kinscherf@staff.uni-marburg.de (R.K.)
[2]   Stable Solutions LLC, Easton Industrial Park, 19 Norfolk Avenue, South Easton, MA 02375, USA; d.driscoll@stablesolns.com
[3]   Department of Medicine, University of Massachusetts Medical School, Worcester, MA 01655, USA
*   Correspondence: gabriel.bonaterra@staff.uni-marburg.de; Tel.: +49-6421-286-4097; Fax: +49-6421-286-8983

Academic Editor: Sylvia Urban
Received: 24 October 2016; Accepted: 10 March 2017; Published: 15 March 2017

Abstract: Background: Parenteral nutrition is often a mandatory therapeutic strategy for cases of septicemia. Likewise, therapeutic application of anti-oxidants, anti-inflammatory therapy, and endotoxin lowering, by removal or inactivation, might be beneficial to ameliorate the systemic inflammatory response during the acute phases of critical illness. Concerning anti-inflammatory properties in this setting, omega-3 fatty acids of marine origin have been frequently described. This study investigated the anti-inflammatory and LPS-inactivating properties of krill oil (KO)-in-water emulsion in human macrophages in vitro. Materials and Methods: Differentiated THP-1 macrophages were activated using specific ultrapure-LPS that binds only on the toll-like receptor 4 (TLR4) in order to determine the inhibitory properties of the KO emulsion on the LPS-binding capacity, and the subsequent release of TNF-α. Results: KO emulsion inhibited the macrophage binding of LPS to the TLR4 by 50% (at 12.5 μg/mL) and 75% (at 25 μg/mL), whereas, at 50 μg/mL, completely abolished the LPS binding. Moreover, KO (12.5 μg/mL, 25 μg/mL, or 50 μg/mL) also inhibited (30%, 40%, or 75%, respectively) the TNF-α release after activation with 0.01 μg/mL LPS in comparison with LPS treatment alone. Conclusion: KO emulsion influences the LPS-induced pro-inflammatory activation of macrophages, possibly due to inactivation of the LPS binding capacity.

Keywords: krill oil-in-water emulsion; omega-3 fatty acids; phospholipids; LPS; cytokines; septic shock

---

1. Introduction

Sepsis and septic shock due to Gram-negative pathogens are responsible for significant morbidity and mortality in human populations [1]. LPS binding to phagocytic cells stimulates the synthesis and release of cytokines, such as TNF-α, IL-1β, and IL-6 [2]. Cytokine secretion is an important component of host defense, but when overstimulation occurs, excessive cytokine secretion may lead to the systemic signs and symptoms of sepsis [3]. Exogenous or endogenous stimulation of biological factors that modulate the extent of binding of LPS to monocytes and macrophages may play a pivotal role in determining the outcome of endotoxin exposure [1]. In this context, serum factors that bind LPS may prevent macrophage activation [4]. In vitro and in vivo, HDL binds LPS and neutralizes it and the LPS-induced cytokine response is attenuated [5,6]. However, the phospholipid (PL) content, rather than the cholesterol content, correlates with the effectiveness of LPS neutralization [7]. Additionally, circulating levels of HDL are reduced in sepsis/septic shock, and this reduction is positively correlated with the severity of the illness [8], and decreased LDL levels (≤70 mg/dL) were associated with increased risks of sepsis [9]. In an optimal way, the substance used to neutralize the endotoxin effect during sepsis should be anti-oxidative, anti-inflammatory, and with endotoxin-binding capacity. In this context, omega-3 fatty acids (*n*-3 fatty acids) decrease the production of inflammatory eicosanoids, cytokines, reactive oxygen species (ROS) and adhesion molecules [10]. The key link between PUFAs and inflammation is that eicosanoids, which are among the primary mediators and regulators of inflammation during acute metabolic stress, are generated from 20-carbon PUFAs [10]. The three types of omega-3 fatty acids involved in human physiology are α-eicosapentaenoic acid (EPA) and docosahexaenoic acid (DHA), both of which are usually found in marine fish oils and linolenic acid (ALA), commonly found in plant oils. With respect to the precursor fatty acid ALA, in human it has poor bioconversion to the essential omega-3 fatty acids EPA and DHA and, therefore, it is an unreliable source for these bioactive fatty acids [11]. In this context, fish oil dietary supplements play a role of increasing the strategic importance in meeting daily requirements of essential nutrients [12].

Applications of intravenous lipid emulsions containing fish oil reduce the length of stay in hospital [13], as well as antibiotic use and mortality [14]. Moreover, fish oil with parenteral nutrition provided to septic intensive care patients increases plasma EPA, modifies inflammatory cytokine, improves gas exchange [15], and may exert profound influence on the status of immunocompetence and inflammation [16,17] The anti-inflammatory properties of marine omega-3 fatty acids have already been described [18,19]. In addition to triglycerides, marine *n*-3 fatty acids are also available in other forms, such as in crude krill oil (KO), which provides EPA and DHA, mainly in the form of PLs, and as ethyl esters of pharmaceutical grade, highly-concentrated preparations [18]. In this context, most recently, KO, which contains a significant portion of its *n*-3 LC-PUFA in PLs, is also increasingly found on the market, and is promoted as being of "higher efficacy" [12,20]. Additionally, most recently a new product category, derived from Antarctic krill (*Euphausia superba* Dana), has been brought onto the omega-3 market, characterized by a greater ease of absorption due to higher PL content [12,20].

KO comes from sustainable fisheries and is nearly at the beginning of a food chain, compared with fish sources that are more affected by environmental pollutants [12,20]. In addition, a higher fraction of omega-3 LC-PUFA is associated with PLs in KO, compared to triacylglycerol in fish oils, and this property may improve gastrointestinal absorption and bioavailability of omega-3 LC-PUFA [21]. KO contains PUFAs, including the bioactive EPA and DHA, (up to 35% *w/w* of the fatty acids profile), with up to 95% *w/w* PLs and up to 45% triglycerides [22]. According to these characteristics, we hypothesize that an injectable KO emulsion might in vitro exert anti-inflammatory properties from the presence of omega-3 fatty acids, and also bind endotoxin, thereby inhibiting LPS mediated effects, i.e., LPS is less able to stimulate and activate macrophages to release pro-inflammatory cytokines.

2. Results

2.1. Effect of KO Emulsion or LPS on the Viability of Differentiated Human THP-1 Macrophages

As shown in Figure 1A, we found that, after 24 h, treatment with 5–250 µg/mL KO did not display any cytotoxicity. Glycerol used as the vehicle was not cytotoxic (Figure 1A). Incubation of differentiated human THP-1 macrophages for 4 h with LPS did not show cytotoxicity (Figure 1B).

2.2. Effect of KO Emulsion on the LPS Binding

We used two binding assays to evaluate the interaction of LPS with macrophage-TLR4 and the inhibitory effect of KO. As shown in Figure 2, macrophages incubated 24 h with 1 µg or 5 µg/mL LPS-EB-biotin displays positive binding, detected by fluorescence (Figure 2), compared with controls without LPS.

(A)

(B)

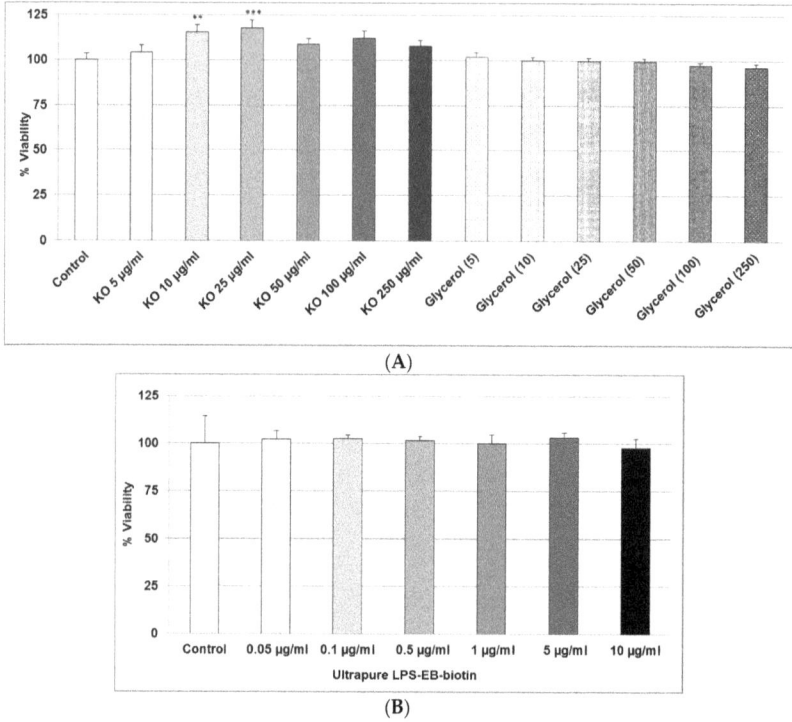

Figure 1. (**A**) Effects of 24 h treatment with KO emulsion or the glycerol vehicle (numbers in brackets indicate the volume of glycerol used in the corresponding KO concentration), on the viability of THP-1 macrophages. Values (in % viability of cells without treatment (control = 100%)) are given as the mean + SEM; ANOVA test, significance vs. negative control, ** $p \leq 0.01$, *** $p \leq 0.001$; $n = 7$ independent experiments; and (**B**) the effects of 4 h treatment with ultrapure LPS-EB-biotin on the viability of differentiated human THP-1 macrophages. Values (in % viability of cells without treatment (control = 100%)) are given as the mean + SEM; $n = 4$ independent experiments.

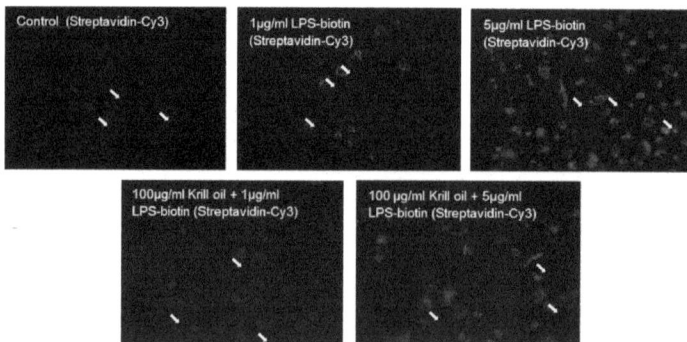

Figure 2. Effect of the KO emulsion on the LPS binding by differentiated human THP-1 macrophages. Photographs of the inhibitory effect of KO on the binding of ultrapure LPS-EB-biotin after 24 h incubation, was detected with streptavidin-Cy3-conjugated as fluorochrome. White arrows: nuclei. Magnification: 200×.

LPS (1 µg/mL) pre-incubated with KO (100 µg/mL) inhibited the binding, but not when using LPS at a concentration of 5 µg/mL (Figure 2). The LPS binding was increased at 0.1 µg/mL (6.5%), 1 µg/mL (20.3%, $p \leq 0.05$), and 5 µg/mL (100%, $p \leq 0.05$) when compared with the negative control (Figure 3).

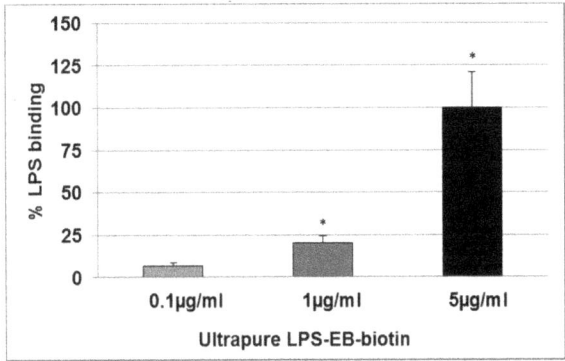

Figure 3. Effect of the KO emulsion on the LPS binding on macrophage-TLR4. Ultrapure LPS-EB-biotin binding assay was performed by using PMA-differentiated THP-1 macrophages after 3 h incubation, including streptavidin-HRP-OPD system for detection and colorimetric quantification. Values (binding relative to negative control (value = 1) without LPS-EB-biotin) are given as the mean + SEM; ANOVA test, significance vs. negative control without LPS, * $p \leq 0.05$; $n = 4$ independent experiments.

After co-incubation of 0.1 µg/mL LPS-EB with KO, the macrophages were treated for 3 h with different concentrations of LPS + KO: at 12.5 µg/mL LPS + KO, LPS-binding decreased by 50% (not significant); at 25 µg/mL LPS + KO LPS binding significantly decreased by 75% ($p \leq 0.05$) and, at 50 µg/mL LPS + KO, it was abolished ($p \leq 0.01$), as shown in Figure 4.

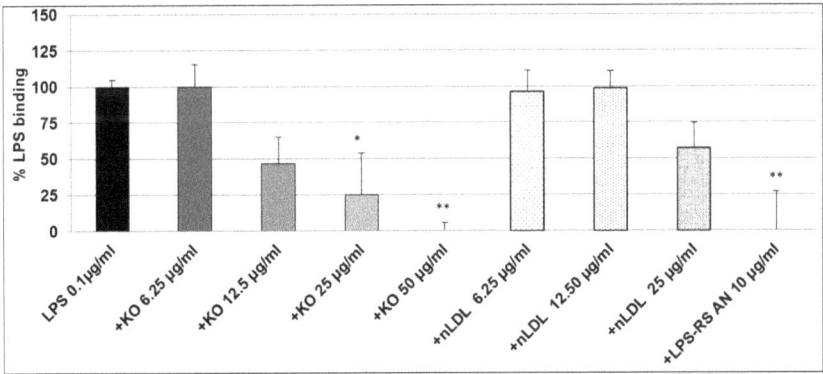

Figure 4. Ultrapure LPS-EB-biotin binding assay on macrophage-TLR4. LPS-EB-biotin (0.1 µg/mL) was co-incubated overnight with different concentrations of KO emulsion or LPS antagonist (LPS-RS AN), added to differentiated THP-1 macrophages (3 h). LPS binding was spectrophotometrically quantified using streptavidin-HRP-OPD system. Values (in % binding relative to positive control (100% binding) with LPS-EB-biotin) are given as the mean + SEM; ANOVA test, significance vs. positive control (=LPS-EB-Biotin), * $p \leq 0.05$, ** $p \leq 0.01$; $n = 6$ independent experiments.

As the LPS-binding control, 25 µg/mL nLDL showed a similar effect as KO at 12.5 µg/mL, however, this was not significant. LPS-RS-AN antagonist significantly ($p \leq 0.01$) prevented the binding of LPS-biotin and, thus, mimicked the effect of KO (Figure 4). A total and significant inhibition of the binding was obtained when 1 µg/mL LPS-EB was co-incubated with 6.25, 12.5 µg/mL ($p \leq 0.01$), 25 µg/mL ($p \leq 0.01$), or 50 µg/mL ($p \leq 0.001$) KO when compared with the control. Furthermore, LDL showed a similar inhibition pattern as KO. Additionally, the LPS-RS antagonist completely inhibited the LPS binding on the TLR4 (Figure 5). Moreover, we have performed this experiment using 5 µg/mL LPS-EB-biotin, however, without a statistically significant difference between LPS/KO co-incubation and LPS alone (data not shown).

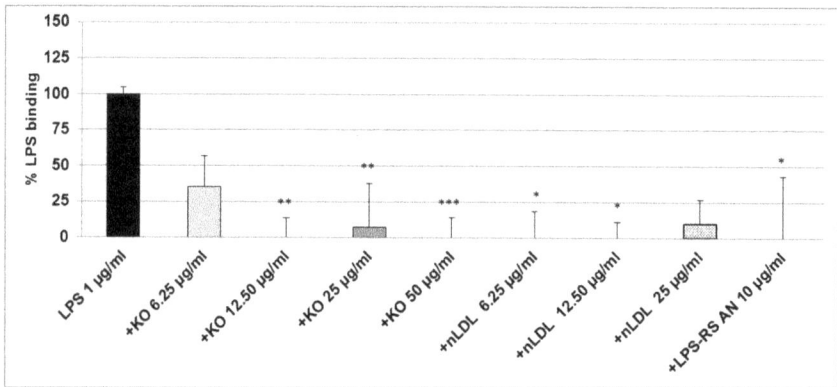

Figure 5. Ultrapure LPS-EB-biotin binding assay on macrophage-TLR4. LPS-EB-biotin (1 µg/mL) was co-incubated overnight together with different concentrations of KO emulsion or 100 µg/mL LPS antagonist (LPS-RS AN) added to differentiated human THP-1 macrophages (3 h). LPS binding was spectrophotometrically quantified using streptavidin-HRP-OPD system. Values (in % binding relative to positive control (100% binding) with LPS-EB-biotin) are given as the mean + SEM; ANOVA test, significance vs. positive control (=LPS-EB-Biotin), * $p \leq 0.05$, ** $p \leq 0.01$, *** $p \leq 0.001$; $n = 6$ independent experiments using four wells per treatment and experiment.

2.3. Effect of KO Emulsion on TNF-α Release of Differentiated Human THP-1 Macrophages Stimulated with LPS-EB

The TNF-α release in cell supernatants of differentiated human THP-1 macrophages treated with LPS-EB (pre-incubated with KO) was markedly reduced when 0.01 µg/mL or 0.1 µg/mL LPS was used. A significant inhibition of the TNF-α release was obtained after incubation of 0.01 µg/mL LPS with 12.5 µg/mL (-30%, $p \leq 0.01$), 25 µg/mL (-40%, $p \leq 0.001$) or 50 µg/mL (-75%, $p \leq 0.001$) KO in comparison with the control LPS (Figure 6A).

Treatment with 0.1 µg/mL LPS pre-incubated with 25 µg/mL or 50 µg/mL KO showed a 50% ($p \leq 0.05$) or 60% ($p \leq 0.01$) inhibition in comparison with control LPS (Figure 6B). Treatment with KO alone had no effect on the TNF-α production (Figure 6B). Incubation with antagonist LPS-RS abolished the TNF-α release (Figure 6B).

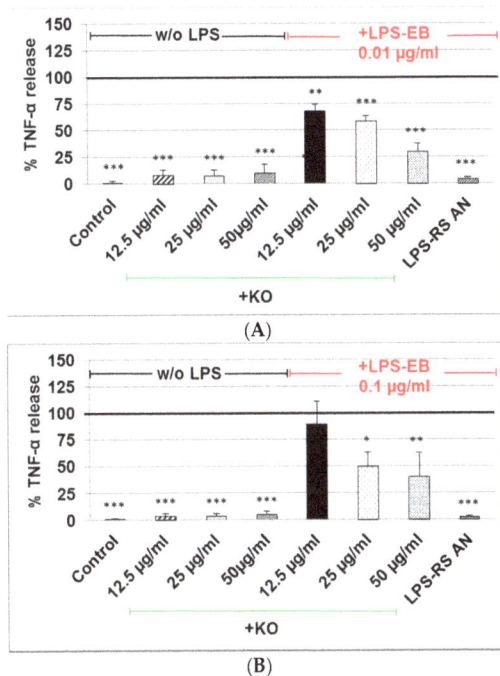

Figure 6. Analyses (by ELISA) of the inhibitory effect of KO emulsion on the TNF-α release by LPS-EB differentiated human THP-1 macrophages. (**A**) LPS-EB 0.01 μg/mL, or (**B**) 0.1 μg/mL was co-incubated overnight together with different concentrations of KO or 10 μg/mL LPS antagonist (LPS-RS-AN) and afterwards for 3 h with the differentiated THP-1 macrophages. Values (in %TNF-α release relative to positive (LPS-EB) control (=100% release)) are given as the mean + SEM; ANOVA test, significance vs. positive control (=LPS-EB), * $p \leq 0.05$, ** $p \leq 0.01$, *** $p \leq 0.001$; $n = 3$–5 independent experiments.

3. Discussion

Sepsis has been defined as a systemic immune activation in patients with infection, characterized by low rates of survival [23]. Bacterial endotoxemia is considered one of the major causes of sepsis, resulting from the release of LPS by microorganisms, e.g., within the colon, that translocate across a compromised intestinal wall [17,24]. In this context, we performed in vitro experiments to test the efficacy of a KO emulsion against endotoxin-triggered pro-inflammatory effects.

Death rates in patients caused by sepsis reach about 30%–80% [25], especially in oncologic patients [26]. Sepsis-associated organ failure and death result from an overwhelming inflammatory immune response that culminates in a generalized autodestructive process [27] and development of multi-organ dysfunction syndrome, or MODS [28]. Under normal physiological conditions, in which ROS levels are controlled by endogenous anti-oxidant systems, sepsis induces an imbalance between pro- and antioxidant systems, which leads to oxidative stress [29]. Moreover, long-chain *n*-3 PUFAs decrease the production of inflammatory mediators (eicosanoids, cytokines, and ROS) [22,30]. Once incorporated into cell membranes, a key step to exert cytoprotection, *n*-3-FAs dramatically modulate the body's response to inflammation, oxidative stress, ischemia, and immune function through several downstream bioactive mediators, (e.g., cytokines, prostaglandins, thromboxanes, leukotrienes, resolvins, protectins, etc.) [22]. The combination of "cytoprotective excipients" [31], together with nutritional, anti-oxidant, and anti-inflammatory properties, make the omega-3 therapy a successful candidate for the management of sepsis in patients under intensive care. By using EPA or fish oil, inhibition of endotoxin-induced TNF-α production by monocytes [32], may exert effects on both, the generation of inflammatory mediators, and on the resolution of inflammatory processes [15]. These observations suggest direct effects of long-chain *n*-3 PUFA on inflammatory gene expression via inhibition of activation of the transcription factor NF-κB [33]. Several studies in healthy human volunteers involving supplementation of the diet with fish oil have demonstrated decreased production of TNF-α, IL-1β, and IL-6 after endotoxin-stimulation of monocytes or mononuclear cells [34,35]. The benefits of fish oil in animal models of experimental endotoxemia have been clearly demonstrated when dietary fish oil or fish oil infused intravenously enhances the survival of guinea pigs after intraperitoneal endotoxin injection [36]. In this context, strategies to attenuate the immune response and prevent organ failure could help patients with sepsis or septic shock. Given that ultrapure LPS-EB specifically binds to TLR4 and the LPS antagonist LPS-RS, leading to an almost complete suppression of the TNF-α release, confirms the specific binding of LPS-EB to TLR4. The same effect observed with KO, can be explained by an inactivation of LPS during the co-incubation with KO or an inhibition of the binding to TLR4. Nevertheless, the present study clearly demonstrates that KO inhibited the production of TNF-α after stimulation with LPS. Available data have confirmed an effective reduction in the LPS level in the patients' blood after this procedure [37], and could effectively eliminate a wide range of the factors as LPS, cytokines, etc., from peripheral blood.

Our findings suggest that KO has properties to inactivate and bind LPS, leading to the inhibition of activation of macrophages. In this context, we found a reduction of LPS binding capacity on differentiated human THP-1 macrophages after 3 h treatment with LPS that has been co-incubated (24 h) with 100 µg/mL KO (Figure 2), by reduction of the fluorescence intensity at a concentration of 1 µg/mL LPS (Figure 2). These data support studies showing less LPS internalization into murine macrophages after 24 h treatment, when LPS was pre-exposed to high-density lipoprotein (HDL) [4]. In agreement with these studies, we used human LDL as binding control and we found that LDL reduced the binding of LPS on macrophages (Figures 4 and 5). Moreover, KO significantly reduced the LPS-binding on macrophages over a range of LPS concentrations when both had been previously co-incubated and, afterwards, were applied to the macrophages. Human serum or LDL inactivate endotoxins and inhibit the IL-1β release in LPS-activated monocytes [38]. Consequently, this observed KO characteristic may be a sign of a possible LPS adsorption, inactivation, or hiding of the binding sites. Using the LPS antagonist *Rhodobacter sphaeroides*, which binds to the TLR4 but does not induce TLR4 signaling, we can confirm the specificity of the activation (Figures 4 and 5). After co-incubation (24 h) of LPS and KO (12.5, 25 or 50 µg/mL), LPS lost its pro-inflammatory capacity and inhibited the TNF-α release. KO alone or LPS antagonist did not affect the TNF-α production, and confirmed that KO has beneficial, and not detrimental, effects. Therefore, our results impressively indicate that KO can efficiently prevent macrophages from being activated by LPS, including suppression of negative consequences. like the release of pro-inflammatory cytokines, such as TNF-α.

The effects we have observed with a KO emulsion are likely the result of two main mechanisms of action. First, the phopsholipids present in KO bind and neutralize LPS endotoxin. This effect has been shown previously in a study of normal human volunteers receiving an intravenous dose of *Escherichia coli* endotoxin during a 6-h PL infusion, and attenuation of the clinical and laboratory responses were directly related to PL levels in the bloodstream [39]. However, when this formulation was tested in the critical care setting in patients with severe Gram-negative sepsis, the high dose arm (1350 mg/kg by continuous infusion) had to be stopped because of the increased incidence of life-threatening significant adverse events and obvious futilty to show a survival advantage [40]. The PL emulsion used in this investigation consisted of 92.5% soy-based, PL, and 7.5% soy triglycerides. Although this formulation contained PLs capable of binding endotoxin, the dosing was probably excessive. Additionally, the fatty acid profile of soybean oil-derived triacylglycerols and phosphoglycerides mainly contain pro-inflammatory omega-6 fatty acids (i.e., linoleic acid). In contrast, the KO-based PL emulsion in the present study contained both the necessary PL to bind the endotoxin, plus the anti-inflammatory omega-3 fatty acids to modulate eicosanoid metabolism, and they clearly exhibit positive synergistic effects.

Finally, the KO emulsion used in this study was a crude formulation made from an unrefined natural source that is widely found in oral supplements and, thus, would be unsuitable for intravenous administration. Therefore, our study provided an indication of a proof-of-concept with this phospholipid-omega-3 combination in a cell culture model. We clearly recognize that for such a product to be a potentially viable injectable emulsion in the clinical setting, the crude KO would need to undergo refinement steps (similar to that applied to fish oil) [41] to concentrate the PL in amounts and levels of purity similar to currently available and widely used egg PL which, for example, contain at least 80% phosphatides (versus ~40%, and about equal amounts of triglycerides). In addition, since the triglycerides present in crude KO are essentially devoid of omega-3 fatty acids, the ideal injectable product would also include fish oil triglcyerides enriched with omega-3 fatty acids for maximal therapeutic efficacy.

We conclude that KO emulsion inhibits the LPS-binding on macrophage-TLR4 and, thus, the TNF-α release induced by LPS in vitro. The addition of omega-3 fatty acids potentiates the therapeutic actions by reducing the intensity of the systemic inflammatory response. These properties may be beneficial for patients under intensive care with septicemia.

4. Materials and Methods

4.1. Krill Oil-In-Water Emulsion

Three separate batches of a 5% KO-in-water emulsion were aseptically prepared in the laboratory and sterilized prior to use. This was done to be sure the emulsions could be successfully made using a crude source of KO, since a pharmaceutical-quality grade, suitable for parenteral administration, does not exist. The mean droplet size of all the emulsions was approximately 190 nm and, thus, they were considered pharmaceutically equivalent. The composition of the final emulsions are shown in Table 1.

Table 1. Krill oil-in-water emulsion.

Ingredient	Amount/1000 mL
Glycerol	25.0 g
Krill Oil *	50.0 g
Sterile Water for Injection	1000 mL
pH	6.89 ± 0.06

* Virgin Krill Oil, LLC, Braintree, MA, USA, Lot #1107091, Key Ingredient Totals (w/w): n-3-FAs = 26%; PLs = 41.2%.

4.2. Cells and Culture Conditions

The in vitro experiments were performed using the THP-1 (human acute monocytic leukemia) cell line (DSMZ GmbH, Braunschweig, Germany), cultured in 90% RPMI-1640 (PAA GmbH, Cölbe, Germany), 10% FBS (PAA GmbH); 100 U/mL penicillin; and 0.1 mg/mL streptomycin (PAA GmbH). All experiments were carried out in medium with 10% FBS.

4.3. Determination of LPS and KO Emulsion Cytotoxicity

THP-1 (5×10^4) cells were seeded in 96-well plates (BD Falcon™, Becton Dickinson GmbH, Heidelberg, Germany). After differentiation with 0.1 µg/mL phorbol-12-myristate-13-acetate (PMA) Sigma-Aldrich, St. Louis, MO, USA), the medium was changed and the macrophages treated with 0.05 µg/mL-10 µg/mL ultrapure biotinylated lipopolysaccharide (LPS-EB) from *E. coli* O111:B4 (Cayla-InvivoGen Europe, Toulouse, France), which is recognized only by toll-like receptor-4 (TLR4); KO 5 µg/mL to 250 µg/mL, or glycerol. After 24 h KO or 4 h LPS treatment, viability was assessed using PrestoBlue™ reagent (Invitrogen-Life Technologies GmbH, Darmstadt, Germany). After 1 h the optical density (OD) was measured at 570 nm/600 nm with a SUNRISE ELISA-reader (Tecan Salzburg, Austria). Results are expressed as the % of viability/survival (OD570 nm/600 nm of samples × 100/OD570 nm/600 nm of control without substances).

4.4. Determination of the LPS Binding on Differentiated THP-1 Macrophages, Effect of the Treatment with KO Emulsion, Detected by Fluorescence Microscopy

THP-1 (5 × 10^4) cells were seeded in 100 μL medium/well in 96-well plates (BD Falcon™). After eight days of differentiation into macrophages using 0.1 μg/mL PMA, the medium was changed and the macrophages were treated with 1 μg/mL or 5 μg/mL ultrapure LPS-EB-biotin alone or with KO. After 3 h the cells were fixed with 1% paraformaldehyde (PFA/PBS) for 20 min and incubated afterwards with streptavidin-Cy3 (Dianova, Hamburg, Germany). Digitalized images were obtained using an inverted microscope Eclipse-TS100 (Nikon GmbH, Düsseldorf, Germany) and an AxioCamMRc/AxioVision digital imaging system (Carl Zeiss GmbH, Jena, Germany).

4.5. Determination of the LPS Binding on Differentiated THP-1 Macrophages Detected by Spectrophotometry

THP-1 (5 × 10^4) cells were seeded as described above and treated with 0.1 μg/mL, 1 μg/mL, or 5 μg/mL ultrapure LPS-EB-biotin. After 3 h the macrophages were fixed with 1% PFA/PBS for 20 min, then incubated with streptavidin-biotinylated horseradish peroxidase (HRP-streptavidin-biotin) complex Amersham (GE Healthcare Europe GmbH, Freiburg, Germany). Subsequently, the macrophages were incubated with 50 μL peroxidase substrate Sigma Fast™ (OPD) (Sigma-Aldrich, St. Louis, MO, USA) for 30 min at RT. The reaction was stopped with 25 μL 3N HCl, and the absorbance was measured at 490 nm/655 nm. Afterwards, the cells were stained with crystal violet solution (0.04% paraformaldehyde crystal violet in 4% (*v/v*)). The OPD absorbance was normalized against the crystal violet absorbance measured at 595 nm/660 nm.

4.6. Inhibitory Effects of the KO Emulsion on the LPS Binding Capacity

PMA-differentiated THP-1 macrophages were treated for 3 h with the LPS-EB-biotin and KO mixture (both previously incubated together at 4 °C overnight), afterwards the cells were fixed with 1% PFA-PBS for 20 min, washed with PBS, incubated with streptavidin-biotin-complex-HRP (Amersham, GE Healthcare), afterwards with OPD and measured as described above. The positive control was performed by incubation of macrophages with ultrapure LPS-EB-biotin alone and the negative controls incubated only with medium or with KO. In humans, low-density lipoproteins (LDLs) may bind LPS and inactivate it; we used native LDL (nLDL, HoelzelDiagnostika GmbH, Cologne, Germany) as a control to compare the adsorption properties of KO [38]. The antagonist LPS from Rhodobactersphaeroides (LPS-RS) was used as control of the TLR4 specific binding by competitive inhibition at 100-fold excess of the agonist LPS-EB.

4.7. Effects of the KO Emulsion on TNF-α Release

The release of TNF-α was determined using ELISA. Therefore, 5 × 10^5 THP-1 cells were seeded in 24-well plates (BD Falcon™). After five days of PMA-induced differentiation (0.1 μg/mL), the macrophages were treated 3 h with LPS-EB and KO (previously incubated as described). After the treatment, the culture medium was harvested and centrifuged at 500× *g* (5 min). The cells were homogenized in RIPA buffer (Cell Signaling Technology, Inc., Danvers, MA, USA) for protein quantification using the bicinchoninic acid assay (Thermo Fisher Scientific, Bonn, Germany). Human TNF-α was determined in the supernatant using the DuoSet-ELISA kit (R&D Systems Europe, Ltd., Abingdon, UK) according to the manufacturer's instructions; 96-well NUNC MaxiSorp™ (Thermo Fisher Scientific) were used. The amount of TNF-α was normalized with the protein content.

4.8. Statistical Analyses

The SigmaPlot®-12 software (Systat Software GmbH, Erkrath, Germany) was used to carry out statistical analyses by one-way analysis of variance test (ANOVA) using Dunnett's method appropriate for multiple comparisons versus the control group. Data are shown as mean + SEM.

Acknowledgments: The authors thank Andrea Cordes for the excellent technical assistance, as well as Ellen Essen and Gabriella Stauch for preparation of the manuscript.

Author Contributions: G.A. Bonaterra, D. Driscoll and R. Kinscherf contributed to the conception and design of the research. G.A. Bonaterra contributed to the acquisition, analysis, interpretation of the data, and drafted the manuscript. H. Schwarzbach analyzed the data and image processing. D. Driscoll and R. Kinscherf critically revised the manuscript, agreed to be fully accountable for ensuring the integrity and accuracy of the work, and read and approved the final manuscript.

Conflicts of Interest: D. Driscoll has been awarded a patent, assigned to Stable Solutions LLC, entitled: Therapeutic application of parenteral krill oil. US 8,895,074 B2. Financial disclosure: This work was supported by B. Braun Melsungen AG, Germany. The founding sponsors had no role in the design of the study; in the collection, analyses, or interpretation of data; in the writing of the manuscript, and in the decision to publish the results

References

1. Bochsler, P.N.; Maddux, J.M.; Neilsen, N.R.; Slauson, D.O. Differential binding of bacterial lipopolysaccharide to bovine peripheral-blood leukocytes. *Inflammation* **1993**, *17*, 47–56. [CrossRef] [PubMed]
2. Adams, J.L.; Czuprynski, C.J. Bacterial lipopolysaccharide induces release of tumor necrosis factor-alpha from bovine peripheral blood monocytes and alveolar macrophages in vitro. *J. Leukoc. Biol.* **1990**, *48*, 549–556. [PubMed]
3. Sprague, A.H.; Khalil, R.A. Inflammatory cytokines in vascular dysfunction and vasculardisease. *Biochem. Pharmacol.* **2009**, *78*, 539–552. [CrossRef] [PubMed]
4. Cavaillon, J.M.; Fitting, C.; Haeffner-Cavaillon, N.; Kirsch, S.J.; Warren, H.S. Cytokine response by monocytes and macrophages to free and lipoprotein-bound lipopolysaccharide. *Infect. Immun.* **1990**, *58*, 2375–2382. [PubMed]
5. Emancipator, K.; Csako, G.; Elin, R.J. In vitro inactivation of bacterial endotoxin by human lipoproteins and apolipoproteins. *Infect. Immunol.* **1992**, *60*, 596–601.
6. Eggesbo, J.B.; Hjermann, I.; Hostmark, A.T.; Kierulf, P. LPS induced release of IL-1 beta, IL-6, IL-8 and TNF-alpha in EDTA or heparin anticoagulated whole blood from persons with high or low levels of serum HDL. *Cytokine* **1996**, *8*, 152–160. [CrossRef] [PubMed]
7. Baumberger, C.; Ulevitch, R.J.; Dayer, J.M. Modulation of endotoxic activity oflipopolysaccharide by high-density lipoprotein. *Pathobiology* **1991**, *59*, 378–383. [CrossRef] [PubMed]
8. Wu, A.; Hinds, C.J.; Thiemermann, C. High-density lipoproteins in sepsis and septic shock: Metabolism, actions, and therapeutic applications. *Shock* **2004**, *21*, 210–221. [CrossRef] [PubMed]
9. Shor, R.; Wainstein, J.; Oz, D.; Boaz, M.; Matas, Z.; Fux, A.; Halabe, A. Low serum LDL cholesterol levels and the risk of fever, sepsis, and malignancy. *Ann. Clin. Lab. Sci.* **2007**, *37*, 343–348. [PubMed]
10. Calder, P.C. *n*-3 polyunsaturated fatty acids, inflammation, and inflammatory diseases. *Am. J. Clin. Nutr.* **2006**, *83* (Suppl. 6), 1505S–1519S. [PubMed]
11. Arterburn, L.M.; Hall, E.B.; Ojen, H. Distribution, interconversion and dose response of *n*-3 fatty acids in humans. *Am. J. Clin. Nutr.* **2006**, *83* (Suppl. 6), 1467S–1476S. [PubMed]
12. Nash, S.M.B.; Schlabach, M.; Nichols, P.D. A nutritional-toxicological assessment of antarctic krill oil versus fish oil dietary supplements. *Nutrients* **2014**, *6*, 3382–3402. [CrossRef] [PubMed]
13. Wichmann, M.W.; Thul, P.; Czarnetski, H.D.; Morlion, B.J.; Kemen, M.; Jauch, K.W. Evaluation of clinical safety and beneficial effects of a fish oil containing lipid emulsion (Lipoplus, MLF541): Data from a prospective, randomized, multicenter trial. *Crit. Care Med.* **2007**, *35*, 700–706. [CrossRef] [PubMed]
14. Heller, A.R.; Rössler, S.; Litz, R.J.; Stehr, S.N.; Heller, S.C.; Koch, R.; Koch, T. Omega-3 fatty acids improve diagnosis-related clinical outcome. *Crit. Care Med.* **2006**, *34*, 972–979. [CrossRef] [PubMed]
15. Barbosa, V.M.; Miles, E.A.; Calhau, C.; Lafuente, E.; Calder, P.C. Effects of a fish oil containing lipid emulsion on plasma phospholipid fatty acids, inflammatory markers, and clinical outcomes in septic patients: A randomized, controlled clinical trial. *Crit. Care* **2010**, *14*, R5. [CrossRef] [PubMed]

16. Mayer, K.; Fegbeutel, C.; Hattar, K.; Sibelius, U.; Krämer, H.J.; Heuer, K.U.; Temmesfeld-Wollbrück, B.; Gokorsch, S.; Grimminger, F.; Seeger, W. Omega-3 vs. omega-6 lipid emulsions exert differential influence on neutrophils in septic shock patients: Impact on plasma fatty acids and lipid mediator generation. *Intensive Care Med.* **2003**, *29*, 1472–1481. [CrossRef] [PubMed]

17. Tsutsumi, R.; Horikawa, Y.T.; Kume, K.; Tanaka, K.; Kasai, A.; Kadota, T.; Tsutsumi, Y.M. Peptide-Based Formulas With ω-3 Fatty Acids Are Protective in LPS-Mediated Sepsis. *JPEN J. Parenter. Enter. Nutr.* **2015**, *39*, 552–561. [CrossRef] [PubMed]

18. Calder, P.C. Marine omega-3 fatty acids and inflammatory processes: Effects, mechanisms and clinical relevance. *Biochim. Biophys. Acta* **2015**, *1851*, 469–484. [CrossRef] [PubMed]

19. Ramsvik, M.S.; Bjørndal, B.; Bruheim, I.; Bohov, P.; Berge, R.K. A Phospholipid-Protein Complex from Krill with Antioxidative and Immunomodulating Properties Reduced Plasma Triacylglycerol and Hepatic Lipogenesis in Rats. *Mar. Drugs* **2015**, *13*, 4375–4397. [CrossRef] [PubMed]

20. Ghasemifard, S.; Turchini, G.M.; Sinclair, A.J. Omega-3 long chain fatty acid "bioavailability": A review of evidence and methodological considerations. *Prog. Lipid Res.* **2014**, *56C*, 92–108. [CrossRef] [PubMed]

21. Cansell, M.; Nacka, F.; Combe, N. Marine lipid-based liposomes increase in vivo FA bioavailability. *Lipids* **2003**, *38*, 551–559. [CrossRef] [PubMed]

22. Bistrian, B.R. Clinical aspects of essential fatty acid metabolism: Jonathan Rhoads Lecture. *JPEN J. Parenter. Enter. Nutr.* **2003**, *27*, 168–175. [CrossRef] [PubMed]

23. Wohlmuth, C.; Dünser, M.W.; Wurzinger, B.; Deutinger, M.; Ulmer, H.; Torgersen, C.; Schmittinger, C.A.; Grander, W.; Hasibeder, W.R. Early fish oil supplementation and organ failure in patients with septic shock from abdominal infections: A propensitymatchedcohort study. *JPEN J. Parenter. Enter. Nutr.* **2010**, *34*, 431–437. [CrossRef] [PubMed]

24. Su, G.L. Lipopolysaccharides in liver injury: Molecular mechanisms of Kupffer cell activation. *Am. J. Physiol. Gastrointest. Liver Physiol.* **2002**, *283*, G256–G265. [CrossRef] [PubMed]

25. Yegenaga, I.; Hoste, E.; Van Biesen, W.; Vanholder, R.; Benoit, D.; Kantarci, G.; Dhondt, A.; Colardyn, F.; Lameire, N. Clinical characteristics of patients developing ARF due to sepsis/systemic inflammatory response syndrome: Results of a prospective study. *Am. J. Kidney Dis.* **2004**, *43*, 817–824. [CrossRef] [PubMed]

26. Angus, D.C.; Linde-Zwirble, W.T.; Lidicker, J.; Clermont, G.; Carcillo, J.; Pinsky, M.R. Epidemiology of severe sepsis in the United States: Analysis of incidence, outcome, and associated costs of care. *Crit. Care Med.* **2001**, *29*, 1303–1310. [CrossRef] [PubMed]

27. Annane, D.; Belissant, E.; Cavaillon, J.M. Septic shock. *Lancet* **2005**, *365*, 63–78. [CrossRef]

28. Thomas, S.; Balasubramanian, K.A. Role of intestine in postsurgical complications: Involvement of free radicals. *Free Radic. Biol. Med.* **2004**, *36*, 745–756. [CrossRef] [PubMed]

29. Coquerel, D.; Kušíková, E.; Mulder, P.; Coëffier, M.; Renet, S.; Dechelotte, P.; Richard, V.; Thuillez, C.; Tamion, F. Omega-3 polyunsaturated fatty acids delay the progression of endotoxic shock-induced myocardial dysfunction. *Inflammation* **2013**, *36*, 932–940. [CrossRef] [PubMed]

30. Yates, C.M.; Calder, P.C.; Ed Rainger, G. Pharmacology and therapeutics of omega-3 polyunsaturated fatty acids in chronic inflammatory disease. *Pharmacol. Ther.* **2014**, *14*, 272–782. [CrossRef] [PubMed]

31. Bonaterra, G.A.; Wakenhut, F.; Röthlein, D.; Wolf, M.; Bistrian, B.R.; Driscoll, D.; Kinscherf, R. Cytoprotection by omega-3 fatty acids as a therapeutic drug vehicle when combined with nephrotoxic drugs in an intravenous emulsion: Effects on intraglomerular mesangial cells. *Toxicol. Rep.* **2014**, *1*, 843–857. [CrossRef]

32. Zhao, Y.; Joshi-Barve, S.; Barve, S.; Chen, L.H. Eicosapentaenoic acid prevents LPS induced TNF-α expression by preventing NF-κB activation. *J. Am. Coll. Nutr.* **2004**, *23*, 71–78. [CrossRef] [PubMed]

33. Calder, P.C. Use of fish oil in parenteral nutrition: Rationale and reality. *Proc. Nutr. Soc.* **2006**, *65*, 264–277. [CrossRef] [PubMed]

34. Trebble, T.; Arden, N.K.; Stroud, M.A.; Wootton, S.A.; Burdge, G.C.; Miles, E.A.; Ballinger, A.B.; Thompson, R.L.; Calder, P.C. Inhibition of tumour necrosis factor-α and interleukin-6 production by mononuclear cells following dietary fish-oil supplementation in healthy men and response to antioxidant co-supplementation. *Br. J. Nutr.* **2003**, *90*, 405–412. [CrossRef] [PubMed]

35. Wallace, F.A.; Miles, E.A.; Calder, P.C. Comparison of the effects of linseed oil and different doses of fish oil on mononuclear cell function in healthy human subjects. *Br. J. Nutr.* **2003**, *89*, 679–689. [CrossRef] [PubMed]

36. Mascioli, E.A.; Leader, L.; Flores, E.; Trimbo, S.; Bistrian, B.; Blackburn, G. Enhanced survival to endotoxin in guinea pigs fed iv fish oil emulsion. *Lipids* **1988**, *23*, 623–625. [CrossRef] [PubMed]

37. Kulabukhov, V.V. Use of an endotoxin adsorber in the treatment of severe abdominal sepsis. *Acta Anaesthesiol. Scand.* **2008**, *52*, 1024–1025. [CrossRef] [PubMed]

38. Weinstock, C.; Ullrich, H.; Hohe, R.; Berg, A.; Baumstark, M.W.; Frey, I.; Northoff, H.; Flegel, W.A. Low density lipoproteins inhibit endotoxin activation of monocytes. *Arterioscler. Thromb.* **1992**, *12*, 341–347. [CrossRef] [PubMed]

39. Gordon, B.R.; Parker, T.S.; Levine, D.M.; Feuerbach, F.; Saal, S.D.; Sloan, B.J.; Chu, C.; Stenzel, K.H.; Parrillo, J.E.; Rubin, A.L. Neutralization of endotoxin by a phospholipid emulsion in healthy volunteers. *J. Infect. Dis.* **2006**, *191*, 1515–1522. [CrossRef] [PubMed]

40. Dellenger, R.P.; Tomayko, J.F.; Angus, D.C.; Opal, S.; Cupo, M.A.; McDermott, S.; Ducher, A.; Calandra, T.; Cohen, J. Lipid Infusion and Patient Outcomes in Sepsis (LIPOS) Investigators. Efficacy and safety of a phospholipid emulsion (GR270773) in gram-negative severe sepsis: Results of a phase II multicenter, randomized, placebo-controlled, dose finding trial. *Crit. Care* **2009**, *37*, 2929–2938. [CrossRef] [PubMed]

41. European Food Safety Authority (EFSA). Panel on Biological Hazards. Scientific opinión on fish oil for human consumption. *EFSA J.* **2010**, *8*, 1874.

marine drugs

MDPI

Article

Biosynthesis of Polyunsaturated Fatty Acids in *Octopus vulgaris*: Molecular Cloning and Functional Characterisation of a Stearoyl-CoA Desaturase and an Elongation of Very Long-Chain Fatty Acid 4 Protein

Óscar Monroig [1,*], Rosa de Llanos [2,3], Inmaculada Varó [4], Francisco Hontoria [4], Douglas R. Tocher [1], Sergi Puig [3] and Juan C. Navarro [4]

[1]   Institute of Aquaculture, Faculty of Natural Sciences, University of Stirling, Stirling FK9 4LA, Scotland, UK; d.r.tocher@stir.ac.uk
[2]   School of Applied Sciences, Edinburgh Napier University, Edinburgh EH11 4BN, Scotland, UK; R.DeLlanos@napier.ac.uk
[3]   Instituto de Agroquímica y Tecnología de Alimentos (IATA-CSIC), Paterna, Valencia 46980, Spain; spuig@iata.csic.es
[4]   Instituto de Acuicultura Torre de la Sal (IATS-CSIC), Ribera de Cabanes, Castellón 12595, Spain; inma@iats.csic.es (I.V.); hontoria@iats.csic.es (F.H.); jcnavarro@iats.csic.es (J.C.N.)
*   Correspondence: oscar.monroig@stir.ac.uk; Tel.: +44-1786-467892

Academic Editors: Rosário Domingues, Ricardo Calado and Pedro Domingues
Received: 14 February 2017; Accepted: 16 March 2017; Published: 21 March 2017

Abstract: Polyunsaturated fatty acids (PUFAs) have been acknowledged as essential nutrients for cephalopods but the specific PUFAs that satisfy the physiological requirements are unknown. To expand our previous investigations on characterisation of desaturases and elongases involved in the biosynthesis of PUFAs and hence determine the dietary PUFA requirements in cephalopods, this study aimed to investigate the roles that a stearoyl-CoA desaturase (Scd) and an elongation of very long-chain fatty acid 4 (Elovl4) protein play in the biosynthesis of essential fatty acids (FAs). Our results confirmed the *Octopus vulgaris* Scd is a $\Delta 9$ desaturase with relatively high affinity towards saturated FAs with $\geq C_{18}$ chain lengths. Scd was unable to desaturate 20:1n-15 ($^{\Delta 5}$20:1) suggesting that its role in the biosynthesis of non-methylene interrupted FAs (NMI FAs) is limited to the introduction of the first unsaturation at $\Delta 9$ position. Interestingly, the previously characterised $\Delta 5$ fatty acyl desaturase was indeed able to convert 20:1n-9 ($^{\Delta 11}$20:1) to $^{\Delta 5,11}$20:2, an NMI FA previously detected in octopus nephridium. Additionally, Elovl4 was able to mediate the production of 24:5n-3 and thus can contribute to docosahexaenoic acid (DHA) biosynthesis through the Sprecher pathway. Moreover, the octopus Elovl4 was confirmed to play a key role in the biosynthesis of very long-chain (>C_{24}) PUFAs.

Keywords: biosynthesis; elongation of very long-chain fatty acids 4 protein; non-methylene-interrupted fatty acids; polyunsaturated fatty acids; *Octopus vulgaris*; stearoyl-CoA desaturase

1. Introduction

Cephalopods have been regarded as promising candidates for the diversification of marine aquaculture due to their great commercial interest [1]. Despite significant progress made over the last decade, culture of cephalopod species with pelagic paralarval stages like the common octopus *Octopus vulgaris* is still challenging due to the massive mortalities occurring upon the settlement phase [2]. The specific factors causing such mortalities of paralarvae remain unclear, although it has become increasingly obvious that nutritional issues associated with inadequate supply of essential nutrients such as lipids are crucial to ensure normal growth and development of *O. vulgaris* paralarvae and ultimately improve their viability [3].

Previous investigations postulated that polyunsaturated fatty acids (PUFAs) are essential nutrients for the common octopus [4,5]. However, the specific PUFAs that satisfy the physiological requirements were not determined, partly due to the difficulties in running nutritional trials on octopus paralarvae. In order to provide insights to the endogenous capability for PUFA biosynthesis in *O. vulgaris*, we have recently conducted a series of studies aiming to identify and characterise the function of genes encoding enzymes that mediate the conversions in the PUFA biosynthetic pathways. First, we identified a fatty acyl desaturase (Fad) cDNA sequence with homology to the vertebrate Fads family [6], enzymes that participate in long-chain (C_{20-24}) PUFA (LC-PUFA) biosynthetic pathways [7–9]. The expression of the common octopus Fad in yeast demonstrated that the enzyme was a $\Delta 5$ desaturase ($\Delta 5$ Fad) with high efficiency towards both saturated and polyunsaturated fatty acid (FA) substrates [6]. Thus, the *O. vulgaris* $\Delta 5$ Fad was able to desaturate the yeast endogenous saturated FAs 16:0 and 18:0 to the corresponding monoenes 16:1n-11 ($^{\Delta 5}$16:1) and 18:1n-13 ($^{\Delta 5}$18:1), respectively. Furthermore, the *O. vulgaris* $\Delta 5$ Fad efficiently desaturated the PUFA 20:4n-3 and 20:3n-6 to the $\Delta 5$ desaturation products eicosapentaenoic acid (EPA, 20:5n-3) and arachidonic acid (ARA, 20:4n-6), respectively (Figure 1).

A second study provided further evidence of the existence of an active PUFA biosynthetic system in the common octopus [10]. Thus, a cDNA encoding a protein with high homology to an elongation of very long-chain fatty acids (Elovl) protein was isolated [10]. Phylogenetic analysis comparing the amino acid (aa) sequence of the *O. vulgaris* Elovl with other elongases from molluscs and vertebrates clearly showed that the common octopus Elovl, as well as other putative elongases from molluscs, was grouped as a basal cluster of the vertebrate Elovl2 and Elovl5 families [10]. Consequently, such an elongase has been termed "Elovl5/2" [11] or "Elovl2/5" [12,13]. Regarding its function, the common octopus Elovl2/5 exhibited substrate specificities resembling those of vertebrate Elovl5 but not Elovl2, as it efficiently elongated C_{18-20} PUFAs [10] but had no activity towards C_{22} substrates. This was hypothesised as one of the reasons accounting for the inability of cephalopods to biosynthesise docosahexaenoic acid (DHA; 22:6n-3) (Figure 1) [10,11] through the so-called "Sprecher pathway" requiring the production of 24:5n-3 as an intermediate in the pathway [14]. Among alternative Elovl-like enzymes with a role in the biosynthesis of LC-PUFAs such as DHA in the common octopus, the Elovl4 elongase is an interesting candidate, as studies on teleost orthologues have shown that Elovl4 can efficiently catalyse the elongation of 22:5n-3 to 24:5n-3 [15–19], and recent studies have further demonstrated a similar elongation ability in molluscs [12].

The above studies on *O. vulgaris* [6,10], as well as those on homologous genes from the common cuttlefish *Sepia officinalis* [20], have enabled us to predict the biosynthetic pathways of PUFAs in cephalopods (Figure 1). Beyond the biosynthesis of standard PUFAs, i.e., FAs whose double bonds are always separated by a methylene group (-CH_2-) [9], one can predict that some pathways involving the $\Delta 5$ Fad and Elovl2/5 lead to the production of so-called "non-methylene-interrupted FAs" (NMI FAs), a particular type of PUFA that had been previously reported in other molluscan classes (bivalves, gastropods), as well in sponges, echinoderms and other phyla [21–23]. Analyses performed in wild-caught specimens of *O. vulgaris* confirmed that the polar lipid fractions of nephridium, male gonad, eye and caecum contained NMI FAs identified as $^{\Delta 5,11}$20:2, $^{\Delta 7,13}$20:2, $^{\Delta 5,11,14}$20:3 and $^{\Delta 7,13}$22:2 [10]. From the unsaturation pattern of these compounds, it became clear that, in addition to $\Delta 5$ Fad, a further desaturase with $\Delta 9$ activity was likely involved in the NMI FA biosynthetic pathways accounting for the $\Delta 5,9$ unsaturation patterns typically found among these compounds [22,23]. The stearoyl-CoA desaturase (Scd), an enzyme that is expressed in virtually all living organisms [24], has $\Delta 9$ desaturation capability and thus appears to play a role in NMI FA biosynthesis [7].

Our overall aim is to characterise the biosynthetic pathways of PUFAs including NMI FAs in cephalopods. Using the common octopus *O. vulgaris* as model species, we herein isolated two cDNAs, namely Scd and Elovl4 sequences, and characterised their functions by heterologous expression in yeast. In order to establish the mechanisms accounting for biosynthesis of Δ5,9 dienes (NMI FA) we further investigated the roles that the herein characterised Scd and the previously reported Δ5 Fad [6] play within these pathways.

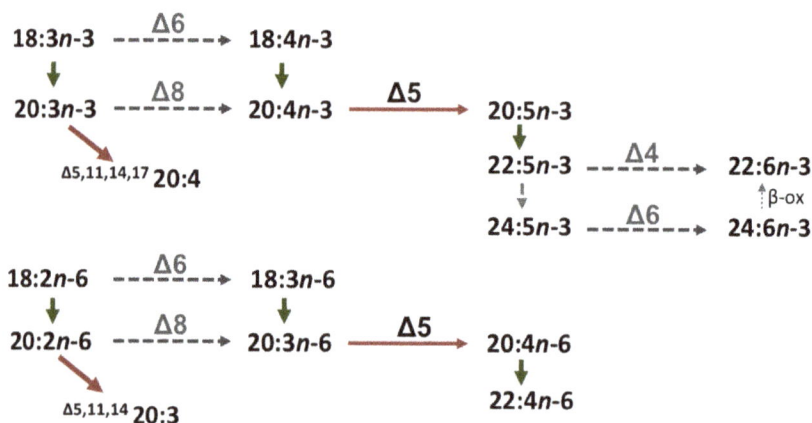

Figure 1. Model of biosynthetic pathways of polyunsaturated fatty acids in cephalopods. Enzymatic activities shown in the diagram are predicted from heterologous expression in yeast (*Saccharomyces cerevisiae*) of fatty acyl desaturases (red arrows) and elongation of very long-chain fatty acid (Elovl) proteins (green arrows) from *Octopus vulgaris* [6,10] and *Sepia officinalis* [20]. Dotted arrows indicate reactions that have not yet been demonstrated prior to the present study. β-ox, partial β-oxidation.

2. Results

2.1. Octopus vulgaris Scd Sequence

The *O. vulgaris* Scd-like cDNA consisted of a 981-bp open reading frame (ORF) encoding a putative protein of 326 amino acids (aa) with a predicted molecular weight of 37.9 kDa. Its sequence was deposited in the GenBank database with the accession number JX310655. In common with other Scd proteins, the *O. vulgaris* putative Scd possessed three histidine boxes (HXXXH, HXXHH and HXXHH), four membrane-spanning regions rich in hydrophobic aa, and lacked the cytochrome b5 domain characteristic of Fads (Figure 2).

The deduced aa sequence from the common octopus Scd was 50.8%–53.4% identical to Scd sequences from vertebrates including *Homo sapiens* (NP_005054.3), *Gallus gallus* (NP_990221.1) and *Xenopus laevis* (NP_001087809.1), and 40.7% and 43.9% identical to Scd from the nematode *Caenorhabditis elegans* FAT-5 (NP_507482.1) and FAT-6 (NP_001255595.1), respectively. When compared to mollusc Scd-like protein sequences, the *O. vulgaris* Scd showed relatively high identity scores with orthologues from *Octopus bimaculoides* (99.0%) (XP_014788510.1), *Crassostrea gigas* (63.0%) (XP_011452904.1) and *Lottia gigantea* (60.7%). Importantly, identities between the newly cloned Scd and several Fads desaturases including the Δ5-like desaturase identified in *O. vulgaris* [6] were below 16.0%.

```
Octopus vulgaris Scd          ------MSPRNLVTEPPP---AEDELHPGELNIEPVITEPTHTQKRPPKIVWRNVILMT 51
Octopus bimaculoides Scd      ------MSPRNLVTEPPP---AEDELHPGELNIEPVITEPTHTQKRPPKIVWRNVILMT 51
Caenorhabditis elegans FAT-5  -----------MTQKVDAIISKQFLAADLNEIRQMQGSKKQ-VIKQEIVWRNVALFV 47
Caenorhabditis elegans FAT-6  ---MTVKTRSNIAKKPEKDGGPETQYLAVDPNEIIQLQEESKKI-PYKQSIVWRNALFA 56
Saccharomyces cerevisiae Ole1p MVSVEFDKKGNEKKSNLDRLLEKDNQEKEEAKTKIHISEQPWTLNNWHQHLNWLNMVLVC 60
                                                                             -------
                                                                                   I

                                          HXXXH
Octopus vulgaris Scd          VLFLSPLYSIT-ILPLAHFYTLIWSLCVYMYAFGIGITAGSHRLWAHRAYKAKLPLEALLA 110
Octopus bimaculoides Scd      VLHLSALYSIT-ILPLAHCYTLIWSLCVYMYAIGIGITAGSHRLWAHRAYKAKLPLEALLA 110
Caenorhabditis elegans FAT-5  ALHFGALVGGYQLVFQRKWAQVGWVFLLHTLGSMEVTGARHLWAHRAYKATLSWMVFLM 107
Caenorhabditis elegans FAT-6  ALFFAAAIGHYQLIFEAKWQVIFTFLLYVFGIFGITAGAHRLWSLKSYKATTPLEGLFLM 116
Saccharomyces cerevisiae Ole1p GMPMLGWYFALSGKVPLHLNVFLFSVFYYAVGGVSITAGYHRLWSHRSYSAHWPLRLFYA 120
                                                                                        II

                              HXXHH
Octopus vulgaris Scd          SMQSAAFQNDIYDWSRDHHVHHKYSETDADPHNAKRGFFSHVGWLLVRKHPDVTARGKL 170
Octopus bimaculoides Scd      SMQSAAFQNDIYDWSRDHHVHHKYSETDADPHNAKRGFFSHVGWLLVRKHPDVTARGKL 170
Caenorhabditis elegans FAT-5  LINSIAFQNDIIDWARDHRCHHKWTDTDADPHSTNGKMPFHWGWLLVRKIDQLKIQGGK 167
Caenorhabditis elegans FAT-6  ILNNIALQNDVIEWARDHHCHHKWTDTDADPHNTTIGFFHSGWLLVRKHPQVKEQGAK 176
Saccharomyces cerevisiae Ole1p IFGSASVEGSAKWWGHSHRIHHRYTDILRDPYDARGLWYSHMGWMLLKPNGKYKAR--- 177

Octopus vulgaris Scd          LLTNDILNDPVHFQRKYYALSWVVFCLAIPTLVGYIEWNENLWNAYFLAGVLRYCGGLN 230
Octopus bimaculoides Scd      LDTNDILNDPVHFQRKYYALSWVIFCPALPTLVGYIEWNENLWNAYFLAGVLRYCGGLN 230
Caenorhabditis elegans FAT-5  LDLSDIYEDPVLMFQRNHLPLYGIFCFALPTFIPYVVLGGSAFIGFYTAALFRYGFTLH 227
Caenorhabditis elegans FAT-6  LDMSDLLSDFVGVFQRSHVFPLYILCCPILPTIIGVYFWKDTAFIGFYTAGTFRYGFTLH 236
Saccharomyces cerevisiae Ole1p ABITEMTDEDWTIRFQHRHYILLMLLTABVPTLICGYFFNDYMG--GLIYAGFIRVFVIQQ 236
                                                        III                        IV

                                                           HXXHH
Octopus vulgaris Scd          ATWLVNSAAHMWGYRPYDKRINEAENICVSLGSMGEGFHNYHHIFPQDYATSEYGWRINL 290
Octopus bimaculoides Scd      ATWLVNSAAHMWGYRPYDKRINEAENICVSLGSMGEGFHNHHHIFPQDYATSEYGWRINL 290
Caenorhabditis elegans FAT-5  ATWCINSVSHWVGWCPFDHQASSVDALWTSIAAVGEGGHNFHHTFPQDYRTSEHAEFLWN 287
Caenorhabditis elegans FAT-6  ATWCINSAAHYFCWNPYDSSITQVDNVFTTIAAVGEGGHNFHHTFPQDYSEYSLKYNW 296
Saccharomyces cerevisiae Ole1p ATFCINSMAHYIGTQPFDDRRTPRDNWITAIVTFGEGGHNFHHEFPTDYRNAIKWYQYDP 296
                              ------

Octopus vulgaris Scd          TTFPIDFMAFLEQAYDRKTIDRDTIRRRDRTSTHT-------------------- 326
Octopus bimaculoides Scd      TTFPIDFMAFLEQAYDRKTIDRDTIRRRDRTSSHT-------------------- 326
Caenorhabditis elegans FAT-5  GRVLLDFGASISMVVDRKTTPEEVQFQOCKKFSCETEREKMLHKLG----------- 333
Caenorhabditis elegans FAT-6  GRVLITAGALGLVVDRKDKACDEIHGRQVSNHSCDIQRGKSIM------------- 339
Saccharomyces cerevisiae Ole1p LKVIHYLTSLVGLADLKKFSQNAHEEALIQQEQKKINKKKAKINWGPVLTDLPMWDKQT 356
```

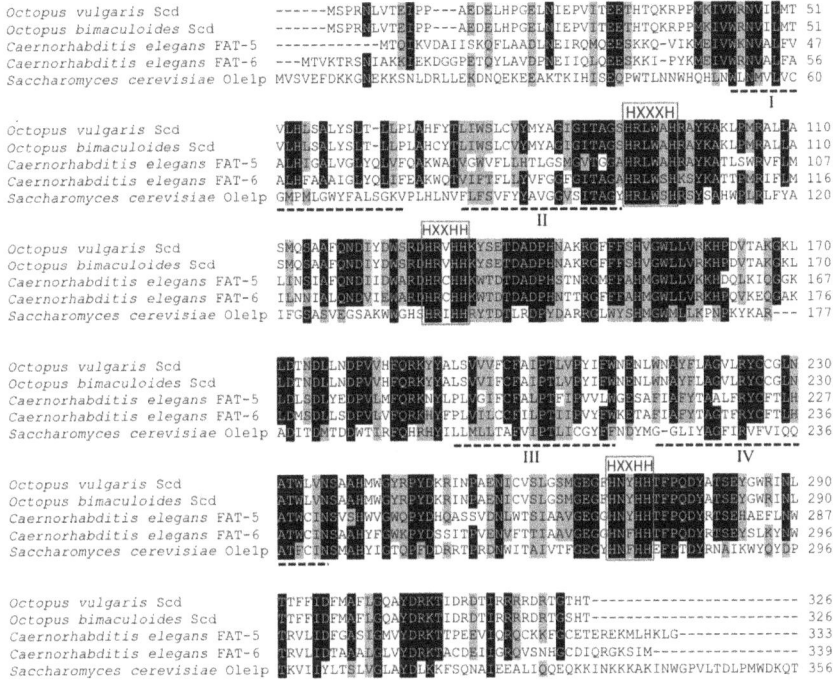

Figure 2. ClustalW amino acid alignment comparing the *Octopus vulgaris* stearoyl-CoA desaturase (Scd) with homologous sequences from the *Octopus bimaculoides* (XP_014788514.1), the *Caenorabditis elegans* FAT-5 (NP_507482.1) and the *C. elegans* FAT-6 (NP_001255595.1), and a portion of the *Saccharomyces cerevisiae* Ole1p sequence. Identical residues are shaded black and similar residues (using ClustalW2 default parameters) are shaded grey. The three histidine boxes (HXXXH, HXXHH and QXXHH) are highlighted with grey squares. Four (I–IV) trans-membrane domains predicted by Watts and Browse [25] are underlined with dashed lines.

2.2. Octopus Elovl4-Like Sequence

The Elovl4-like cDNA consisted of an ORF of 930 bp whose deduced protein had 309 aa with a predicted molecular weight of 37.9 kDa (deposited in GenBank database with accession number KJ590963). From analogy to vertebrate orthologues [15–18], five putative transmembrane domains containing hydrophobic aa stretches can be predicted (Figure 3). The common octopus putative Elovl4 contained the histidine dideoxy-binding motif HXXHH, and the putative endoplasmic reticulum (ER) retrieval signal with a histidine (H) and lysine (K) residues at the carboxyl terminus, HXKXX (Figure 3) [26].

Comparison of the deduced aa sequence of the *O. vulgaris* Elovl4 with other orthologues revealed high identity with the *O. bimaculoides* Elovl4 (93.5%) (XP_014784234.1), with remarkable lower identity scores obtained when compared with Elovl4 sequences from *L. gigantea* (57.9%) (XP_009051096.1), *C. gigas* (41.6%) (XP_011450778.1), the sea squirt *Ciona intestinalis* (48.9%) (AAV67802.1) [27], *H. sapiens* (50.8%) (NP_073563.1), *G. gallus* (49.1%) (NP_001184238.1) and *Anolis carolinensis* (47.6%) (XP_003215742.1). Identity scores of 36.6% and 36.2% were obtained by comparing the *O. vulgaris* Elovl4 with the previously characterised Elovl2/5 from *O. vulgaris* (AFM93779.1) [10] and *S. officinalis* (AKE92956.1) [20], respectively.

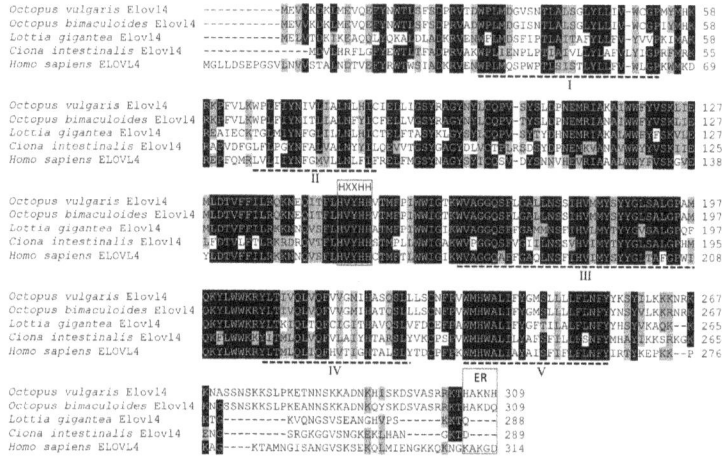

Figure 3. ClustalW amino acid alignment of the *Octopus vulgaris* Elovl4 with homologous sequences from *Octopus bimaculoides* (XP_014784234.1), *Lottia gigantea* (XP_009051096.1), *Ciona intestinalis* (AAV67802.1) and *Homo sapiens* (NP_073563.1). Identical residues are shaded black and similar residues (using ClustalW default parameters) are shaded grey. Indicated are the conserved histidine box motif HXXHH, five (I–V) putative membrane-spanning domains, and the putative endoplasmic reticulum (ER) retrieval signal.

2.3. Functional Characterisation of the Octopus Scd

Yeast *Saccharomyces cerevisiae* cells lacking the *OLE1* gene are unable to synthesise $\Delta 9$-monounsaturated FAs including palmitoleic acid (16:1n-7) and oleic acid (18:1n-9), which are essential for growth [28]. To address whether the *O. vulgaris* Scd was able to complement yeast *OLE1* function, an *S. cerevisiae ole1Δ* mutant strain (L8-14C) was transformed with a plasmid expressing the octopus Scd (p416OLE1-Scd) under the control of yeast *OLE1* promoter. The *ole1Δ* mutant yeast were also transformed with a plasmid expressing the *S. cerevisiae OLE1* (Δ9 desaturase) under the control of its own promoter (p416OLE1-OLE1) as a positive control, and with empty vector (pRS416) as a negative control. All three yeast transformants were able to grow in media supplemented with at least one supplemented FA (16:1n-7 and/or 18:1n-9) (Figure 4a–c). However, only yeast cells expressing either the *S. cerevisiae OLE1* or the *O. vulgaris* Scd grew on medium that was not supplemented with monounsaturated FAs (Figure 4d). These results indicated that the common octopus Scd complemented the function of the yeast *OLE1* desaturase.

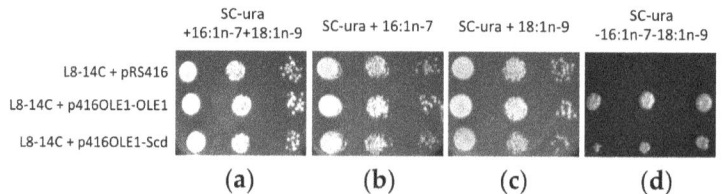

Figure 4. Complementation of the *Saccharomyces cerevisiae ole1Δ* mutant strain L8-14C with the *Octopus vulgaris* stearoyl-CoA desaturase (Scd) coding region. Yeast were transformed with the pRS416 empty vector (negative control), p416OLE1-OLE1 (expressing the *S. cerevisiae OLE1* under the control of its own promoter) and p416OLE1-Scd (expressing the *O. vulgaris* Scd under the control of *S. cerevisiae OLE1* promoter) and grown in SC medium lacking uracil (SC-ura) supplemented with 16:1n-7 and 18:1n-9 (**a**), 16:1n-7 (**b**) or 18:1n-9 (**c**), or in the absence of these fatty acids (**d**).

To analyse in more detail the complemention of yeast *ole1Δ* mutants by the *O. vulgaris* Scd, we determined the FA composition of *ole1Δ* yeast cells transformed with either p416OLE1-OLE1 (i.e., expressing the *S. cerevisiae* OLE1) or p416OLE1-Scd (i.e., expressing the *O. vulgaris* Scd), and grown in liquid medium with no exogenously supplemented FAs. Yeast expressing the *O. vulgaris* Scd grew notably slower compared to controls and formed clumps as previously described [29]. Moreover, the p416OLE1-Scd yeast contained significantly ($p \leq 0.05$) less 16:1*n*-7 and more 16:0 than control yeast expressing the *S. cerevisiae* OLE1, suggesting a low affinity of the *O. vulgaris* Scd towards 16:0. On the contrary, contents of 18:1*n*-9 showed no statistical differences between yeast cells expressing Scd or OLE1, indicative of the octopus Scd participating in the biosynthesis of 18:1*n*-9 similar to *S. cerevisiae* Ole1p. The distinctive substrate specificities of the *O. vulgaris* Scd were further emphasised by the higher 18:1*n*-9/16:1*n*-7 ratio of yeast complemented with the *O. vulgaris* Scd compared to that of *OLE1* control yeast (Table 1).

The substrate specificities of the *O. vulgaris* Scd were further investigated through an overexpression assay using the *S. cerevisiae* strain InvSc1 (Invitrogen, Paisley, UK), possessing the endogenous Ole1p. Comparison of the FA profiles between the control yeast (i.e., transformed with the empty pYES2) and yeast expressing the *O. vulgaris* Scd (i.e., transformed with pYES2-Scd) confirmed a role of the common octopus Scd in the biosynthesis of monounsaturated FAs (Table 2). Thus, significant ($p \leq 0.05$) increases in the contents of 18:1*n*-9 (oleic acid) in pYES2-Scd yeast compared to control yeast were observed. Parallel decreased levels of the saturated FA precursor 18:0 were detected in pYES2-Scd yeast ($p \leq 0.05$). Additionally, other monoenes corresponding to 20:1*n*-11 and 22:1*n*-13 were detected in total lipids of InvSc1 yeast expressing the *O. vulgaris* Scd. These results confirmed that the *O. vulgaris* Scd is a Δ9 desaturase with activity towards saturated FAs with chain-lengths $\geq C_{18}$. In contrast, shorter FAs ($\leq C_{16}$) did not appear to be adequate substrates for the octopus Scd and, for instance, the content of 16:1*n*-7 in yeast expressing the common octopus Scd was significantly lower than that of control yeast ($p \leq 0.05$) (Table 2).

Table 1. Fatty acid (FA) composition of the *S. cerevisiae* strain L8-14C transformed with the yeast-endogenous *OLE1* (SC OLE1) or stearoyl-CoA desaturase from *O. vulgaris* (OV Scd). Results are expressed as an area percentage of total fatty acids (FAs) found in transformed yeast. Different letters for each FA or ratio indicate significant differences among treatments (*t*-test, $p \leq 0.05$).

Fatty Acid	SC OLE1	OV Scd
16:0	23.5 ± 3.6 [b]	59.3 ± 4.9 [a]
16:1*n*-9	0.5 ± 0.4	0.7 ± 0.7
16:1*n*-7	35.0 ± 2.2 [a]	4.5 ± 1.0 [b]
18:0	8.0 ± 1.7	5.6 ± 1.3
18:1*n*-9	31.8 ± 2.5	29.8 ± 6.0
18:1*n*-7	1.1 ± 0.2 [a]	0.0 ± 0.0 [b]
18:1*n*-9/16:1*n*-7	0.9 ± 0.1 [b]	6.8 ± 2.0 [a]

Table 2. Fatty acid (FA) composition of the *S. cerevisiae* InvSc1 transformed with either empty pYES2 vector (Control) or the common octopus Scd open reading frame (ORF). Different letters for each FA indicate significant differences among treatments (*t*-test, $p \leq 0.05$).

Fatty Acid	Control	OV Scd
14:0	1.6 ± 0.2 [a]	1.0 ± 0.1 [b]
14:1*n*-5	0.5 ± 0.1	0.7 ± 0.1
16:0	25.9 ± 0.3 [a]	22.2 ± 1.2 [b]
16:1*n*-7	37.7 ± 0.8 [a]	28.6 ± 3.3 [b]
18:0	8.5 ± 0.3 [a]	4.4 ± 0.9 [b]
18:1*n*-9	24.1 ± 0.7 [b]	40.9 ± 3.6 [a]
18:1*n*-7	1.1 ± 0.1	1.1 ± 0.2
20:0	0.1 ± 0.0	0.1 ± 0.0
20:1*n*-11	N.D. [b]	0.4 ± 0.2 [a]
22:0	0.1 ± 0.0	0.1 ± 0.0
22:1*n*-13	N.D. [b]	0.1 ± 0.0 [a]

In order to establish the biosynthetic pathways of NMI FAs with Δ5,9 unsaturation patterns (or their derivatives such as Δ5,11) found in the common octopus lipids [10], InvSc1 yeast transformed with pYES2-Scd were grown in the presence of the monoene 20:1*n*-15 ($^{\Delta 5}$20:1), while InvSc1 yeast expressing the Δ5 Fad (i.e., transformed with pYES2-Fad) [6] were grown in the presence of 20:1*n*-9 ($^{\Delta 11}$20:1) (Figure 5). Our results showed that the *O. vulgaris* Scd was unable to desaturate the substrate $^{\Delta 5}$20:1 (20:1*n*-15) to $^{\Delta 5,9}$20:2 (Figure 5a), although yeast transformed with pYES2-Fad were able to produce $^{\Delta 5,11}$20:2 (2.7% ± 0.2% conversion), thus confirming activity as Δ5 desaturase on $^{\Delta 11}$20:1) (Figure 5b).

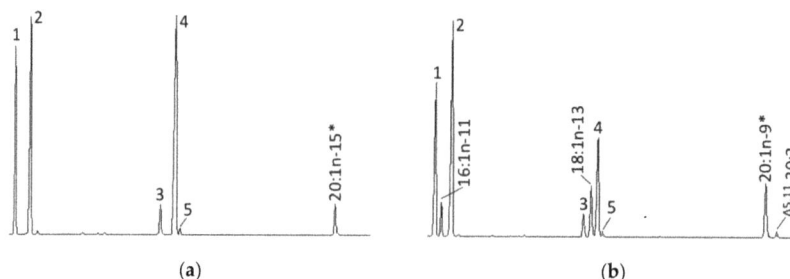

(a)   (b)

Figure 5. Biosynthesis of non-methylene interrupted fatty acids (FAs) in cephalopods: (**a**) Gas chromatography (GC) trace of yeast *Saccharomyces cerevisiae* expressing the *Octopus vulgaris* Scd and grown in the presence of 20:1*n*-15 ($^{\Delta 5}$20:1); (**b**) GC trace of yeast *S. cerevisiae* expressing the previously characterised *O. vulgaris* Δ5 Fad [6] and grown in the presence of 20:1*n*-9 ($^{\Delta 11}$20:1). Peaks 1-5 represent *S. cerevisiae* endogenous FAs, namely 16:0 (1), 16:1 isomers (2), 18:0 (3), 18:1*n*-9 (4) and 18:1*n*-7 (5). Peaks derived from exogenously added substrates (*) and the desaturation product $^{\Delta 5,11}$20:2 (**b**) are indicated accordingly.

2.4. Functional Characterisation of the Octopus Elovl4

The role of the *O. vulgaris* Elovl4 in the biosynthesis of very long-chain (>C$_{24}$) PUFAs (VLC-PUFAs) was investigated in yeast *S. cerevisiae* (strain InvSc1) expressing the Elovl4 and grown in the presence of one of either C$_{18}$ (18:3*n*-3, 18:2*n*-6, 18:4*n*-3 and 18:3*n*-6), C$_{20}$ (20:5*n*-3 and 20:4*n*-6), C$_{22}$ (22:5*n*-3, 22:4*n*-6, 22:6*n*-3) or C$_{24}$ (24:5*n*-3) PUFA substrates (Table 3). Gas Chromatography-Mass Spectrometry (GC–MS) analyses confirmed that the control yeast did not have the ability to elongate PUFAs, consistent with the previously reported lack of a PUFA elongase in *S. cerevisiae* strain InvSc1 [30]. However, the common octopus Elovl4 conferred the yeast the ability to elongate PUFAs to the corresponding elongated polyenoic products of both *n*-3 and *n*-6 series (Table 3). With the exception of DHA (22:6*n*-3), the addition of C$_{22}$ and C$_{24}$ PUFA substrates resulted in the production of polyenes of C$_{32}$ products or even C$_{34}$ when 24:5*n*-3 was used as substrate (Table 3). Indeed, the endogenous production of PUFAs with chain lengths ≥ C$_{26}$ in yeast supplemented with exogenously supplemented C$_{22}$ and C$_{24}$ PUFAs allowed us to estimate the Elovl4 efficiency (as % conversion) towards potential VLC-PUFA substrates that are not commercially available. The results showed that the highest % conversions (often over 88.0%) were consistently detected on C$_{28}$, C$_{30}$ and C$_{32}$ substrates (Table 3). It is noteworthy that the octopus Elovl4 was able to convert the exogenously added 20:5*n*-3 and 22:5*n*-3 to 24:5*n*-3, the substrate for DHA biosynthesis via the Sprecher pathway [14], although it showed relative low elongase activity towards DHA itself, which was marginally elongated (0.7%) to 24:6*n*-3.

Table 3. Role of the *Octopus vulgaris* Elovl4 in the biosynthesis of very long-chain (>C$_{24}$) polyunsaturated fatty acids (FAs). Conversions were calculated for each stepwise elongation according to the formula (areas of first product and longer chain products/(areas of all products with longer chain than substrate + substrate area)) × 100. The substrate FA varies as indicated in each step-wise elongation.

FA Substrate	Product	% Conversion	Elongation
18:3*n*-3	20:3*n*-3	2.8	C18→22
	22:3*n*-3	3.0	C20→22
18:2*n*-6	20:2*n*-6	1.1	C18→22
	22:2*n*-6	26.2	C20→22
18:4*n*-3	20:4*n*-3	1.0	C18→20
18:3*n*-6	20:3*n*-6	0.8	C18→20
20:5*n*-3	22:5*n*-3	1.9	C20→24
	24:5*n*-3	8.1	C22→24
20:4*n*-6	22:4*n*-6	1.0	C20→24
	24:4*n*-6	6.2	C22→24
22:5*n*-3	24:5*n*-3	4.8	C22→32
	26:5*n*-3	37.9	C24→32
	28:5*n*-3	95.0	C26→32
	30:5*n*-3	93.5	C28→32
	32:5*n*-3	51.9	C30→32
22:4*n*-6	24:4*n*-6	3.6	C22→32
	26:4*n*-6	44.5	C24→32
	28:4*n*-6	94.1	C26→32
	30:4*n*-6	91.5	C28→32
	32:4*n*-6	46.3	C30→32
22:6*n*-3	24:6*n*-3	0.7	C22→24
24:5*n*-3	26:5*n*-3	1.4	C24→34
	28:5*n*-3	88.8	C26→34
	30:5*n*-3	88.6	C28→34
	32:5*n*-3	63.2	C30→34
	34:5*n*-3	17.3	C32→34

3. Discussion

Fish and seafood are the primary sources of omega-3 long-chain (C$_{20-24}$) PUFAs for humans [31] and this partly explains the considerable interest in elucidating the PUFA biosynthetic pathways in aquatic and marine species, particularly in farmed fish for which current trends in feed formulation are impacting the nutritional quality for human consumers [32]. Previous investigations on *O. vulgaris* [6,10,33] and *S. officinalis* [20] revealed that cephalopods possess active desaturase and elongase enzymes involved in the biosynthesis of PUFAs including NMI FAs. Using the common octopus *O. vulgaris* as model species, the present study aimed to expand our knowledge of roles that further desaturases and elongases could have on PUFA biosynthesis in cephalopods. Consequently, we characterised a stearoyl-CoA desaturase (Scd) with a putative role in the biosynthesis of NMI FAs, and an Elovl4 elongase that, in addition to its participation in the biosynthesis of VLC-PUFAs [15–19,34], could potentially catalyse the elongation of 22:5*n*-3 to 24:5*n*-3 required for DHA biosynthesis through the Sprecher pathway [14].

The Scd, present in virtually all living organisms [24], is an enzyme with Δ9 desaturase activity. Consistently, the newly-cloned *O. vulgaris* Scd cDNA was confirmed to encode a Δ9 desaturase that was able to operate on a range of saturated FA substrates with different chain lengths, particularly ≥ C$_{18}$. These functions are similar to those described for two homologous sequences found in the nematode

C. elegans, namely FAT-6 and FAT-7, which can efficiently desaturate 18:0 to 18:1*n*-9 but have lower activity towards 16:0 [25]. Interestingly, FAT-5, another Scd-like sequence existing in *C. elegans*, efficiently desaturated 16:0 to 16:1*n*-7 but had nearly undetectable activity on 18:0, and it is thus regarded as a palmitoyl-CoA-specific desaturase [25]. The functional characterisation of the *O. vulgaris* Scd and its particularly high substrate affinity towards 18:0 is consistent with 18:1*n*-9 typically appearing several folds above the levels of shorter monoenes such as 16:1*n*-7 in lipids across the Mollusca phylum [21]. Beyond its role in biosynthesis of monounsaturated FA, occurrence of NMI FAs with $\Delta5,9$ unsaturation patterns in lipids from molluscs [22] suggested a role of Scd in the biosynthesis of this particular type of PUFA. Our results revealed that the *O. vulgaris* Scd was not able to introduce a second double bond on a $\Delta5$ monoene such as 20:1*n*-15 ($^{\Delta5}$20:1). While the possibility that cephalopod Scd can introduce $\Delta9$ desaturations on other monoenes cannot be ruled out, these results strongly suggest that the role of Scd in the biosynthesis of NMI FAs with $\Delta5,9$ unsaturation patterns is limited to the insertion of the first unsaturation at $\Delta9$ position on saturated FAs. Indeed, the *O. vulgaris* $\Delta5$ Fad, previously characterised as $\Delta5$ desaturase [6], was able to introduce a $\Delta5$ unsaturation on 20:1*n*-9 ($^{\Delta11}$20:1) producing the NMI FA $^{\Delta5,11}$20:2. Overall, these results allow us to hypothesise that the insertion of double bonds required for NMI FA biosynthesis is initiated with a $\Delta9$ desaturation by Scd and a subsequent $\Delta5$ desaturation catalysed by $\Delta5$ Fad. Compared to other molluscan classes, the occurrence of NMI FAs in cephalopods appears to be rather limited and hence the contribution of endogenous production of NMI FAs vs. dietary input is difficult to establish. Nevertheless, the mechanism of $^{\Delta5,11}$20:2 biosynthesis postulated herein aligns well with the occurrence of $^{\Delta5,11}$20:2 in the polar lipids of nephridia of adult *O. vulgaris* [10], suggesting that NMI FA biosynthesis is possible in cephalopods and likely other molluscs.

In addition to desaturases, certain Elovl enzymes play crucial roles in the biosynthesis of essential PUFAs such as EPA, ARA and DHA [9,11,26]. Whereas the previously characterised cephalopod Elovl2/5 demonstrated an ability to efficiently elongate C_{18} and C_{20} PUFA substrates [6,20], lack of elongase capability towards 22:5*n*-3 was hypothesised as a major limiting factor for DHA biosynthesis through the Sprecher pathway [6,20]. Functional characterisation of the *O. vulgaris* Elovl4 demonstrated the ability of this enzyme to contribute potentially to DHA biosynthesis by producing 24:5*n*-3, which is the substrate for $\Delta6$ desaturation and chain-shortening to produce DHA in the Sprecher pathway [14]. Such an ability of Elovl4 has been reported in fish orthologues [15–19]. Despite the elongation capability of *O. vulgaris* herein revealed, our overall observations on gene repertoire and function [6,10,20,33] strongly suggest that DHA is still an essential FA for cephalopods. The limited ability to biosynthesise DHA is related, rather than to an issue with elongase activity, to the apparent absence of key desaturation abilities ($\Delta6$ or $\Delta4$) required for DHA biosynthesis according to the two aerobic pathways known in vertebrates (Figure 1). In addition to the abovementioned $\Delta6$ desaturation required in the Sprecher pathway [14], an alternative pathway described in some teleost fish involves a $\Delta4$ desaturation from 22:5*n*-3 to produce DHA directly (Figure 1) [9]. Molecular evidence provided by the recently published *Octopus bimaculoides* genome project [35] suggests that cephalopods possess one single Fad-like desaturase in their genome and this is likely to be a $\Delta5$ desaturase since all the Fad-like desaturase functionally characterised not only from cephalopods such as *O. vulgaris* and *S. officinalis* [6,10] but also from bivalves [36] and gastropods [37], have $\Delta5$ desaturase activities. In non-cephalopod molluscs, further Fad-like desaturases exist [38] but, with the exception of the $\Delta8$ desaturase found in the scallop *Chlamys nobilis* [39], their functions have not been determined and therefore it is yet not possible to predict whether $\Delta6$ and/or $\Delta4$ desaturase activities exist to enable DHA biosynthesis in those species.

The *O. vulgaris* Elovl4 is also involved in the biosynthesis of VLC-PUFAs as it was able to elongate exogenously supplemented PUFAs ranging from C_{18-24} to PUFA products with chain lengths of $\geq C_{26}$. Such conversions are largely consistent with those previously exhibited by human [34] and fish [15–19] Elovl4 proteins. Surprisingly, no elongation products beyond C_{24} were reported in the functional characterisation of the bivalve *C. nobilis* Elovl4 [39]. The elongation activities of the *O. vulgaris* Elovl4 estimated in the yeast expression system strongly suggested that this enzyme is particularly efficient towards C_{26-30} substrates, for which the highest % conversions were observed. Similarly, the human ELOVL4 had PUFAs of $\geq C_{26}$ as preferred substrates for elongation, although the human ELOVL4 was able to produce up to C_{38} PUFA elongation products in mammalian cell lines [34]. Due to technical challenges in their analysis and relatively low abundance, the functions of VLC-PUFAs are not fully understood [40]. Nevertheless, the structural features of VLC-PUFAs combining those from saturated FAs at one end and those from PUFAs at the other allow unique membrane lipid conformations in photoreceptors and spermatozoa, thus suggesting important roles of VLC-PUFAs in vision and reproduction of vertebrates [40–43]. Investigations of VLC-PUFAs in cephalopods and, therefore, the herein reported activities of the *O. vulgaris* Elovl4 cannot be correlated with the presence of VLC-PUFAs in vivo. Interestingly, some very long-chain NMI FAs have been reported in nudibranchs [44,45] but, unfortunately, the potential role of Elovl4 in the biosynthesis of such compounds was not determined. Interestingly, unpublished data on tissue distribution analysis of the common octopus Elovl4 mRNA indicated that gonads and, to a lesser extent eye, were also major sites for VLC-PUFA biosynthesis in cephalopods.

In conclusion, the present study demonstrated that the common octopus *O. vulgaris* possesses an Scd with high affinity for saturated FA substrates with chain lengths of 18 carbons or longer. Its inability to introduce a double bond into Δ5 monoene strongly suggested that the role of the Scd in NMI FA biosynthesis was restricted to the introduction of the first double bond at the Δ9 position, whereas the *O. vulgaris* Δ5 Fad can introduce the second unsaturation at the Δ5 position according to the herein confirmed ability to convert 20:1n-9 to $^{\Delta 5,11}$20:2, an NMI FA previously reported in nephridia of *O. vulgaris* adult specimens. Beyond desaturases, we could also demonstrate that *O. vulgaris* possesses an Elovl4 responsible for the biosynthesis of VLC-PUFAs.

4. Materials and Methods

4.1. Tissue Samples

Tissue samples including brain, nerve, muscle, heart, hepatopancreas, gill, caecum, eye, nephridium and gonads were obtained from the dissection of two (male and female) common octopus adult specimens as previously described [6,10]. Briefly, two (male and female) wild *O. vulgaris* adults (~1.5 kg) were maintained in seawater tanks at the facilities of the Instituto de Acuicultura Torre de la Sal; before they were cold, they were anesthetised and sacrificed by direct brain puncture. After collection, samples were immediately frozen at −80 °C until further analysis. Total RNA was extracted from octopus tissues using TriReagent® (Sigma-Aldrich, Alcobendas, Spain) according to manufacturer's instructions. Two μg of total RNA tissue samples were used for synthesis of first strand cDNA using M-MLV reverse transcriptase (Promega, Southampton, UK) primed with random hexamers.

4.2. Molecular Cloning of the Scd and Elovl4 cDNA Sequences

The full-length sequences of the Scd and Elovl4 cDNA were obtained as follows. The deduced aa sequences of Scd proteins from *Homo sapiens* (NP_005054.3), *Gallus gallus* (NP_990221.1), *Anolis carolinensis* (XP_003226591.1), *Caenorhabditis elegans* (FAT-7) (NP_504814.1), *Acheta domesticus* (AAK25796.1) and *Pediculus humanus* (XP_002424386.1) were aligned using BioEdit v5.0.6 (Tom Hall, Department of Microbiology, North Carolina State University, Raleigh, NC, USA). Conserved regions were used for in silico searches of mollusc expressed sequence tags (EST) using The National Center for Biotechnology Information (NCBI) tblastn tool (http://www.ncbi.nlm.nih.gov/). Processed Scd-like Expressed Sequence Tags (ESTs) from the molluscs *Lottia gigantea* (FC767047.1 and FC644199.1), *Aplysia californica* (FF074235.1, FF076278.1, EB307164.1, EB253812.1 and EB281044.1), *Limnaea stagnalis* (ES580199.1 and ES579822.1), *Crassostrea gigas* (CU997931.1 and FP006991.1), *Euprymna scolopes* (DW280836.1, DW269578.1 and DW273589.1) and *Ruditapes decussatus* (AM870139.1), were aligned

(Bioedit) and conserved regions used for the design of the degenerate primers UNID9F (5′-ATCACAGCTGGWGCTCAYCG-3′) and UNID9R (5′-TGGCATTGTGWGGGTCWGCATC-3′).

To clone the first fragment of the octopus Elovl4, blastn searches of mollusc ESTs were performed using the so-called "transcript 2" from *L. gigantea* that was previously identified by [6] (gi | Lotgi1 | 178149 |) as query. Thus, additional Elovl4-like consensus sequences derived from ESTs from *A. californica* (EB285681.1, GD233360.1, EB316848.1, GD212825.1, EB325217.1, EB345626.1, EB345430.1), *Saccostrea kagaki* (AB375033.1) and *C. gigas* (HS215834.1, AM866458.1) were obtained and aligned with *L. gigantea* Elovl4-like to design the degenerate primers UNIE4F (5′-GCCAAGGCATTRTGGTGGTT-3′) and UNIE4R (5′-GTSAGRTATCKYTTCCACCA-3′).

Polymerase chain reactions (PCR) were performed with the GoTaq® Green Master Mix (Promega) and using a mixture of cDNAs from brain, nerve and hepatopancreas as template. The PCR consisted of an initial denaturing step at 95 °C for 2 min, followed by 35 cycles of denaturation at 95 °C for 30 s, annealing at 50 °C for 30 s, extension at 72 °C for 40 s, followed by a final extension at 72 °C for 5 min. PCR products of approximately 180 bp (Scd) and 290 bp (Elovl4) were obtained and thereafter confirmed as positive by sequencing (DNA Sequencing Service, IBMCP-UPV, Valencia, Spain). Full-length cDNA of the octopus Scd and Elovl4 were completed by 5′ and 3′ rapid amplification of cDNA ends (RACE) PCR (FirstChoice RLM-RACE kit, Ambion, Applied Biosystems, Warrington, UK) with gene-specific primers shown in Table S1.

4.3. Sequence Analysis

The deduced aa sequences of the *O. vulgaris* Scd and Elovl4 ORF were compared to corresponding orthologues from other invertebrate and vertebrate species and sequence identity scores were calculated using the EMBOSS Needle Pairwise Sequence Alignment tool (http://www.ebi.ac.uk/Tools/psa/emboss_needle/). Deduced aa sequence alignments were carried out using the built-in ClustalW tool (BioEdit v7.0.9, Tom Hall, Department of Microbiology, North Carolina State University, Raleigh, NC, USA).

4.4. Functional Characterisation of Octopus Scd: Complementation Assay of the S. cerevisiae ole1Δ Mutant Strain L8-14C

The open reading frame (ORF) of the common octopus Scd was amplified from a mixture of cDNAs (brain, nerve and hepatopancreas) using the high fidelity *Pfu* Turbo DNA polymerase (Promega) and the primers OVD9VF and OVD9VR, containing restriction sites *Hind*III and *Xho*I, respectively (underlined in Table S1). PCR conditions consisted of an initial denaturing step at 95 °C for 2 min, followed by 32 cycles of denaturation at 95 °C for 30 s, annealing at 58 °C for 30 s, extension at 72 °C for 2.5 min, followed by a final extension at 72 °C for 5 min. After restriction of the PCR product, the *O. vulgaris* Scd ORF was cloned into the yeast expression vector p416TEF (a centromeric plasmid with a *URA3* selectable marker) to produce the construct p416TEF-Scd, in which octopus Scd was under the control of the yeast *TEF1* promoter. Subsequently, the promoter region of the *S. cerevisiae OLE1* Δ9-fatty acid desaturase gene amplified from yeast genomic DNA with the primers SCPromOLE1F and SCPromOLE1R containing *Sac*I and *Hind*III sites, respectively, was cloned upstream of the *O. vulgaris* Scd, replacing the *TEF1* promoter and producing the construct p416OLE1-Scd, in which octopus Scd is expressed under the control of the yeast *OLE1* promoter. As a positive control, the promoter and ORF of *S. cerevisiae OLE1* were amplified by PCR from yeast genomic DNA using the primers SCPromOLE1F and SCOLE1R that contained restriction sites for *Sac*I and *Xho*I, respectively (Table S1). To obtain the p416OLE1-OLE1 plasmid, the *OLE1* PCR fragment was cloned into p416TEF *Sac*I and *Xho*I restriction sites, allowing *TEF1* promoter replacement by *OLE1* promoter and ORF.

To address the common octopus Scd function we used *S. cerevisiae* L8-14C (MATa *ole1Δ::LEU2, leu2-3,112, trp1-1, ura3-52, his4*; kindly donated by Dr. Charles E. Martin), a strain whose *OLE1* gene has been deleted. Therefore, L8-14C cells lack the only yeast Δ9-fatty acid desaturase and require the supply of monounsaturated FAs including 0.5 mM palmitoleic (16:1n-7) and 0.5 mM oleic (18:1n-9) acids in the medium for growth. To address whether the octopus Scd complemented the growth defect displayed by L8-14C cells in media lacking FAs, yeast mutants transformed with pRS416 (empty vector as negative control), p416OLE1-OLE1 (positive control) or p416OLE1-Scd plasmids were grown in liquid SC medium lacking uracil (SC-ura) with supplemented FAs until exponential phase, and then assayed for growth on plates in 10-fold serial dilution drops starting at OD$_{600}$ = 0.1. Plates were incubated for 3 days at 30 °C and photographed. To determine FA composition, four L8-14C colonies from yeast transformed with either p416OLE1-OLE1 or p416OLE1-Scd were grown in 10 mL of yeast extract peptone dextrose (YPD) broth lacking 18:1n-9 and 16:1n-7. After incubation at 30 °C for 48 h, a 5 mL aliquot of yeast culture was pelleted (1000 g for 2 min), washed twice with ddH$_2$O, homogenised in chloroform/methanol (2:1, v/v) containing 0.01% butylated hydroxy toluene (BHT) [30,46], and kept at −20 °C until further analysis.

4.5. Functional Characterisation of Octopus Scd: Overexpression in the S. cerevisiae Strain InvSc1

Primers containing *Hind*III (forward) and *Xho*I (reverse) restriction sites (underlined in Table S1) OVD9VF and OVD9VR (Scd) were used to amplify the ORF of the *O. vulgaris* Scd, using the high fidelity *Pfu* Turbo DNA polymerase (Promega). Further cloning into the yeast expression vector pYES2 (Invitrogen), which contains a *GAL1* promoter that is inducible by galactose and has *URA3* as selective marker, was achieved after ligation of restricted ORF amplicons and plasmid pYES2 to produce the construct pYES2-Scd. The recombinant plasmids pYES2-Scd or pYES2 empty (negative control) were transformed into *S. cerevisiae* competent cells InvSc1 (S.c. EasyComp Transformation Kit, Invitrogen). Yeast were grown in SC-ura for 3 days.

One single yeast colony containing the pYES2-Scd or pYES2 was grown overnight at 30 °C in 5 mL of liquid SC-ura. Cell cultures were then used to inoculate 10 mL of fresh SC-ura for a final OD$_{600}$ of 0.4. Four replicates for each construct (pYES2-Scd or pYES2) were run. Cells were grown at 30 °C for 5 h before the expression of the transgene was induced by the addition of galactose to 2% (w/v) [46]. After 48 h of galactose induction, yeast samples were collected, washed and homogenised in chloroform/methanol (2:1, v/v) containing 0.01% BHT. Samples were kept at −20 °C until further analysis.

4.6. Role of the Octopus Scd and Δ5 Fad in the Biosynthesis of Non-Methylene-Interrupted FAs

In order to establish the role of Scd and the previously characterised Δ5 Fad [6] in the biosynthetic pathways of Δ5,9 (or Δ5,11) non-methylene interrupted FAs (NMI FAs) we conducted the following experiment. First, yeast InvSc1 transformed with pYES2-Scd, i.e., expressing the octopus Scd, were grown in the presence of exogenously added 5-eicosenoic acid (20:1n-15 or ^{Δ5}20:1). Second, yeast InvSc1 transformed with pYES2-Fad, i.e., expressing the octopus Fads that was previously reported to exhibit Δ5-desaturase [6], were grown in the presence of 11-eicosenoic acid (20:1n-9 or ^{Δ11}20:1). Final concentration of exogenously-added substrates (^{Δ5}20:1 or ^{Δ11}20:1) was 0.75 mM. Culture conditions and yeast sample collection were performed as described above. Yeast transformed with empty pYES2 were also grown in presence of ^{Δ5}20:1 and ^{Δ11}20:1 as control treatments.

4.7. Functional Characterisation of Octopus Elovl4: Expression in the S. cerevisiae Strain InvSc1

The octopus Elovl4 was functionally characterised by heterologous expression in yeast *S. cerevisiae* (strain InvSc1, Invitrogen). Similarly, as described above for the Scd overexpression experiment, the octopus Elovl4 ORF was amplified with primers OVE4VF and OVE4VR containing restriction sites for *Hind*III and *Xho*I, respectively (Table S1) which allowed its cloning into pYES2 to produce the construct pYES2-Elovl4. Yeast InvSc1 transformed with pYES2-Elovl4 were grown in SC-ura plates. One colony was subsequently grown in SC-ura broth to produce a bulk culture that allowed us to establish subcultures of OD$_{600}$ 0.4 in Erlenmeyer flasks that contained 5 mL of SC-ura broth and were in some cases supplemented with potential PUFA substrates for fatty acyl elongases. In order to assess the ability of the octopus Elovl4 to elongate PUFA substrates, yeast transformed with pYES2-Elovl4 were grown in the presence of one of the following PUFA substrates: 18:3*n*-3, 18:2*n*-6, 18:4*n*-3, 18:3*n*-6, 20:5*n*-3, 20:4*n*-6, 22:5*n*-3, 22:4*n*-6, 22:6*n*-3 and 24:5*n*-3. The FA substrates were added to the yeast cultures at final concentrations of 0.5 (C_{18}), 0.75 (C_{20}), 1.0 (C_{22}) and 1.2 (C_{24}) mM to compensate for decreased efficiency of uptake with increased chain length [47].

To test the ability of the octopus Elovl4 to elongate saturated FAs, yeast transformed with pYES2-Elovl4 or pYES2 (negative control) were grown in the absence of exogenously added substrates and the capability of the Elovl4 to elongate saturated FAs was estimated by comparing the saturated FA profiles of both transformants. Three different replicates for each treatment were established.

4.8. Fatty Acid Analysis by GC-MS

Lipid extracts [48] from the transgenic yeast were utilised for preparing fatty acid methyl esters (FAME) as previously described [6,10]. Briefly, FAME were identified and quantified using an Agilent 6850 Gas Chromatograph coupled to a 5975 series Mass Selective Detector (MSD, Agilent Technologies, Santa Clara, CA, USA). The activity of the newly cloned octopus Scd was estimated by comparing the FA profiles (expressed as % of total FAs) of control yeast with those from yeast transformed with p416OLE1-Scd (complementation experiment) or pYES2-Scd (overexpression experiment). The efficiency of the *O. vulgaris* Δ5 Fad to desaturate $^{\Delta 11}$20:1 into the NMI FAs $^{\Delta 5,11}$20:2 was calculated as % conversion according to the formula: (area of product/(area of product + area of substrate)) $\times$ 100. The double bond positions in FA products derived from conversion by transgenic yeast towards $^{\Delta 5}$20:1 and $^{\Delta 11}$20:1 was confirmed by preparing FA picolinyl ester derivatives according to the methodology described by [49] and modified according to [10]. Finally, the ability of the *O. vulgaris* Elovl4 to elongate the exogenously-added PUFA substrates (18:3*n*-3, 18:2*n*-6, 18:4*n*-3, 18:3*n*-6, 20:5*n*-3, 20:4*n*-6, 22:5*n*-3, 22:4*n*-6, 22:6*n*-3 and 24:5*n*-3) was calculated by the step-wise proportion of substrate FAs converted to elongated product as (areas of first product and longer chain products/(areas of all products with longer chain than substrate + substrate area)) $\times$ 100 [50].

4.9. Statistical Analyses

For the functional characterisation experiments (complementation and overexpression) of the octopus Scd, FA analyses from yeast samples were expressed as mean values $\pm$ standard deviation ($n = 4$). Similarly, the assay aiming to determine the ability of the octopus Elovl4 for elongation saturated FA was run in replicates ($n = 3$) and FA contents expressed as mean values $\pm$ standard deviation. Homogeneity of variances was checked by Barlett's test. Comparison of FA profiles from control and yeast expressing the *O. vulgaris* Scd (complementation and overexpression experiments) or Elovl4 were compared with a Student's *t*-test. Comparisons of the means with *p* values less or equal than 0.05 were considered significantly different. All the statistical analyses were carried out using the SPSS statistical package (SPSS Inc., Chicago, IL, USA).

4.10. Materials

All PUFA substrates were purchased from Nu-Chek Prep, Inc. (Elysian, MN, USA), except stearidonic acid (18:4*n*-3) from Sigma-Aldrich (Alcobendas, Spain) and tetracosapentaenoic acid (24:5*n*-3) from Larodan (Larodan Fine Chemicals AB, Malmö, Sweden). All chemicals used to prepare the *S. cerevisiae* media were from Sigma-Aldrich, except for the bacteriological agar obtained from Oxoid Ltd. (Hants, UK).

Mar. Drugs **2017**, *15*, 82

Supplementary Materials: The following are available online at www.mdpi.com/1660-3397/15/3/82/s1, Table S1: Primer sequences used in the present study.

Acknowledgments: This research and OM were supported by a Marie Curie Reintegration Grant within the 7th European Community Framework Programme (PERG08-GA-2010-276916, LONGFA), with additional support from "Ministerio de Ciencia e Innovación" through the OCTOPHYS Project (AGL-2010-22120-C03-02) and a Juan de la Cierva postdoctoral contract to OM. Further support was obtained from the Generalitat Valenciana through grants within the PROMETEO (2010/006), "Grupos Emergentes" (GV/2013/123) programes, and AGL2011-29099 and BIO2014-56298-P grants from the Spanish Ministry of Economy and Competitiveness. R.d.L. was supported by a postdoctoral JAE-Doc contract from the Spanish Research Council (CSIC) and the European Social Fund.

Author Contributions: O.M., S.P. and J.C.N. conceived and designed the experiments; O.M., R.d.L. and F.H. performed the experiments; O.M., R.d.L. and I.V. analysed the data; I.V., S.P. and J.C.N. contributed reagents/materials/analysis tools; O.M., D.R.T. and J.C.N. wrote the paper.

Conflicts of Interest: The authors declare no conflict of interest. The founding sponsors had no role in the design of the study; in the collection, analyses, or interpretation of data; in the writing of the manuscript, and in the decision to publish the results.

References

1. Iglesias, J.; Fuentes, L.; Villanueva, R. *Cephalopod Culture*; Springer: Dordrecht, The Netherlands, 2014; p. 494.

2. Iglesias, J.; Sánchez, F.J.; Bersano, J.G.F.; Carrasco, J.F.; Dhont, J.; Fuentes, L.; Linares, F.; Muñoz, J.L.; Okumura, S.; Roo, J.; et al. Rearing of *Octopus vulgaris* paralarvae: Present status, bottlenecks and trends. *Aquaculture* **2007**, *266*, 1–15. [CrossRef]

3. Navarro, J.C.; Monroig, Ó.; Sykes, A.V. Nutrition as a key factor for cephalopod aquaculture. In *Cephalopod Culture*; Iglesias, J., Fuentes, L., Villanueva, R., Eds.; Springer: Dordrecht, The Netherlands, 2014; pp. 77–96.

4. Navarro, J.C.; Villanueva, R. Lipid and fatty acid composition of early stages of cephalopods: An approach to their lipid requirements. *Aquaculture* **2000**, *183*, 161–177. [CrossRef]

5. Navarro, J.C.; Villanueva, R. The fatty acid composition of *Octopus vulgaris* paralarvae reared with live and inert food: Deviation from their natural fatty acid profile. *Aquaculture* **2003**, *219*, 613–631. [CrossRef]

6. Monroig, Ó.; Navarro, J.C.; Dick, J.R.; Alemany, F.; Tocher, D.R. Identification of a Δ5-like fatty acyl desaturase from the cephalopod *Octopus vulgaris* (Cuvier 1797) involved in the biosynthesis of essential fatty acids. *Mar. Biotechnol.* **2012**, *14*, 411–422. [CrossRef] [PubMed]

7. Guillou, H.; Zadravec, D.; Martin, P.G.P.; Jacobsson, A. The key roles of elongases and desaturases in mammalian fatty acid metabolism: Insights from transgenic mice. *Prog. Lipid Res.* **2010**, *49*, 186–199. [CrossRef] [PubMed]

8. Monroig, Ó.; Navarro, J.C.; Tocher, D.R. Long-chain polyunsaturated fatty acids in fish: Recent advances on desaturases and elongases involved in their biosynthesis. In *Proceedings of the XI International Symposium on Aquaculture Nutrition*; Cruz-Suarez, L.E., Ricque-Marie, D., Tapia-Salazar, M., Nieto-López, M.G., Villarreal-Cavazos, D.A., Gamboa-Delgado, J., Hernández-Hernández, L.H., Eds.; Universidad Autónoma de Nuevo León: Monterrey, Mexico, 2011; pp. 257–282.

9. Castro, L.F.C.; Tocher, D.R.; Monroig, O. Long-chain polyunsaturated fatty acid biosynthesis in chordates: Insights into the evolution of Fads and Elovl gene repertoire. *Prog. Lipid Res.* **2016**, *62*, 25–40. [CrossRef] [PubMed]

10. Monroig, Ó.; Guinot, D.; Hontoria, F.; Tocher, D.R.; Navarro, J.C. Biosynthesis of essential fatty acids in *Octopus vulgaris* (Cuvier, 1797): Molecular cloning, functional characterisation and tissue distribution of a fatty acyl elongase. *Aquaculture* **2012**, *360–361*, 45–53. [CrossRef]

11. Monroig, Ó.; Tocher, D.R.; Navarro, J.C. Biosynthesis of polyunsaturated fatty acids in marine invertebrates: Recent advances in molecular mechanisms. *Mar. Drugs* **2013**, *11*, 3998–4018. [CrossRef] [PubMed]

12. Liu, H.; Zheng, H.; Wang, S.; Wang, Y.; Li, S.; Liu, W.; Zhang, G. Cloning and functional characterization of a polyunsaturated fatty acid elongase in a marine bivalve noble scallop *Chlamys nobilis* Reeve. *Aquaculture* **2013**, *416*, 146–151. [CrossRef]

13. Monroig, Ó.; Lopes-Marques, M.; Navarro, J.C.; Hontoria, F.; Ruivo, R.; Santos, M.M.; Venkatesh, B.; Tocher, D.R.; Castro, L.F.C. Evolutionary wiring of the polyunsaturated fatty acid biosynthetic pathway. *Sci. Rep.* **2016**, *6*, 20510. [CrossRef] [PubMed]

14. Sprecher, H. Metabolism of highly unsaturated *n*-3 and *n*-6 fatty acids. *Biochim. Biophys. Acta* **2000**, *1486*, 219–231. [CrossRef]

15. Monroig, Ó.; Rotllant, J.; Cerdá-Reverter, J.M.; Dick, J.R.; Figueras, A.; Tocher, D.R. Expression and role of Elovl4 elongases in biosynthesis of very long-chain fatty acids during zebrafish *Danio rerio* early embryonic development. *Biochim. Biophys. Acta* **2010**, *1801*, 1145–1154. [CrossRef] [PubMed]

16. Monroig, Ó.; Webb, K.; Ibarra-Castro, L.; Holt, G.J.; Tocher, D.R. Biosynthesis of long-chain polyunsaturated fatty acids in marine fish: Characterization of an Elovl4-like elongase from cobia *Rachycentron canadum* and activation of the pathway during early life stages. *Aquaculture* **2011**, *312*, 145–153. [CrossRef]

17. Carmona-Antoñanzas, G.; Monroig, Ó.; Dick, J.R.; Davie, A.; Tocher, D.R. Biosynthesis of very long-chain fatty acids (C>24) in Atlantic salmon: Cloning, functional characterisation, and tissue distribution of an Elovl4 elongase. *Comp. Biochem. Physiol. B* **2011**, *159*, 122–129. [CrossRef] [PubMed]

18. Monroig, Ó.; Wang, S.; Zhang, L.; You, C.; Tocher, D.R.; Li, Y. Elongation of long-chain fatty acids in rabbitfish *Siganus canaliculatus*: Cloning, functional characterisation and tissue distribution of Elovl5- and Elovl4-like elongases. *Aquaculture* **2012**, *350–353*, 63–70. [CrossRef]

19. Li, S.; Monroig, Ó.; Navarro, J.C.; Yuan, Y.; Xu, W.; Mai, K.; Tocher, D.R.; Ai, Q. Molecular Cloning, functional characterization and nutritional regulation by dietary fatty acid profiles of a putative Elovl4 gene in orange-spotted grouper *Epinephelus coioides*. *Aquac. Res.* **2017**, *48*, 537–552. [CrossRef]

20. Monroig, Ó.; Hontoria, F.; Varó, I.; Tocher, D.R.; Navarro, J.C. Investigating the essential fatty acids in the common cuttlefish *Sepia officinalis* (Mollusca, Cephalopoda): Molecular cloning and functional characterisation of fatty acyl desaturase and elongase. *Aquaculture* **2016**, *450*, 38–47. [CrossRef]

21. Joseph, J.D. Lipid composition of marine and estuarine invertebrates. Part II: Mollusca. *Prog. Lipid Res.* **1982**, *21*, 109–153. [CrossRef]

22. Barnathan, G. Non-methylene-interrupted fatty acids from marine invertebrates: Occurrence, characterization and biological properties. *Biochimie* **2009**, *91*, 671–678. [CrossRef] [PubMed]

23. Kornprobst, J.M.; Barnathan, G. Demospongic acids revisited. *Mar. Drugs* **2010**, *8*, 2569–2577. [CrossRef] [PubMed]

24. Castro, L.F.C.; Wilson, J.M.; Gonçalves, O.; Galante-Oliveira, S.; Rocha, E.; Cunha, I. The evolutionary history of the stearoyl-CoA desaturase gene family in vertebrates. *BMC Evol.* **2011**, *11*, 132. [CrossRef] [PubMed]

25. Watts, J.L.; Browse, J. A palmitoyl-CoA-specific Δ9 fatty acid desaturase from *Caenorhabditis elegans*. *Biochem. Biophys. Res. Commun.* **2000**, *272*, 263–269. [CrossRef] [PubMed]

26. Jakobsson, A.; Westerberg, R.; Jacobsson, A. Fatty acid elongases in mammals: Their regulation and roles in metabolism. *Prog. Lipid Res.* **2006**, *45*, 237–249. [CrossRef] [PubMed]

27. Meyer, A.; Kirsch, H.; Domergue, F.; Abbadi, A.; Sperling, P.; Bauer, J.; Cirpus, P.; Zank, T.K.; Moreau, H.; Roscoe, T.J.; et al. Novel fatty acid elongases and their use for the reconstitution of docosahexaenoic acid biosynthesis. *J. Lipid Res.* **2004**, *45*, 1899–1909. [CrossRef] [PubMed]

28. Stukey, J.; McDonough, V.; Martin, C. Isolation and characterization of OLE1, a gene affecting fatty acid desaturation from *Saccharomyces cerevisiae*. *J. Biol. Chem.* **1989**, *264*, 16537–16544. [PubMed]

29. Porta, A.; Fortino, V.; Armenante, A.; Maresca, B. Cloning and characterization of a D9-desaturase gene of the Antarctic fish *Chionodraco hamatus* and *Trematomus bernacchii*. *J. Comp. Physiol. B* **2013**, *183*, 379–392. [CrossRef] [PubMed]

30. Agaba, M.; Tocher, D.R.; Dickson, C.; Dick, J.R.; Teale, A.J. Zebrafish cDNA encoding multifunctional fatty acid elongase involved in production of eicosapentaenoic (20:5*n*-3) and docosahexaenoic (22:6*n*-3) acids. *Mar. Biotechnol.* **2004**, *6*, 251–261. [CrossRef] [PubMed]

31. Bell, M.V.; Tocher, D.R. Biosynthesis of polyunsaturated fatty acids in aquatic ecosystems: General pathways and new directions. In *Lipids in Aquatic Ecosystems*; Arts, M.T., Brett, M., Kainz, M., Eds.; Springer: New York, NY, USA, 2009; pp. 211–236.

32. Tocher, D.R. Omega-3 long-chain polyunsaturated fatty acids and aquaculture in perspective. *Aquaculture* **2015**, *449*, 94–107. [CrossRef]

33. Reis, D.B.; Acosta, N.G.; Almansa, E.; Navarro, J.C.; Tocher, D.R.; Monroig, O.; Andrade, J.P.; Sykes, A.V.; Rodríguez, C. In vivo metabolism of unsaturated fatty acids in *Octopus vulgaris* hatchlings determined by incubation with ^{14}C-labelled fatty acids added directly to seawater as protein complexes. *Aquaculture* **2014**, *431*, 28–33. [CrossRef]

34. Agbaga, M.-P.; Brush, R.S.; Mandal, M.N.A.; Henry, K.; Elliott, M.H.; Anderson, R.E. Role of Stargardt-3 macular dystrophy protein (ELOVL4) in the biosynthesis of very long chain fatty acids. *Proc. Natl. Acad. Sci. USA* **2008**, *105*, 12843–12848. [CrossRef] [PubMed]

35. Albertin, C.B.; Simakov, O.; Mitros, T.; Wang, Z.Y.; Pungor, J.R.; Edsinger-Gonzales, E.; Brenner, S.; Ragsdale, C.W.; Rokhsar, D.S. The octopus genome and the evolution of cephalopod neural and morphological novelties. *Nature* **2015**, *524*, 220–224. [CrossRef] [PubMed]

36. Liu, H.; Guo, Z.; Zheng, H.; Wang, S.; Wang, Y.; Liu, W.; Zhang, G. Functional characterization of a Δ5-like fatty acyl desaturase and its expression during early embryogenesis in the noble scallop *Chlamys nobilis* Reeve. *Mol. Biol. Rep.* **2014**, *41*, 7437–7445. [CrossRef] [PubMed]

37. Li, M.; Mai, K.; He, G.; Ai, Q.; Zhang, W.; Xu, W.; Wang, J.; Liufu, Z.; Zhang, Y.; Zhou, H. Characterization of Δ5 fatty acyl desaturase in abalone *Haliotis discus hannai* Ino. *Aquaculture* **2013**, *416–417*, 48–56. [CrossRef]

38. Surm, J.M.; Prentis, P.J.; Pavasovic, A. Comparative analysis and distribution of Omega-3 LC-PUFA biosynthesis genes in marine molluscs. *PLoS ONE* **2015**, *10*, e0136301. [CrossRef] [PubMed]

39. Liu, H.; Zang, H.; Zheng, H.; Wang, S.; Guo, Z.; Zhang, G. PUFA Biosynthesis pathway in marine scallop *Chlamys nobilis* Reeve. *J. Agric. Food Chem.* **2014**, *62*, 12384–12391. [CrossRef] [PubMed]

40. Agbaga, M.-P.; Mandal, M.N.A.; Anderson, R.E. Retinal very long-chain PUFAs: New insights from studies on ELOVL4 protein. *J. Lipid Res.* **2010**, *51*, 1624–1642. [CrossRef] [PubMed]

41. McMahon, A.; Jackson, S.N.; Woods, A.S.; Kedzierski, W. A Stargardt disease-3 mutation in the mouse Elovl4 gene causes retinal deficiency of C_{32}–C_{36} acyl phosphatidylcholines. *FEBS Lett.* **2007**, *581*, 5459–5463. [CrossRef] [PubMed]

42. Furland, N.E.; Oresti, G.M.; Antollini, S.S.; Venturino, A.; Maldonado, E.N.; Aveldaño, M.I. Very long-chain polyunsaturated fatty acids are the major acyl groups of sphingomyelins and ceramides in the head of mammalian spermatozoa. *J. Biol. Chem.* **2007**, *282*, 18151–18161. [CrossRef] [PubMed]

43. Zadravec, D.; Tvrdik, P.; Guillou, H.; Haslam, R.; Kobayashi, T.; Napier, J.A.; Capecchi, M.R.; Jacobsson, A. ELOVL2 controls the level of *n*-6 28:5 and 30:5 fatty acids in testis, a prerequisite for male fertility and sperm maturation in mice. *J. Lipid Res.* **2011**, *52*, 245–255. [CrossRef] [PubMed]

44. Zhukova, N.V. Lipid classes and fatty acid composition of the tropical nudibranch mollusks *Chromodoris* sp. and *Phyllidia coelestis*. *Lipids* **2007**, *42*, 1169–1175. [CrossRef] [PubMed]

45. Zhukova, N.V. Lipids and fatty acids of nudibranch mollusks: Potential sources of bioactive compounds. *Mar. Drugs* **2014**, *12*, 4578–4592. [CrossRef] [PubMed]

46. Hastings, N.; Agaba, M.; Tocher, D.R.; Leaver, M.J.; Dick, J.R.; Sargent, J.R.; Teale, A.J. A vertebrate fatty acid desaturase with Δ5 and Δ6 activities. *Proc. Natl. Acad. Sci. USA* **2001**, *98*, 14304–14309. [CrossRef] [PubMed]

47. Lopes-Marques, M.; Ozório, R.; Amaral, R.; Tocher, D.R.; Monroig, Ó.; Castro, L.F.C. Molecular and functional characterization of a *fads2* orthologue in the Amazonian teleost, *Arapaima gigas*. *Comp. Biochem. Physiol. B* **2017**, *203*, 84–91. [CrossRef] [PubMed]

48. Folch, J.; Lees, N.; Sloane-Stanley, G.H. A simple method for the isolation and purification of total lipids from animal tissues. *J. Biol. Chem.* **1957**, *226*, 497–509. [PubMed]

49. Destaillats, F.; Angers, P. One-step methodology for the synthesis of FA picolinyl esters from intact lipids. *J. Am. Oil Chem. Soc.* **2002**, *79*, 253–256. [CrossRef]

50. Li, Y.; Monroig, Ó.; Zhang, L.; Wang, S.; Zheng, X.; Dick, J.R.; You, C.; Tocher, D.R. Vertebrate fatty acyl desaturase with Δ4 activity. *Proc. Natl. Acad. Sci. USA* **2010**, *107*, 16840–16845. [CrossRef] [PubMed]

marine drugs

MDPI

Article

An Extract from Shrimp Processing By-Products Protects SH-SY5Y Cells from Neurotoxicity Induced by Aβ25–35

Yongping Zhang [1,2], Guangling Jiao [3,4], Cai Song [1,2,5,*], Shelly Gu [2], Richard E. Brown [2], Junzeng Zhang [4], Pingcheng Zhang [2], Jacques Gagnon [3], Steven Locke [6], Roumiana Stefanova [4], Claude Pelletier [3], Yi Zhang [1] and Hongyu Lu [1]

[1] Research Institute for Marine Drugs and Nutrition, College of Food Science and Technology, Guangdong Ocean University, Zhanjiang 524088, China; zhangyp2015@163.com (Y.Z.); hubeizhangyi@163.com (Y.Z.); irislhy@126.com (H.L.)

[2] Department of Psychology and Neuroscience, Dalhousie University, Halifax, NS B3H 4R2, Canada; xiaomei.gu@dal.ca (S.G.); Richard.Brown@dal.ca (R.E.B.); pczhang2000@yahoo.com (P.Z.)

[3] Coastal Zones Research Institute Inc., 232B, avenue de l'Église, Shippagan, NB E8S 1J2, Canada; sebrina2006@gmail.com (G.J.); jacques.gagnon@umoncton.ca (J.G.); Claude.pelletier@umoncton.ca (C.P.)

[4] Aquatic and Crop Resource Development, National Research Council of Canada, 1411 Oxford Street, Halifax, NS B3H 3Z1, Canada; Junzeng.Zhang@nrc-cnrc.gc.ca (J.Z.); Roumiana.Stefanova@nrc-cnrc.gc.ca (R.S.)

[5] Graduate Institute of Neural and Cognitive Sciences, China Medical University Hospital, Taichung 40402, Taiwan

[6] Aquatic and Crop Resource Development, National Research Council of Canada, 550 University Avenue, Charlottetown, PE C1A 4P3, Canada; Steven.Locke@nrc-cnrc.gc.ca

* Correspondence: cai.song@dal.ca; Tel.: +86-759-239-6047

Academic Editors: Rosário Domingues, Ricardo Calado and Pedro Domingues
Received: 8 December 2016; Accepted: 15 March 2017; Published: 22 March 2017

Abstract: Increased evidence suggests that marine unsaturated fatty acids (FAs) can protect neurons from amyloid-β (Aβ)-induced neurodegeneration. Nuclear magnetic resonance (NMR), high performance liquid chromatography (HPLC) and gas chromatography (GC) assays showed that the acetone extract 4-2A obtained from shrimp *Pandalus borealis* industry processing wastes contained 67.19% monounsaturated FAs and 16.84% polyunsaturated FAs. The present study evaluated the anti-oxidative and anti-inflammatory effects of 4-2A in Aβ25–35-insulted differentiated SH-SY5Y cells. Cell viability and cytotoxicity were measured by using 3-(4,5-Dimethylthiazol-2-yl)-2,5-diphenyltetrazolium bromide (MTT) and lactate dehydrogenase (LDH) assays. Quantitative PCR and Western blotting were used to study the expression of neurotrophins, pro-inflammatory cytokines and apoptosis-related genes. Administration of 20 μM Aβ25–35 significantly reduced SH-SY5Y cell viability, the expression of nerve growth factor (NGF) and its tyrosine kinase TrkA receptor, as well as the level of glutathione, while increased reactive oxygen species (ROS), nitric oxide, tumor necrosis factor (TNF)-α, brain derived neurotrophic factor (BDNF) and its TrkB receptor. Aβ25–35 also increased the Bax/Bcl-2 ratio and Caspase-3 expression. Treatment with 4-2A significantly attenuated the Aβ25–35-induced changes in cell viability, ROS, GSH, NGF, TrkA, TNF-α, the Bax/Bcl-2 ratio and Caspase-3, except for nitric oxide, BDNF and TrkB. In conclusion, 4-2A effectively protected SH-SY5Y cells against Aβ-induced neuronal apoptosis/death by suppressing inflammation and oxidative stress and up-regulating NGF and TrkA expression.

Keywords: Acetone extract from shrimp processing by-product (4-2A); polyunsaturated fatty acids; Aβ25–35; human neuroblastoma cell (SH-SY5Y); neuroprotection; Alzheimer's disease (AD)

1. Introduction

The neurodegenerative process in Alzheimer's disease (AD) is associated with progressive accumulation of intracellular and extracellular neurotoxic amyloid-β (Aβ) oligomers in the brain [1–3]. Excessive Aβ-deposition may induce AD through oxidative stress and neuroinflammation. Amyloid-beta oligomers can activate microglia in vitro and in vivo [4], resulting in the production and release of reactive oxygen species (ROS) and pro-inflammatory cytokines such as tumor necrosis factor (TNF)-α, both of which can cause neural degeneration. Elevated levels of ROS interfere with the actions of many key molecules including enzymes, membrane lipids and DNA, which leads to cell apoptosis or death [5,6]. Increased pro-inflammatory cytokine release may stimulate neurons to produce increased amounts of Aβ oligomers and cause neuronal dysfunction and apoptosis [7,8]. Another hallmark of AD is decreased neurogenesis due to the dysfunction in neurotrophic signaling mechanisms [9]. In particular, nerve growth factor (NGF) and brain-derived neurotrophic factor (BDNF) and their receptors in the brain are disrupted. Reduced BDNF expression in the brain is a common feature of AD and cognitive dysfunction [10]. In addition, Aβ peptides are able to interfere with BDNF signal transduction pathways involved in neuronal survival and synaptic plasticity, hampering the transmission of neurotrophic responses [11].

Because the etiology of AD remain unknown, treatments that target AD are ineffective and often cause severe side-effects [12]. Most neurodegenerative diseases, including AD, are irreversible because the failure of neurogenesis and the increase in neuron death occurs before the clinical symptoms appear [13]. Thus, much effort is directed towards the discovery of neural pathways and their molecular mechanism that can be targeted by novel therapeutics to prevent AD. Natural substances with anti-oxidative and/or anti-inflammatory activity could provide effective treatments for the prevention of AD.

In the past decade, many studies have demonstrated that unsaturated fatty acids of marine origin, such as omega (*n*)-3 and *n*-9 fatty acids, could play a beneficial role in brain functions. Our previous studies have highlighted the effectiveness of dietary *n*-3 polyunsaturated fatty acids (PUFAs) as a potential treatment strategy for affective diseases [14,15]. Recent studies have demonstrated that eicosapentaenoic acid (EPA) and docosahexaenoic acid (DHA) possess neuroprotective properties because of their anti-oxidant and anti-inflammatory functions [14–16]. There are also data showing neuroprotective potential of monounsaturated fatty acids (MUFAs). For example, oleic acid, which belongs to *n*-9 MUFAs, has been found to modulate mitochondrial dysfunction, insulin resistance and inflammatory signaling [17–19] and act as neurotrophic factors for neurons [20]. Palmitoleic acid, a naturally occurring 16-carbon *n*-7 MUFA, and one of the most abundant fatty acids in the serum and tissues, was considered to be a lipokine and has been found to benefit some physiological function, such as regulating cell proliferation, and decreasing the expression of pro-inflammatory mediators and adipokines [21–23].

Tchoukanova and Benoit [24] developed a method to recover organic solids and oils from marine by-products. Using this method, shrimp oil was produced and found to be rich in long-chain unsaturated fatty acids [25]. The solid residue may also contain nutritional and bioactive ingredients that can be exploited. In the present study, an acetone extract 4-2A obtained from the solid phase was found to be rich in *n*-3 and *n*-9 unsaturated fatty acids. Because of the above mentioned neuroprotective function of unsaturated fatty acids, the present study aimed to determine whether the 4-2A extract from shrimp by-products would protect neurons from Aβ-induced neurotoxicity via its regulation of neurotrophic function, anti-inflammatory and anti-oxidative effects. To carry out this experiment, a cellular model of AD was set up using $A\beta_{25-35}$-insulted differentiated SH-SY5Y cells. Following this, the effects of 4-2A on $A\beta_{25-35}$-induced changes in cell viability, oxidative stress (ROS, NO, and GSH) and neurotrophins (NGF, BDNF and their TrkA and TrkB receptors) were measured. Since TNF-α is a key "pro-neuropathic" cytokine [26] and can activate a pro-apoptotic factor JNK pathway and trigger cellular death signaling [27], TNF-α expression was measured to test 4-2A anti-inflammatory effect in the present study. Then, the ability of 4-2A to regulate the expression of apoptosis-related genes (Bcl-2,

Bax and Caspase-3) in the model was explored. The experimental design and content is presented in Figure 1.

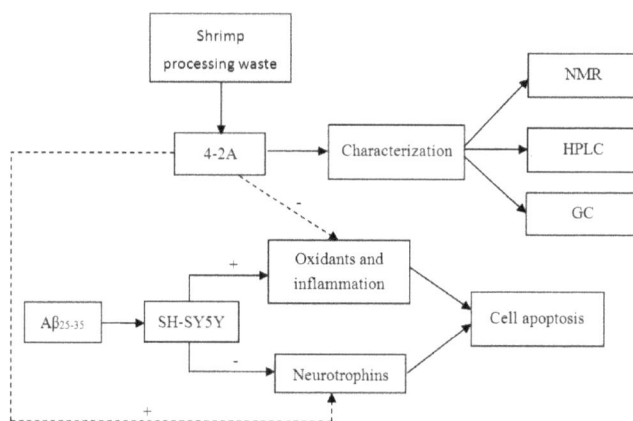

Figure 1. Experimental design and contents. NMR: Nuclear Magnetic Resonance; HPLC: High Performance Liquid Chromatography; GC: Gas Chromatograph. "+": strengthening; "−": weakening.

2. Results

2.1. Characterization of Shrimp Extract 4-2A

Reddish and oily sample 4-2A was extracted from the solid residue of shrimp processing waste using acetone and yielded 2.55% of original dry mass. Both ^{1}H-NMR and ^{13}C-NMR spectra indicated that this shrimp extract was rich in lipids. The detailed assignments of the most common proton and carbon signals are shown in Figure 2. They have been compiled in the online resources of AOCS Lipid Library by a wide range of fatty compounds [28]. The spectra ranging from 0 to 6 ppm in the ^{1}H-NMR spectrum covered the proton chemical shifts of most lipids (Figure 2A). To be noted, the signals around 2.8 ppm indicated the presence of methylenes between two double bonds –CH=CH-CH$_2$-CH=CH-, a typical feature of polyunsaturated fatty acids. The shifts at 0.93–1.01 ppm suggested the protons of ω-3 terminal methyl groups. The intensity of these signals indicated that 4-2A contained relatively high level of ω-3 PUFAs. The chemical shifts between 3.5 and 4.0 ppm showed the presence of small amount of monoglycerides.

In the ^{13}C-NMR spectrum (Figure 2B), the chemical shifts at 180.3 ppm indicated the carbons of C=O groups. The shifts around 130 ppm showed the two olefinic carbon atoms of double bonds. The carbons of the terminal methyl groups of lipids were identified by the chemical shifts at 15.3–15.6 ppm. Collectively, the major components of 4-2A were identified by NMR as lipids containing large amount of unsaturated fatty acids.

Lipid standards were subjected on HPLC to qualify the separations. As indicated in Figure 3, the major components of 4-2A are free fatty acids and monoglycerides with elution window from 16 min to 36 min by charged aerosol detector (CAD). Based on the chromatogram, this extract does not contain much triglycerides and phospholipids. This is consistent with the result obtained from NMR analyses.

In order to understand the fatty acids profiles in this shrimp extract, the total fatty acids were released from 4-2A and analyzed on GC instrument. As a result shown in Table 1, the major FAME profiles of 4-2A was 18:1n-9 (20.65%), 16:1n-7 (14.75%), 22:1n-7 (11.04%), 20:5n-3 (8.45%), 20:1n-9 (7.97%), 18:1n-7 (7.07%), 16:0 (6.56%) and 22:6n-3 (6.54%). Most fatty acids (over 93%) are unsaturated, including 67.19% of monounsaturated fatty acids (MUFAs) and 16.84% of polyunsaturated fatty acids (PUFAs). To be mentioned, 4-2A contains 14.99% of omega-3 fatty acids (20:5n-3, Eicosapentaenoic acid, EPA and 22:6n-3, Docosahexaenoic acid, DHA).

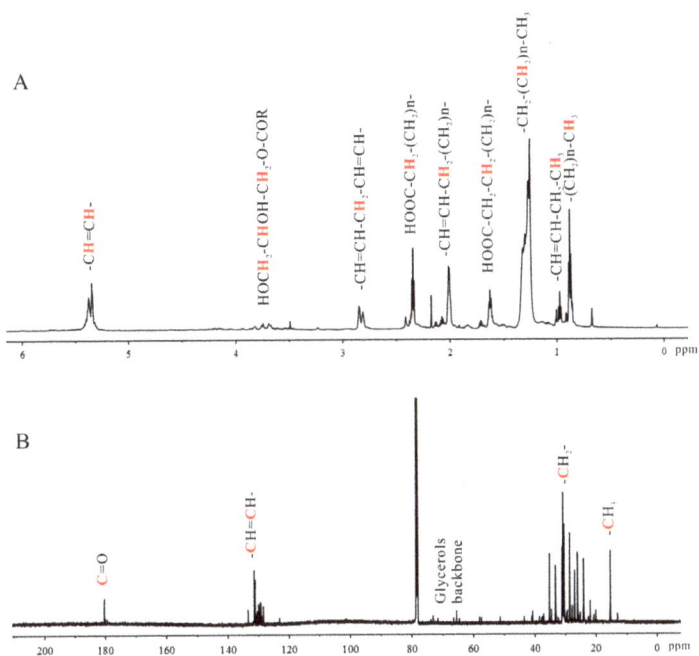

Figure 2. Characterization of 4-2A by NMR: (**A**) ^{1}H-NMR spectrum ($n \geq 1$); and (**B**) ^{13}C-NMR spectrum.

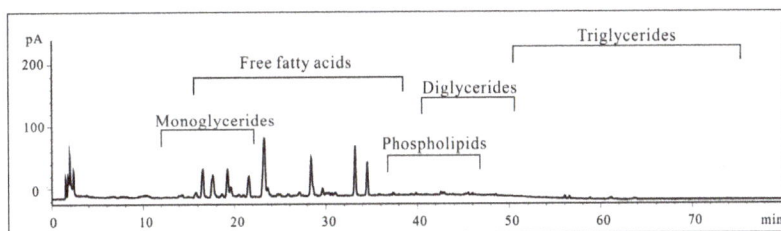

Figure 3. Characterization of 4-2A by HPLC with CAD detector.

Table 1. Major fatty acid methylated esters (FAME) profiles of 4-2A by GC analysis.

FAME	Intensity (%)	Stdev ($n = 4$, %)
14:0	2.10	0.25
16:0	6.56	0.58
18:0	0.70	0.08
16:1*n*-7	14.75	0.21
18:1*n*-5	0.77	0.16
18:1*n*-7	7.07	0.31
18:1*n*-9	20.65	2.62
20:1*n*-7	1.08	0.03
20:1*n*-9	7.97	0.97
20:1*n*-11	1.95	0.01
22:1*n*-7	11.04	0.24
22:1*n*-9	1.91	0.78
18:2*n*-6	0.86	0.10
20:4*n*-6	0.99	0.12

Table 1. *Cont.*

FAME	Intensity (%)	Stdev (n = 4, %)
20:5n-3	8.45	0.83
22:6n-3	6.54	1.02
Total	93.39	
Saturated	9.36	
Monounsaturated	67.19	
Polyunsaturated	16.84	
Omega-3	14.99	
Omega-6	1.85	
Omega-7	33.94	
Omega-9	30.53	

In brief, the acetone extract 4-2A obtained from shrimp processing wastes was identified by NMR, HPLC and GC to contain mainly unsaturated free fatty acids and monoglycerides, including ω-3 PUFAs.

2.2. $A\beta_{25-35}$-Induced Decrease in Neuronal Cell Viability Was Ameliorated by 4-2A

The cell morphology of differentiated SH-SY5Y cells is more like the classic neuron-like cells with long synapse compared to undifferentiated cells (Figure 4). As detected by 3-(4,5-Dimethylthiazol-2-yl)-2,5-diphenyltetrazolium bromide (MTT) method, $A\beta_{25-35}$ significantly decreased cell viability of differentiated SH-SY5Y cells in a time- and dose-dependent manner, such as at 15 μM for 12 h ($p > 0.05$), 24 h ($p < 0.05$) and 48 h ($p < 0.01$); and at 10 μM ($p > 0.05$), 15 μM ($p < 0.05$) and 20 μM ($p < 0.01$) for 24 h (Figure 5A). 4-2A treatment (1–20 μg/mL) alone did not exert any significant influence on the survival rate of SH-SY5Y cells, though 40 μg/mL of 4-2A slightly decreased the cell viability ($p > 0.05$) (Figure 5B). However, pretreated with different concentrations of 4-2A markedly attenuated the reduction of cell viability caused by $A\beta_{25-35}$ in a dose-dependent manner at 1 ($p > 0.05$), 5 ($p < 0.05$), and 10–20 μg/mL (all $p < 0.01$) (Figure 5C). The protective role of 4-2A against $A\beta_{25-35}$-induced insults in SH-SY5Y cells was further confirmed by lactate dehydrogenase (LDH) release assay (Figure 5D), which is an index of cell death. Combining the results from above assays, we could safely draw the conclusion that 4-2A could effectively protect the differentiated SH-SY5Y cells from $A\beta_{25-35}$-induced cellular damage.

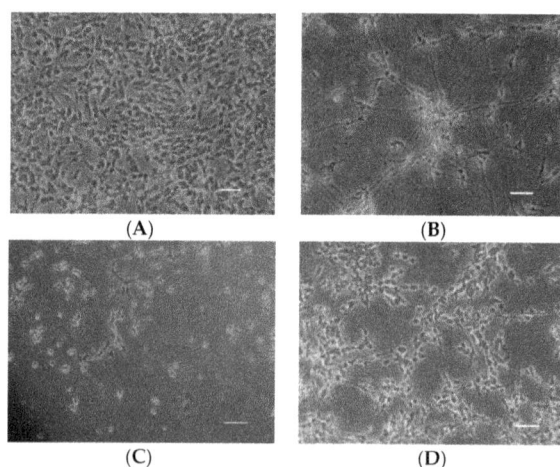

Figure 4. Cell morphology of undifferentiated SH-SY5Y cells (**A**); differentiated SH-SY5Y cells (**B**); differentiated SH-SY5Y cells insulted by $A\beta_{25-35}$ (**C**); and differentiated SH-SY5Y cells treated with 4-2A and $A\beta_{25-35}$ (**D**). Scale bar = 50 μm.

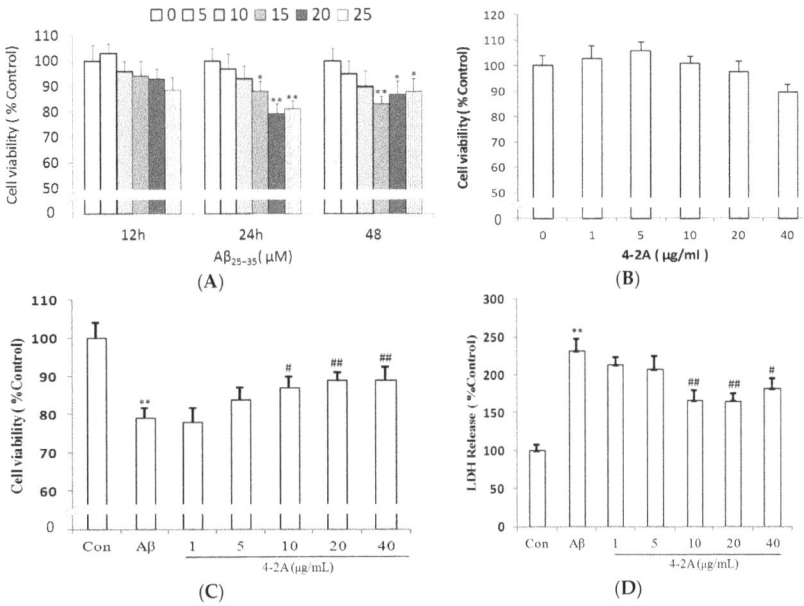

Figure 5. Cell viability and Cytotoxicity: (**A**) Treatment of Aβ$_{25-35}$ at various doses for 12 h, 24 h and 48 h; (**B**) cells treated with 4-2A at indicated doses for 24 h; (**C**) changes in cell survival rate after the treatment with 20 μM Aβ$_{25-35}$ for 24 h in the absence or presence of 4-2A at indicated doses; and (**D**) changes in lactate dehydrogenase (LDH) release after the treatment with 20 μM Aβ$_{25-35}$ for 24 h in the absence or presence of 4-2A at indicated doses. Data represent mean ± SEM of three separate experiments (*n* = 4 in each experiment) and three mean values from independent experiment were used for statistics. * $p < 0.05$, ** $p < 0.01$ vs. control group; # $p < 0.05$, ## $p < 0.01$ vs. Aβ$_{25-35}$ group.

2.3. Treatment with 4-2A Attenuated the Changes in ROS, Nitric Oxide (NO) and Glutathione (GSH) Level Induced by Aβ$_{25-35}$

Figure 6A illustrated that Aβ$_{25-35}$ markedly increased ROS fluorescence when compared to control group ($p < 0.01$). However, cells pretreated with 4-2A (10 μg/mL) showed a partial decrease in mean fluorescence intensities by about 23% when compared to Aβ$_{25-35}$-insulted group ($p < 0.05$). As shown in Figure 6C, a significant increase of about 44.42% in the level of nitrate was observed when the cells were treated with Aβ$_{25-35}$ alone ($p < 0.05$), while 4-2A pre-treatment could not attenuate the Aβ$_{25-35}$-induced change in NO concentration. The results shown in Figure 6C indicate a marked reduction in the GSH content of the Aβ$_{25-35}$ insulted cells ($p < 0.05$), and 4-2A pre-treatment could completely restore this reduction ($p < 0.01$).

2.4. Treatment with 4-2A Attenuated the Aβ$_{25-35}$-Inducued Increased Expression of Pro-Inflammatory Cytokine TNF-α

TNF-α mRNA expression was significantly increased by Aβ$_{25-35}$ administration as early as 4 h of incubation ($p < 0.05$), but the increase in protein expression could not be found until 12 h of incubation with Aβ$_{25-35}$ ($p < 0.01$). Pretreatment with 4-2A alone did not significantly affect TNF-α expression either in mRNA or in protein levels, but it could partially but significantly decrease the effect of Aβ$_{25-35}$ (mRNA expression: $p < 0.05$ and protein expression $p < 0.01$, Figure 7).

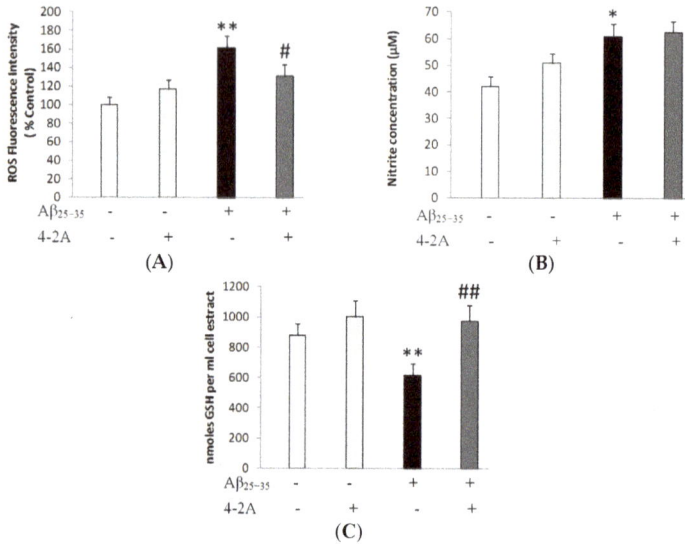

Figure 6. The effect of 4-2A on the changes of oxidative and anti-oxidative response induced by Aβ25-35. (A) ROS production in fully differentiated SH-SY5Y cells treated with Aβ25-35 20 μM and/or 4-2A 10 μM for 24 h. Data represent mean ± SEM percentage of control. (B) NO production and (C) GSH content in fully differentiated SH-SY5Y cells treated with Aβ25-35 20 μM and/or 4-2A 10 μM for 24 h. Data represent mean ± SEM of concentration (μM per mL). The statistics were based on data from six separate experiments, $n = 6$ in each experiment. * $p < 0.05$, ** $p < 0.01$ vs. control group; # $p < 0.05$, ## $p < 0.01$ vs. Aβ25-35 group.

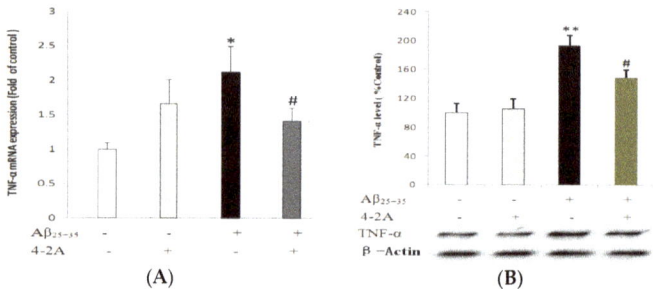

Figure 7. The expression of pro-inflammatory cytokine tumor necrosis factor-alpha (TNF-α). mRNA expression level (A); and protein expression level (B) of TNF-α, normalized to the corresponding level expression of housekeeping gene β-Actin, in differentiated SH-SY5Y cells treated with Aβ25-35 (20 μM) and/or 4-2A (10 μg/mL) for indicated times. The data of both mRNA and protein expression from six independent experiments ($n = 1$ in each experiment) were analyzed and expressed as mean ± SEM. * $p < 0.05$, ** $p < 0.01$ vs. control group; # $p < 0.05$ vs. Aβ25-35 group.

2.5. Effect of 4-2A on the Expression of Neurotrophins: NGF and BDNF and Their TrkA and TrkB Receptors

Similar to TNF-α gene, the obvious change of NGF gene expression appeared at 4 h of Aβ25-35 treatment. Compared to control group, NGF mRNA expression in Aβ25-35-insulted cells was significantly down-regulated ($p < 0.01$, Figure 8A). Meanwhile, BDNF gene mRNA expression in Aβ25-35-insulted cells was also down-regulated at both 4 and 8 h ($p < 0.01$, Figure 8B). Compared to the control, a significantly decreased protein expression of NGF was found until 12 h incubation ($p < 0.01$, Figure 8C), whereas BDNF protein was significantly increased in $A\beta_{25-35}$-insulted cells ($p < 0.01$, Figure 8D). However, in cells treated with 4-2A only, either NGF or BDNF gene was not affected at both mRNA and protein level compared to the control ($p > 0.05$). 4-2A pre-treatment could partially but significantly attenuate the $A\beta_{25-35}$-induced change in NGF mRNA ($p < 0.05$) and protein ($p < 0.01$) expression (Figure 8A,C) and BDNF mRNA expression (both $p < 0.01$, Figure 8D), but not in BDNF protein expression ($p > 0.05$, Figure 8D).

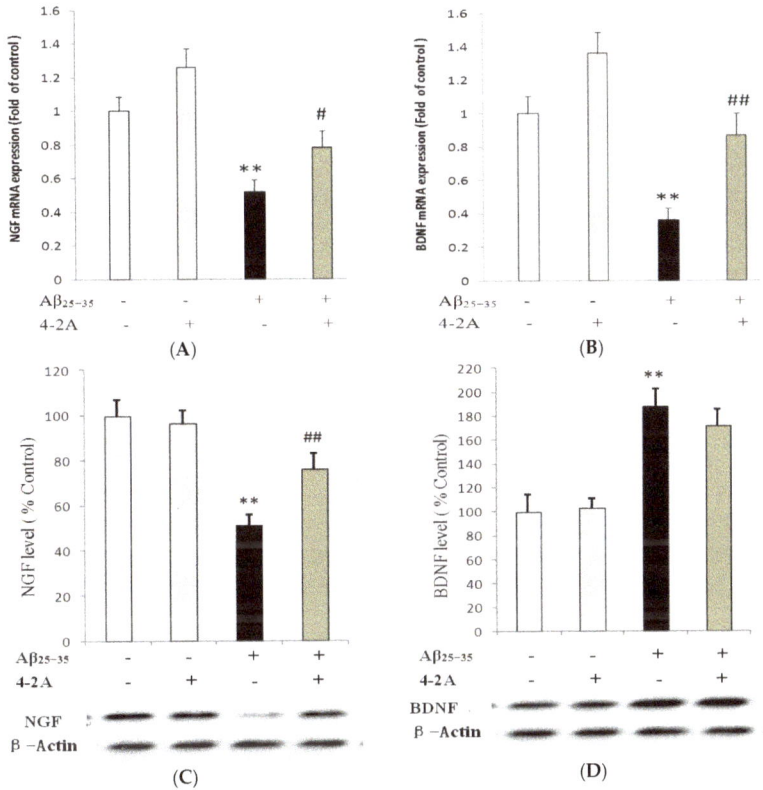

Figure 8. The effect of 4-2A on the expression of neurotrophin genes: NGF and BDNF. NGF mRNA expression (**A**); and protein expression (**C**); and BDNF mRNA expression (**B**); and protein expression (**D**), normalized to the expression of housekeeping gene β-Actin, in differentiated SH-SY5Y cells treated with $A\beta_{25-35}$ (20 μM) and/or 4-2A (10 μg/mL) for 4, 8, 12, 24 h. The data of both mRNA and protein expression from six independent experiments ($n = 1$ in each experiment) were analyzed and expressed as mean ± SEM. ** $p < 0.01$ vs. control group; # $p < 0.05$, ## $p < 0.01$ vs. $A\beta_{25-35}$ group.

Unlike previously discussed genes, TrkA and TrkB were only affected at 24 h following $A\beta_{25-35}$ administration. $A\beta_{25-35}$ significantly decreased TrkA protein expression ($p < 0.01$). Pretreatment with 4-2A only had no effect on TrkA protein expression, but it could significantly attenuate the effect of $A\beta_{25-35}$ on the receptor ($p < 0.01$, Figure 9A). In contrast to TrkA changes, TrkB protein in the cells treated with 4-2A only was significantly increased ($p < 0.01$) compared to the control. In the cells administrated with $A\beta_{25-35}$ alone, a less but significantly increased protein of TrkB was also found ($p < 0.05$). 4-2A pretreatment did not reverse $A\beta_{25-35}$-induced change, but further increased in TrkB expression ($p < 0.05$, Figure 9B).

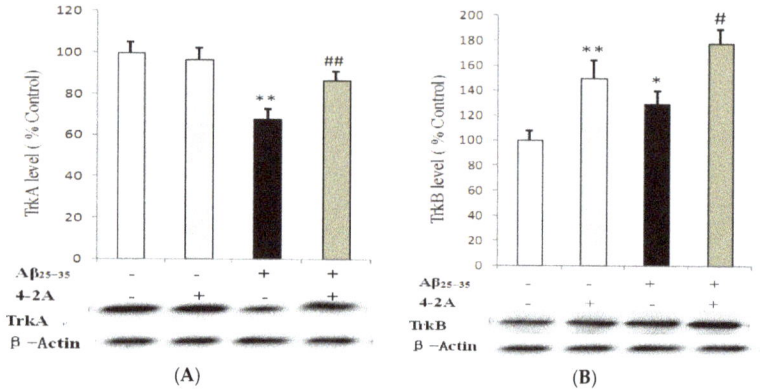

(A)   (B)

Figure 9. The effect of 4-2A on the expression of neurotrophin receptors: TrkA and TrkB. TrkA (**A**); and TrkB (**B**) protein expression, normalized to the protein expression of housekeeping gene β-Actin, in differentiated SH-SY5Y cells treated with Aβ$_{25-35}$ (20 μM) and/or 4-2A (10 μg/mL) for 24 h. The data of protein expression from six independent experiments ($n = 1$ in each experiment) were analyzed and expressed as mean ± SEM. * $p < 0.05$, ** $p < 0.01$ vs. control group; # $p < 0.05$, ## $p < 0.01$ vs. Aβ$_{25-35}$ group.

2.6. Treatment with 4-2A Regulated the Expression of Apoptosis Related Genes: Bax, Bcl-2 and Caspase-3

The genes Bax, Bcl-2 and Caspase-3 protein expression were tested at different incubation times with Aβ$_{25-35}$ (4, 8, 12 and 24 h). The expression of Bcl-2 was strongly decreased by Aβ$_{25-35}$ at 24 h of incubation with Aβ$_{25-35}$, which was significantly increased ($p < 0.01$) by 4-2A. 4-2A by itself had no effect on Bcl-2. With regard to Bax, no obvious change was found after 4–24 h incubation. The Bax: Bcl-2 ratio was significantly increased in cells administrated with Aβ$_{25-35}$ alone ($p < 0.01$). Pretreatment with 4-2A significantly reversed the ratio of Bax: Bcl-2 to the control level ($p < 0.01$, Figure 10A). For Caspase-3, Aβ$_{25-35}$ significantly increased its protein expression ($p < 0.01$), whereas pretreatment with 4-2A only did not affect the expression, but it could partially attenuate the effect of Aβ$_{25-35}$ ($p < 0.05$, Figure 10B).

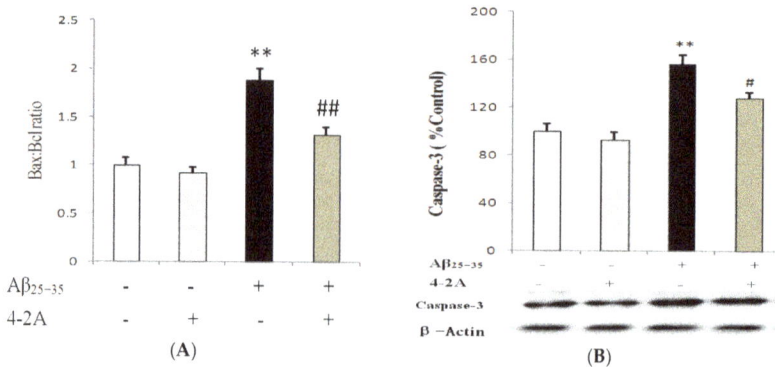

(A)   (B)

Figure 10. 4-2A regulated the expression of apoptosis related genes: Bcl-2, Bax and Caspase-3. Bax:Bcl-2 ratio (**A**); and Caspase-3 (**B**) protein expression (normalized to the protein expression of housekeeping gene β-Actin) in differentiated SH-SY5Y cells treated with Aβ$_{25-35}$ 20 μM and/or 4-2A 10 μg/mL for 24 h. The data of both mRNA and protein expression from six independent experiments ($n = 1$ in each experiment) were analyzed and expressed as mean ± SEM. ** $p < 0.01$ vs. control group; # $p < 0.05$, ## $p < 0.01$ vs. Aβ$_{25-35}$ group.

3. Discussion

Aβ_{25-35}, an active fragment corresponding to amino acids 25–35 in full-length Aβ, possesses the same β-sheet structure and retains full toxicity of full-length Aβ_{1-42} [29]. Many experiments have demonstrated that Aβ_{25-35} can induce neurotoxicity and AD-like pathology, such as activating glial cells, increasing cholinesterase expression [30] and oxidative stress [31], as well as impairing spatial learning and memory [32,33]. Our previous study also showed that Aβ_{25-35} could result in neuroinflammatory response and dysfunction of neurotrophin system [34]. As such, Aβ_{25-35} has been popularly utilized to induce in vitro or in vivo AD models [32,33,35,36]. The present study demonstrates that 4-2A protects the SH-SY5Y cells from Aβ_{25-35}-induced reduction in cell viability by suppressing oxygen stress and inflammation, regulating neurotrophin levels, hence attenuating neuron apoptosis.

Even though the exact pathophysiology of AD is unclear, oxidative stress has been found to play a fatal role in the pathogenic process of AD. Previous studies demonstrated that toxicity of Aβ_{25-35} in models of neurodegenerative diseases in vitro and in vivo was associated with the enhancement of ROS and NO liberation and oxidative damage [37–41], which up-regulated redox-sensitive transcription factors such as NF-κB, an important factor responsible for oxidative and inflammatory reactions in AD [42]. In agreement with these studies, Aβ_{25-35} increased the ROS and NO production from SH-SY5Y cells in the present study. ROS and NO are oxidants in the Alzheimer's brain. However, NO is also a neurotransmitter, which may protect synapses by increasing neuronal excitability [43,44]. Thus, whether the Aβ-induced increase in NO acts as a compensatory and neuroprotective or neurotoxic role is unclear. The administration of 4-2A into Aβ_{25-35}-treated cells partially but significantly decreased ROS production, which means 4-2A may partially protect neurons against free radicals or may improve mitochondrial dysfunction which is the major source of ROS. Furthermore, the decrease in GSH content caused by Aβ_{25-35} was significantly attenuated by 4-2A treatment. These findings indicated that 4-2A could restore the imbalance between oxidative stress factors and antioxidant systems. However, 4-2A could not affect the Aβ_{25-35}-induced NO change, which may be related to a hypothesis that 4-2A may contribute to the self-protective ability of Aβ-insulted cells if this increase in NO is neuroprotective.

Inflammation is another important contributor to neurodegenerative diseases. Experimental and clinical findings provide evidence for the hypothesis that the neuronal degeneration in AD is not simply due to the Aβ deposition, but to neuroinflammation [45,46]. Consistent with the above studies, an increased expression of TNF-α gene was found in the present AD cellular model. Increased TNF-α can activate a pro-apoptotic factor JNK pathway that is involved in cell differentiation and proliferation [47] and trigger cellular death signaling [27]. In the present study, 4-2A showed anti-inflammatory property since it partially down-regulated the expression of TNF-α either in the mRNA or in protein level.

In our previous study, we showed that neuroinflammation could reduce the levels of neurotrophic factors, such as NGF [48] and BDNF [49]. In the present study, Aβ_{25-35} differently regulated the expression of the two neurotrophic factor genes. The decreased expression of NGF and its receptor TrkA seems to be reasonable for the low neuronal viability induced by Aβ. However, a significant increase in the expression of BDNF and its receptor TrkB protein in SH-SY5Y cells exposed to Aβ_{25-35} was unexpected, which was unparalleled by the decreased mRNA level in BDNF. The present data are partially in agreement with a previous in vitro study showing that the exposure of SH-SY5Y cells to Aβ_{25-35} induced a significant increase of BDNF [50]. We speculate that the increase of BDNF levels might act as a compensatory response against amyloid toxicity, while Aβ_{25-35} may trigger distinct effects on BDNF expression in different systems, conditions and incubation time.

Over-production of pro-inflammatory cytokines, oxidative stress and neurotrophin dysfunction may reduce neurogenesis and induce apoptosis [51,52], which result in neuronal death and memory loss in AD [53–56]. In the present study, Aβ_{25-35} treatment increased the pro-apoptotic Bax/anti-apoptotic Bcl-2 ratio and Caspase-3 expression, which were attenuated by pretreatment of 4-2A, suggesting 4-2A can regulate the imbalance between pro- and anti-apoptotic gene expression. The mechanism by which 4-2A attenuated Aβ-induced neuron damage may be through its unique components.

The acetone extract 4-2A from the shrimp by-products consists of lipids containing large amount of unsaturated fatty acids, especially monounsaturated fatty acid (MUFAs, 67.19%), including n-9 MUFAs and n-7 MUFAs. Oleic acid (18:1n-9) and palmitoleic acid (16:1n-7) are the most common MUFAs, which represent n-9 and n-7 MUFAs, respectively. The total fatty acids analysis profiles that 18:1n-9 (oleic acid, 20.65%) is the most abundant fatty acid in 4-2A. As the major n-9 MUFAs, oleic acid is high in olive oil which is the main characteristic of the Mediterranean Style Diet (MSD). Recent data from large epidemiological studies suggest a relationship between MSD adherence and significant reduction in incidence of Parkinson's disease and AD and mild cognitive decline or risk of dementia [57]. Moreover, the protective effect of oleate against palmitate-induced mitochondrial dysfunction, insulin resistance and inflammatory signaling has been evaluated in several cell models [17,58]. Interestingly, in neuronal cells, Kwon et al. demonstrated that oleate preconditioning was superior to DHA or linoleate (18:2n-6) in the protection from the above palmitate-induced insults [58]. These effects may be associated with the neuroprotective ability of 4-2A to exert anti-inflammatory effects and restore the imbalance between oxidative stress factors and antioxidant systems. In addition, oleic acid may behave as a neurotrophic factor for neurons via up-regulation of molecular markers of axonal and dendritic growth, such as GAP-43 and MAP-2 [20]. This may be an important factor related to the ability of 4-2A to modulate Aβ_{25-35}-induced abnormality in neurotrophic systems. These results strongly suggest that the ability of 4-2A to prevent SH-SY5Y cells from neurotoxicity induced by Aβ_{25-35} may be associated with its high content of oleic acid.

In recent years, n-7 palmitoleic acid (16:1n-7, 14.75%) has drawn increasing attention since its characterization as a bioactive lipid that coordinates metabolic crosstalk between the liver and adipose tissue [59]. Studies in cultured hepatocytes and mouse models of diet-induced obesity suggest that palmitoleic acid has anti-inflammatory and insulin-sensitizing effects [60]. Moreover, both in vitro and in vivo studies have demonstrated that palmitoleic acid can decrease the level of pro-inflammatory mediators and reduce the level of C-reactive protein in mice [21,23]. These anti-inflammatory effects may also contribute to the neuro-protective effect of 4-2A in the present study. However, because of rare reports about the effect of other MUFAs in health, the role of other two n-9 MUFAs (20:1n-9 and 22:1n-9) and three n-7 MUFAs (18:1n-7, 20:1n-7 and 22:1n-7) of longer chain length ($\geq$C18) in 4-2A is unclear.

It is not surprising that 4-2A could protect SH-SY5Y neurons-like cells against Aβ_{25-35}-induced apoptosis and death since 4-2A is rich in polyunsaturated n-3 fatty acids, such as EPA and DHA. Our previous studies have confirmed that n-3 fatty acids possess broad spectrum of neuroprotective activities in both in vitro and in vivo experiments because of their anti-oxidant and anti-inflammatory properties [61–66]. Additionally, Wu et al. [67] showed that n-3 PUFAs significantly attenuated the gene expression of pro-inflammatory cytokines, including IL-1β, IL-6, and TNF-α, and the protein levels of NF-κB and iNOS in brain tissues of rats with doxorubicin-induced depressive-like behaviors and neurotoxicity. In another in vivo study, Taepavarapruk and Song [48] revealed that n-3 PUFAs improved memory through IL-1-glucocorticoid-ACh release and IL-1-NGF-ACh release pathways. Moreover, there is evidence that dietary supplementation with DHA reduced the intraneuronal accumulation of not only amyloid-beta, but also tau, another important pathology marker for AD, in the 3xTg-AD mouse model via decreasing steady-state levels of presenilin 1 [68,69]. Based on these findings, there is no doubt that the neuroprotective effect of 4-2A was related to the rich n-3 PUFAs component. As the concentration of n-6 PUFAs are very low (1.85%), their effect in 4-2A can be ignored.

Finally, the saturated fatty acid in 4-2A may have no effect or only serve as an energy component in the cell cultural system because no biological activity was reported.

Taken together, shrimp processing by-product acetone extract 4-2A was prepared and characterized as lipids consisting of large amount of unsaturated fatty acids by NMR, HPLC-CAD and GC analyses. As a multifunctional agent, 4-2A showed potent inhibition against Aβ_{25-35} cytotoxicity, which confirms our hypothesis that 4-2A may exert neuroprotective effect via anti-oxidant, anti-inflammation and increasing neurotrophins. These effects of 4-2A may result from its various FAs components targeting various molecular pathways. The limitations of the present study are firstly that the treatment of 4-2A should be studied in the animal model of AD. Secondly, the effect of each FA component and their combination in different ratios should be determined, which can reveal the exact role of each FA and potential synergistic action of their combination in inflammatory, oxidant and neurotrophic functions in the brain. Thirdly, as the main components of neuronal membranes, the effect of FAs in 4-2A on the function of cell membrane and other mechanisms involved should also be investigated.

4. Materials and Methods

4.1. Preparation of the Shrimp Extract 4-2A

Fatty acids from the solid residue of shrimp *Pandalus borealis* processing waste were extracted with hexane (15 mL/g, *v/w* dry weight) at room temperature for 30 min, followed by 10 min of sonication, filtered, and then extracted with acetone under the same conditions. The liquid acetone extracts were combined, concentrated using rotary evaporator at reduced pressure, and then dried under N_2. The resulting extract was named 4-2A.

4.2. Characterization of Shrimp Extract 4-2A

4.2.1. Characterization of 4-2A by NMR

Both ^{1}H-NMR and ^{13}C-NMR spectra of 4-2A (1 mg, dissolved in 700 μL of 99.8% CDCl$_3$) were recorded at 4 °C on a Bruker AV-III 700 MHz spectrometer (Bruker BioSpin Canada, Milton, ON, Canada), equipped with a 5 mm TCI cryoprobe. The data were processed using the standard TopSpin V 2.1 (Bruker BioSpin Canada, Milton, ON, Canada) software.

4.2.2. Characterization of 4-2A by HPLC

HPLC analysis was carried out on an 1100 series instrument (Agilent Technologies, Santa Clara, CA, USA) with a Thermo Scientific™ Dionex™ Corona™ Charged Aerosol Detector (Burlington, ON, Canada). The method was adopted from dionex.com [70] with modifications. Samples were prepared by diluting 1 mg of analyte in 1 mL of methanol/chloroform (1:1, *v/v*). Ten microliter of each sample solution was injected on a HALO C8 column (2.1 mm × 100 mm, 2.7 μm, Advanced Materials Technology Inc., Wilmington, DE, USA) with a 0.45 mL/min flow rate at 40 °C. A 5 × 2.1 mm C8 cartridge (2.7 μm, Advanced Materials Technology Inc.) was used as a pre-column. Mobile phase A methanol/water/acetic acid (750:250:4) and phase B acetonitrile/methanol/tetrahydrofuran/acetic acid (500:375:125:4) were proceeded from 100% A at the beginning at 0.8 mL/min to 30% A and 70% B at 40 min at 1.0 mL/min, then reached to 20% A and 80% B in 10 min at 1.0 mL/min and finally to 100% B in 10 min at 1.0 mL/min, held for 10 min, and then back to 100% A in 10 min at 0.8 mL/min. The column temperature was at 40 °C.

Free fatty acids (myristic acid, *cis*-palmitoleic acid, *cis*-vaccenic acid, *trans*-vaccenic acid, oleic acid, elaidic acid, pentadecanoic acid, palmitic acid, heptadecanoic acid, *cis*-10-heptadecenoic acid, stearic acid, 9-*cis*,12-*cis*-linoleic acid, *cis*,*cis*,*cis*-9,12,15-octadecatrienoic acid, *cis*,*cis*,*cis*-6,9,12-octadecatrienoic acid, *cis*-11-eicosenoic acid, arachidonic acid, *cis*-5,8,11,14,17-eicosapentaenoic acid, *cis*-4,7,10,13,16,19-docosahexaenoic acid and tricosanoic acid), monoglycerides (DL-α-palmitin, 1-monopalmitoleoyl-*rac*-glycerol and 1-Oleoyl-*rac*-glycerol), diglycerides (1,2-distearoyl-*rac*-glycerol and 1,3-distearoylglycerol), triglycerides (olive oil), and phospholipids mixer (L-α-lysophosphatidylcholine, L-α-phosphatidylcholine, L-α-phosphatidylethanolamine and L-α-phosphatidy- linositol sodium salt) purchased from Sigma-Aldrich Inc. (St. Louis, MO, USA) were applied on the column under the same conditions.

4.2.3. GC Analysis of 4-2A

GC analysis was performed as described by Jiao et al. [25]. Total fatty acids were released from 4-2A by hydrolyzing with 1.5 N NaOH methanol solution under N_2 at 100 °C for 5 min, then methylated with 14% of BF_3 methanol solution at 100 °C for 30 min. Distilled water was then added to stop the reaction. The methylated fatty acids were extracted with hexane and subjected on an Agilent Technologies 7890A GC spectrometer (Agilent Technologies, Santa Clara, CA, USA) using an Omegawax 250 fused silica capillary column (30 m × 0.25 mm × 0.25 μm film thickness, Sigma-Aldrich, St. Louis, MO, USA). Supelco® 37 component fatty acid methylated esters (FAME) mix and PUFA-3 (Supelco, Bellefonte, PA, USA) were used as FAME standards.

4.3. SH-SY5Y Culture and Differentiation

SH-SY5Y cells were obtained from ATCC (CRL-2266, Lot. 61983120) and were maintained in DMEM/F12 medium (Gibco®, Burlington, ON, Canada) supplemented with 10% fetal bovine serum (FBS, Gibco®, Burlington, ON, Canada) and 1% penicillin–streptomycin in a humid atmosphere of 5% CO_2 at 37 °C. SH-SY5Y cells were differentiated into fully human neuron-like cells with treatment all-*trans*-retinoic acid (RA, Sigma Aldrich, Oakville, ON, Canada), at a final concentration of 10 μM in DMEM/F12 with 3% FBS (media changed every 2 days), for 7–8 days.

4.4. Experimental Design

Once the SH-SY5Y cells were differentiated into classic neuron-like cells with long axons compared to undifferentiated cells, they were used in the experiments. $A\beta_{25-35}$ (synthetic, ≥97% HPLC, Sigma Aldrich, Oakville, ON, Canada, A4559) was dissolved in sterile double-distilled water at a concentration of 1 mmol/L stock solution, was aged at 37 °C for 4 day, and then stored at −20 °C before use (This aging procedure was proved to produce birefringent fibril-like structures, globular amorphous aggregates and induce cognitive impairment in rats [34,71]). The 4-2A was dissolved in ethanol at a stock concentration of 100 mg/mL, and then added to DMEM/F12 (final ethanol concentration 0.1%). The effects of $A\beta_{25-35}$ at 5, 10, 15, 20 and 25 μM at 12, 24 and 48 h and the effects of 4-2A at 1, 5, 10, 20 and 40 μg/mL on $A\beta_{25-35}$-induced SH-SY5Y cell injury were then determined separately. The optimal dose and culture duration of $A\beta_{25-35}$ (20 μM, 24 h) and 4-2A (10 μg/mL, pretreated for 12 h before addition of 20 μM $A\beta_{25-35}$) were selected based on when the significant decrease in cell viability or attenuation of this effect appeared respectively. In the present study, four groups of SH-SY5Y cells were used: (i) control (cells in culture media which contained 0.1% *v/v* ethanol); (ii) 4-2A (cells in culture media with 10 μg/mL 4-2A); (iii) $A\beta_{25-35}$ (cells in culture media with 20 μM $A\beta_{25-35}$); and (iv) 4-2A and $A\beta_{25-35}$ (cells pretreated with 10 μg/mL 4-2A for 12 h before addition of 20 μM $A\beta_{25-35}$). After the addition of $A\beta_{25-35}$, the SH-SY5Y cells were incubated for 24 h, after which cell viability, oxidative stress, inflammatory cytokine TNF-a, neurotrophins and apoptosis were measured.

4.5. Measurement of Cell Viability by MTT Assay

Cell viability was measured using MTT assay which measures the cell proliferation rate and the reduction in cell viability. Cells were seeded in 96-well plates and 90 μL of cell suspension added to each well. Following experimental treatment, 10 μL MTT (ATCC) was added to each well and the plate was incubated for additional 4 h at 37 °C. The optical density was measured at 570 nm using a microplate reader (BioTek, Winooski, VT, USA). The absorbance of the control group was considered as 100% of the cell viability.

4.6. Cytotoxicity Assay by LDH Assay

As MTT assay was sensitive to cell numbers which was affected by both cell proliferation and cell viability, it was essential to use another assay to confirm the result. The CytoTox-96 assay kit (Promega, Madison, WI, USA) was employed to evaluate the total release of cytoplasmic lactate dehydrogenase (LDH) into the medium, which is a consequence of cellular integrity damage. The assay is based upon a coupled enzymatic conversion from 2-*p*-(iodophenyl)-3-(*p*-nitrophenyl)-5-phenyltetrazolium chloride (INT, a tetrazolium salt) into a formazan product, and the enzymatic reaction is catalyzed by LDH released from cells and diaphorase in the assay substrate mixture. Absorbance was read at 490 nm by the microplate reader. The mean absorbance of each group was normalized to the percentage of the control value.

4.7. Measurement of Oxidative Stress and Antioxidant Response

The intracellular level of ROS was measured with a fluorometric intracellular ROS kit (Sigma Aldrich) according to manufacturer's instructions. The fluorescence intensity was detected at lex = 650/lem = 675 nm using a fluorescence microplate reader (Reader Synergy HT, BioTek). The intracellular level of NO production was determined by the Griess Reagent System (Promega) according to manufacturer's instructions. The absorbance was measured at 540 nm using a microplate reader.

The level of GSH was determined with a glutathione assay kit (Sigma Aldrich) according to manufacturer's instructions. The fluorescence intensity was measured by a fluorimeter plate reader set at an excitation wavelength of 390 nm and emission wavelength of 478 nm.

4.8. Determination of Gene Expression with Quantitative PCR

SH-SY5Y cells treated as described above were harvested. The total RNA was extracted as recommended by the manufacturer (RNeasy® Lipid Tissue Handbook, Qiagen, Germantown, MD, USA). Complementary DNA (cDNA) was synthesized from 2 μg RNA using the GoScript™ Reverse Transcriptase (Promega, Madison, WI, USA). Primer sequences (Table 2) were obtained from Invitrogen Corporation. PCR reactions were prepared using Quantitect SYBR Green master mix (Qiagen, Germantown, MD, USA) and carried out using a Real-Time PCR Detection Systems (Bio-Rad, Hercules, CA, USA) CFX96™ Real-Time System. The real-time PCR was optimized to run with conditions of the initial incubation at 95 °C for 5 min, denaturation at 94 °C for 15 s, annealing at 59 °C for 30 s, and extension at 72 °C for 30 s with a single fluorescence measurement and up to 38 cycles. Expression levels of target mRNAs were normalized to beta actin (relative quantification) with the ΔΔCT correction.

Table 2. Primer names and sequences.

Primer Names	Sequences
hActin F	GATGAGATTGGCATGGCTTT
hActin R	CACCTTCACCGTTCCAGTTT
hBAX F	GGGGACGAACTGGACAGTAA
hBAX R	CAGTTGAAGTTGCCGTCAGA
hBCL 2 F	TCTAGGGGAGGTGGTAGGCT
hBCL 2 R	CTGAGCAAGTCAGAGACCCC
hCaspase 3 F	GACTCTAGACGGCATCCAGC
hCaspase 3 R	TGACAGCCAGTGAGACTTGG
hNGF F	ACCTTTCTCAGTAGCGGCAA
hNGF R	TGTGTCACCTTGTCAGGGAA
hTNF-α F	AGGTTTGGCCTCACAAGGAC
hTNF-α R	GCGGTAGGGACAGTTCACAG
hBDNF F	GGAGACACATCCAGCAAT
hBDNF R	ACAAGAACGAACACAACAG
hTrkB F	CTATGCTGTGGTGGTGATT
hTrkB R	CCGAAGAAGATGGAGTGTTA
hTrkA F	TACAGCACCGACTATTACC
hTrkA R	ATGATGGCGTAGACCTCT

4.9. Analysis of Protein Expression by Western Blotting

SH-SY5Y cells treated as described above were collected and centrifuged at $10,000 \times g$ for 10 min. The cell pellet was lysed with a RIPA buffer (RIPA, Thermo Fisher Scientific, Hudson, NH, USA) aided by sonication, and then spun down at $10,000 \times g$ for 10 min at 4 °C. The supernatant was collected and aliquots containing 20–40 μg of protein were loaded after boiling and separated on 10% SDS PAGE gels at 100 V for 60 min in electrophoresis buffer. After running the gels, the proteins were transferred to Polyvinylidene difluoride (PVDF) membranes (Millipore, Bellerica, MA, USA). The membranes were blocked with 5% non-fat milk in Tris buffer saline for 1 h at 20 °C. Following blocking, blots were washed with TBST for 5 min and incubated with primary antibodies, including the rabbit origin polyclonal antibodies for Actin (Abcam, Cambridge, MA, USA; ab8227, 42 kDa, 1:4000), NGF (Abcam, ab52918, 27 kDa, 1:500), BDNF (Abcam, ab6201, 28kDa, matured form, 1:200), TrkA (Abcam, ab59272, 85 kDa, 1:800), TrkB (Thermo Fisher Science, Hudson, NH, USA; MA5-14903, 90-140 kDa, 1:1000), TNF-α (Abcam, ab9739, 17 kDa, 1:2500), Bcl-2 (Abcam, ab136285, 26 kDa, 1:5000), Bax (Abcam, ab32503, 21 kDa, 1:2000) and Caspase-3 (Abcam, ab44976, 32 kDa, activated forms, 1:500), overnight at 4 °C, followed by the secondary antibody, peroxidase (HRP)-conjugated anti-rabbit IgG (Abcam, ab6721, 1:5000), for 1 h at 20 °C. The blots were washed in TBS three times. Immunoreactive bands were detected by Clarity™ Western ECL Substrate Kit (Bio-Rad, Hercules, CA, USA) on a ChemiDoc™ MP System with Image Lab™ Software (Bio-Rad, Hercules, CA, USA). All target proteins were quantified by normalizing them to β-Actin re-probed on the same membrane and then calculated as a percentage of the control group.

4.10. Statistics

Results were expressed as mean ± SEM. Statistical evaluation was performed with IBM SPSS Statistics 22. For the dose–response test of $4\text{-}2A/A\beta_{25\text{-}35}$, results were analyzed by one-way ANOVA, followed by Dunnett post-hoc test in case of significant main effects ($p < 0.05$). The possible interaction between $A\beta_{25\text{-}35}$ and 4-2A was measured by two-way ANOVA with Tukey's post-hoc tests. A value of $p < 0.05$ was considered statistically significant.

Acknowledgments: This study was supported by grants to Cai Song by Atlantic Innovation Fund of Canada, National Natural Science Foundation of China (No. 81171118 and No. 81471223), Project of Enhancing School with Innovation of Guangdong Ocean University (GDOU2013050110) and Project of Science and Technology of Zhanjiang of China. Guanglin Jiao and Junzeng Zhang were supported by Natural Health Product (NHP) Program of National Research Council of Canada. We would like to thank Island Fishermen Cooperative Association (IFCA) processing plant for providing the shrimp byproducts.

Author Contributions: C.S., Y.Z., J.Z., J.G., and G.J. conceived and designed the experiments; Y.Z., G.J. and S.G. performed the experiments; Y.Z. and G.J. analyzed the data; R.E.B., J.G., P.Z. Y.Z. and H.L. contributed to materials, analytical methods, data interpretation and discussion; C.P., R.S., and S.L. contributed to sample analysis and data verification; Y.Z., and C.S. wrote the paper with all others contributing in editing and revision.

Conflicts of Interest: The authors declare no conflict of interest.

References

1. Selkoe, D.J. Alzheimer's disease. *Cold Spring Harbor Perspect. Biol.* **2011**, *3*. [CrossRef] [PubMed]
2. Kumar, S.; Okello, E.J.; Harris, J.R. Experimental inhibition of fibrillogenesis and neurotoxicity by amyloid-beta (Abeta) and other disease-related peptides/proteins by plant extracts and herbal compounds. *Sub-Cell. Biochem.* **2012**, *65*, 295–326.
3. Selkoe, D.J.; Hardy, J. The amyloid hypothesis of Alzheimer's disease at 25 years. *EMBO Mol. Med.* **2016**, *8*, 595–608. [CrossRef] [PubMed]
4. Kim, H.G.; Moon, M.; Choi, J.G.; Park, G.; Kim, A.J.; Hur, J.; Lee, K.T.; Oh, M.S. Donepezil inhibits the amyloid-beta oligomer-induced microglial activation in vitro and in vivo. *Neurotoxicology* **2014**, *40*, 23–32. [CrossRef] [PubMed]

5. Tabner, B.J.; Turnbull, S.; King, J.E.; Benson, F.E.; El-Agnaf, O.M.; Allsop, D. A spectroscopic study of some of the peptidyl radicals formed following hydroxyl radical attack on beta-amyloid and alpha-synuclein. *Free Radic. Res.* **2006**, *40*, 731–739. [CrossRef] [PubMed]

6. Choi, D.J.; Cho, S.; Seo, J.Y.; Lee, H.B.; Park, Y.I. Neuroprotective effects of the Phellinus linteus ethyl acetate extract against H2O2-induced apoptotic cell death of SK-N-MC cells. *Nutr. Res.* **2016**, *36*, 31–43. [CrossRef] [PubMed]

7. Sondag, C.M.; Dhawan, G.; Combs, C.K. Beta amyloid oligomers and fibrils stimulate differential activation of primary microglia. *J. Neuroinflamm.* **2009**, *6*, 1. [CrossRef] [PubMed]

8. Lee, M.; McGeer, E.; McGeer, P.L. Activated human microglia stimulate neuroblastoma cells to upregulate production of beta amyloid protein and tau: Implications for Alzheimer's disease pathogenesis. *Neurobiol. Aging* **2015**, *36*, 42–52. [CrossRef] [PubMed]

9. Allen, S.J.; Watson, J.J.; Dawbarn, D. The neurotrophins and their role in Alzheimer's disease. *Curr. Neuropharmacol.* **2011**, *9*, 559–573. [CrossRef] [PubMed]

10. Ciaramella, A.; Salani, F.; Bizzoni, F.; Orfei, M.D.; Langella, R.; Angelucci, F.; Spalletta, G.; Taddei, A.R.; Caltagirone, C.; Bossu, P. The stimulation of dendritic cells by amyloid beta 1-42 reduces BDNF production in Alzheimer's disease patients. *Brain Behav. Immun.* **2013**, *32*, 29–32. [CrossRef] [PubMed]

11. Poon, W.W.; Carlos, A.J.; Aguilar, B.L.; Berchtold, N.C.; Kawano, C.K.; Zograbyan, V.; Yaopruke, T.; Shelanski, M.; Cotman, C.W. beta-Amyloid (Abeta) oligomers impair brain-derived neurotrophic factor retrograde trafficking by down-regulating ubiquitin C-terminal hydrolase, UCH-L1. *J. Biol. Chem.* **2013**, *288*, 16937–16948. [CrossRef] [PubMed]

12. Kumar, A.; Singh, A.; Ekavali. A review on Alzheimer's disease pathophysiology and its management: An update. *Pharmacol. Rep.* **2015**, *67*, 195–203. [CrossRef] [PubMed]

13. Luchtman, D.W.; Meng, Q.; Song, C. Ethyl-eicosapentaenoate (E-EPA) attenuates motor impairments and inflammation in the MPTP-probenecid mouse model of Parkinson's disease. *Behav. Brain Res.* **2012**, *226*, 386–396. [CrossRef] [PubMed]

14. Song, C.; Manku, M.S.; Horrobin, D.F. Long-chain polyunsaturated fatty acids modulate interleukin-1beta-induced changes in behavior, monoaminergic neurotransmitters, and brain inflammation in rats. *J. Nutr.* **2008**, *138*, 954–963. [PubMed]

15. Song, C. Essential fatty acids as potential anti-inflammatory agents in the treatment of affective disorders. *Mod. Trends Pharmacopsychiatry* **2013**, *28*, 75–89.

16. Song, C.; Li, X.; Leonard, B.E.; Horrobin, D.F. Effects of dietary *n*-3 or *n*-6 fatty acids on interleukin-1beta-induced anxiety, stress, and inflammatory responses in rats. *J. Lipid Res.* **2003**, *44*, 1984–1991. [CrossRef] [PubMed]

17. Peng, G.; Li, L.; Liu, Y.; Pu, J.; Zhang, S.; Yu, J.; Zhao, J.; Liu, P. Oleate blocks palmitate-induced abnormal lipid distribution, endoplasmic reticulum expansion and stress, and insulin resistance in skeletal muscle. *Endocrinology* **2011**, *152*, 2206–2218. [CrossRef] [PubMed]

18. Coll, T.; Eyre, E.; Rodriguez-Calvo, R.; Palomer, X.; Sanchez, R.M.; Merlos, M.; Laguna, J.C.; Vazquez-Carrera, M. Oleate reverses palmitate-induced insulin resistance and inflammation in skeletal muscle cells. *J. Biol. Chem.* **2008**, *283*, 11107–11116. [CrossRef] [PubMed]

19. Yuzefovych, L.; Wilson, G.; Rachek, L. Different effects of oleate vs. palmitate on mitochondrial function, apoptosis, and insulin signaling in L6 skeletal muscle cells: Role of oxidative stress. *Am. J. Physiol. Endocrinol. Metab.* **2010**, *299*, E1096–E1105. [CrossRef] [PubMed]

20. Bento-Abreu, A.; Tabernero, A.; Medina, J.M. Peroxisome proliferator-activated receptor-alpha is required for the neurotrophic effect of oleic acid in neurons. *J. Neurochem.* **2007**, *103*, 871–881. [CrossRef] [PubMed]

21. Frigolet, M.E.; Gutierrez-Aguilar, R. The Role of the Novel Lipokine Palmitoleic Acid in Health and Disease. *Adv. Nutr.* **2017**, *8*, 173S–181S. [CrossRef] [PubMed]

22. Gong, J.; Campos, H.; McGarvey, S.; Wu, Z.; Goldberg, R.; Baylin, A. Adipose tissue palmitoleic acid and obesity in humans: Does it behave as a lipokine? *Am. J. Clin. Nutr.* **2011**, *93*, 186–191. [CrossRef] [PubMed]

23. Ouchi, N.; Parker, J.L.; Lugus, J.J.; Walsh, K. Adipokines in inflammation and metabolic disease. *Nat. Rev. Immunol.* **2011**, *11*, 85–97. [CrossRef] [PubMed]

24. Tchoukanova, N.; Benoit, G. Method for extracting organic solids and oil from marine organisms enriched with astaxanthin. U.S. Patent App. 14/776,481, 2014.

25. Jiao, G.; Hui, J.; Burton, I.; Thibault, M.-H.; Pelletier, C.; Boudreau, J.; Tchoukanova, N.; Subramanian, B.; Djaoued, Y.; Ewart, S.; et al. Characterization of Shrimp Oil from Pandalus borealis by High Performance Liquid Chromatography and High Resolution Mass Spectrometry. *Mar. Drugs* **2015**, *13*, 3849–3876. [CrossRef] [PubMed]

26. Morioka, N.; Zhang, F.F.; Nakamura, Y.; Kitamura, T.; Hisaoka-Nakashima, K.; Nakata, Y. Tumor necrosis factor-mediated downregulation of spinal astrocytic connexin43 leads to increased glutamatergic neurotransmission and neuropathic pain in mice. *Brain Behav. Immunity* **2015**, *49*, 293–310. [CrossRef] [PubMed]

27. Tezel, G. TNF-alpha signaling in glaucomatous neurodegeneration. *Prog. Brain Res.* **2008**, *173*, 409–421. [PubMed]

28. Gunstone, F.; Knothe, G. NMR Spectroscopy of Fatty Acids and Their Derivatives. AOCS Lipid Library. Available online: http://lipidlibrary.aocs.org/Analysis/content.cfm?ItemNumber=40256 (accessed on 21 March 2017).

29. Millucci, L.; Ghezzi, L.; Bernardini, G.; Santucci, A. Conformations and biological activities of amyloid beta peptide 25-35. *Curr. Protein Pept. Sci.* **2010**, *11*, 54–67. [CrossRef] [PubMed]

30. Sohanaki, H.; Baluchnejadmojarad, T.; Nikbakht, F.; Roghani, M. Pelargonidin improves memory deficit in amyloid beta25-35 rat model of Alzheimer's disease by inhibition of glial activation, cholinesterase, and oxidative stress. *Biomed. Pharmacother.* **2016**, *83*, 85–91. [CrossRef] [PubMed]

31. Vedagiri, A.; Thangarajan, S. Mitigating effect of chrysin loaded solid lipid nanoparticles against Amyloid beta25-35 induced oxidative stress in rat hippocampal region: An efficient formulation approach for Alzheimer's disease. *Neuropeptides* **2016**, *58*, 111–125. [CrossRef] [PubMed]

32. Mokhtari, Z.; Baluchnejadmojarad, T.; Nikbakht, F.; Mansouri, M.; Roghani, M. Riluzole ameliorates learning and memory deficits in Abeta25-35-induced rat model of Alzheimer's disease and is independent of cholinoceptor activation. *Biomed. Pharmacother.* **2017**, *87*, 135–144. [CrossRef] [PubMed]

33. Fedotova, J.; Soultanov, V.; Nikitina, T.; Roschin, V.; Ordyan, N.; Hritcu, L. Cognitive-enhancing activities of the polyprenol preparation Ropren(R) in gonadectomized beta-amyloid (25-35) rat model of Alzheimer's disease. *Physiol. Behav.* **2016**, *157*, 55–62. [CrossRef] [PubMed]

34. Ji, C.; Song, C.; Zuo, P. The mechanism of memory impairment induced by Abeta chronic administration involves imbalance between cytokines and neurotrophins in the rat hippocampus. *Curr. Alzheimer Res.* **2011**, *8*, 410–420. [CrossRef] [PubMed]

35. Diaz, A.; Rojas, K.; Espinosa, B.; Chavez, R.; Zenteno, E.; Limon, D.; Guevara, J. Aminoguanidine treatment ameliorates inflammatory responses and memory impairment induced by amyloid-beta 25-35 injection in rats. *Neuropeptides* **2014**, *48*, 153–159. [CrossRef] [PubMed]

36. Yu, H.; Yao, L.; Zhou, H.; Qu, S.; Zeng, X.; Zhou, D.; Zhou, Y.; Li, X.; Liu, Z. Neuroprotection against Abeta25-35-induced apoptosis by Salvia miltiorrhiza extract in SH-SY5Y cells. *Neurochem. Int.* **2014**, *75*, 89–95. [CrossRef] [PubMed]

37. Briyal, S.; Shepard, C.; Gulati, A. Endothelin receptor type B agonist, IRL-1620, prevents beta amyloid (Abeta) induced oxidative stress and cognitive impairment in normal and diabetic rats. *Pharmacol. Biochem. Behav.* **2014**, *120*, 65–72. [PubMed]

38. Cioanca, O.; Hritcu, L.; Mihasan, M.; Trifan, A.; Hancianu, M. Inhalation of coriander volatile oil increased anxiolytic-antidepressant-like behaviors and decreased oxidative status in beta-amyloid (1-42) rat model of Alzheimer's disease. *Physiol. Behav.* **2014**, *131*, 68–74. [CrossRef] [PubMed]

39. Suganthy, N.; Devi, K.P. Protective effect of catechin rich extract of Rhizophora mucronata against β-amyloid-induced toxicity in PC12 cells. *J. Appl. Biomed.* **2016**, *14*, 137–146. [CrossRef]

40. Arimon, M.; Takeda, S.; Post, K.L.; Svirsky, S.; Hyman, B.T.; Berezovska, O. Oxidative stress and lipid peroxidation are upstream of amyloid pathology. *Neurobiol. Dis.* **2015**, *84*, 109–119. [CrossRef] [PubMed]

41. Yue, T.; Shanbin, G.; Ling, M.; Yuan, W.; Ying, X.; Ping, Z. Sevoflurane aggregates cognitive dysfunction and hippocampal oxidative stress induced by beta-amyloid in rats. *Life Sci.* **2015**, *143*, 194–201. [CrossRef] [PubMed]

42. Padayachee, E.; Ngqwala, N.; Whiteley, C.G. Association of beta-amyloid peptide fragments with neuronal nitric oxide synthase: Implications in the etiology of Alzheimers disease. *J. Enzyme Inhib. Med. Chem.* **2012**, *27*, 356–364. [CrossRef] [PubMed]

43. Gamper, N.; Ooi, L. Redox and nitric oxide-mediated regulation of sensory neuron ion channel function. *Antioxid. Redox Signal.* **2015**, *22*, 486–504. [CrossRef] [PubMed]

44. Ooi, L.; Gigout, S.; Pettinger, L.; Gamper, N. Triple cysteine module within M-type K+ channels mediates reciprocal channel modulation by nitric oxide and reactive oxygen species. *J. Neurosci.* **2013**, *33*, 6041–6046. [CrossRef] [PubMed]

45. Shadfar, S.; Hwang, C.J.; Lim, M.S.; Choi, D.Y.; Hong, J.T. Involvement of inflammation in Alzheimer's disease pathogenesis and therapeutic potential of anti-inflammatory agents. *Arch. Pharm. Res.* **2015**, *38*, 2106–2119. [CrossRef] [PubMed]

46. Heppner, F.L.; Ransohoff, R.M.; Becher, B. Immune attack: The role of inflammation in Alzheimer disease. *Nat. Rev. Neurosci.* **2015**, *16*, 358–372. [CrossRef] [PubMed]

47. Wang, Y.L.; He, H.; Liu, Z.J.; Cao, Z.G.; Wang, X.Y.; Yang, K.; Fang, Y.; Han, M.; Zhang, C.; Huo, F.Y. Effects of TNF-alpha on Cementoblast Differentiation, Mineralization, and Apoptosis. *J. Dent. Res.* **2015**, *94*, 1225–1232. [CrossRef] [PubMed]

48. Taepavarapruk, P.; Song, C. Reductions of acetylcholine release and nerve growth factor expression are correlated with memory impairment induced by interleukin-1beta administrations: Effects of omega-3 fatty acid EPA treatment. *J. Neurochem.* **2010**, *112*, 1054–1064. [CrossRef] [PubMed]

49. Song, C.; Zhang, Y.; Dong, Y. Acute and subacute IL-1beta administrations differentially modulate neuroimmune and neurotrophic systems: Possible implications for neuroprotection and neurodegeneration. *J. Neuroinflamm.* **2013**, *10*, 59. [CrossRef] [PubMed]

50. Lattanzio, F.; Carboni, L.; Carretta, D.; Candeletti, S.; Romualdi, P. Treatment with the neurotoxic Abeta (25-35) peptide modulates the expression of neuroprotective factors Pin1, Sirtuin 1, and brain-derived neurotrophic factor in SH-SY5Y human neuroblastoma cells. *Exp. Toxicol. Pathol.* **2016**, *68*, 271–276. [CrossRef] [PubMed]

51. Loo, D.T.; Copani, A.; Pike, C.J.; Whittemore, E.R.; Walencewicz, A.J.; Cotman, C.W. Apoptosis is induced by beta-amyloid in cultured central nervous system neurons. *Proc. Natl. Acad. Sci. USA* **1993**, *90*, 7951–7955. [CrossRef] [PubMed]

52. McGeer, E.G.; McGeer, P.L. Neuroinflammation in Alzheimer's disease and mild cognitive impairment: A field in its infancy. *J. Alzheimer's Dis. JAD* **2010**, *19*, 355–361. [PubMed]

53. Asadi, F.; Jamshidi, A.H.; Khodagholi, F.; Yans, A.; Azimi, L.; Faizi, M.; Vali, L.; Abdollahi, M.; Ghahremani, M.H.; Sharifzadeh, M. Reversal effects of crocin on amyloid beta-induced memory deficit: Modification of autophagy or apoptosis markers. *Pharmacol. Biochem. Behav.* **2015**, *139 Pt A*, 47–58. [CrossRef] [PubMed]

54. Pieri, M.; Amadoro, G.; Carunchio, I.; Ciotti, M.T.; Quaresima, S.; Florenzano, F.; Calissano, P.; Possenti, R.; Zona, C.; Severini, C. SP protects cerebellar granule cells against beta-amyloid-induced apoptosis by down-regulation and reduced activity of Kv4 potassium channels. *Neuropharmacology* **2010**, *58*, 268–276. [CrossRef] [PubMed]

55. Meng, P.; Yoshida, H.; Tanji, K.; Matsumiya, T.; Xing, F.; Hayakari, R.; Wang, L.; Tsuruga, K.; Tanaka, H.; Mimura, J.; et al. Carnosic acid attenuates apoptosis induced by amyloid-beta 1–42 or 1–43 in SH-SY5Y human neuroblastoma cells. *Neurosci. Res.* **2015**, *94*, 1–9. [CrossRef] [PubMed]

56. Sharoar, M.G.; Islam, M.I.; Shahnawaz, M.; Shin, S.Y.; Park, I.S. Amyloid beta binds procaspase-9 to inhibit assembly of Apaf-1 apoptosome and intrinsic apoptosis pathway. *Biochim. Biophys. Acta* **2014**, *1843*, 685–693. [CrossRef] [PubMed]

57. Feart, C.; Samieri, C.; Alles, B.; Barberger-Gateau, P. Potential benefits of adherence to the Mediterranean diet on cognitive health. *Proc. Nutr. Soc.* **2013**, *72*, 140–152. [CrossRef] [PubMed]

58. Kwon, B.; Lee, H.K.; Querfurth, H.W. Oleate prevents palmitate-induced mitochondrial dysfunction, insulin resistance and inflammatory signaling in neuronal cells. *Biochim. Biophys. Acta* **2014**, *1843*, 1402–1413. [CrossRef] [PubMed]

59. Cao, H.; Gerhold, K.; Mayers, J.R.; Wiest, M.M.; Watkins, S.M.; Hotamisligil, G.S. Identification of a lipokine, a lipid hormone linking adipose tissue to systemic metabolism. *Cell* **2008**, *134*, 933–944. [CrossRef] [PubMed]

60. Guo, X.; Li, H.; Xu, H.; Halim, V.; Zhang, W.; Wang, H.; Ong, K.T.; Woo, S.L.; Walzem, R.L.; Mashek, D.G.; et al. Palmitoleate induces hepatic steatosis but suppresses liver inflammatory response in mice. *PLoS ONE* **2012**, *7*, e39286. [CrossRef] [PubMed]

61. Orr, S.K.; Trepanier, M.O.; Bazinet, R.P. n-3 Polyunsaturated fatty acids in animal models with neuroinflammation. *Prostaglandins Leukot. Essent. Fat. Acids* **2013**, *88*, 97–103. [CrossRef] [PubMed]

62. Barros, M.; Poppe, S.; Bondan, E. Neuroprotective Properties of the Marine Carotenoid Astaxanthin and Omega-3 Fatty Acids, and Perspectives for the Natural Combination of Both in Krill Oil. *Nutrients* **2014**, *6*, 1293–1317. [CrossRef] [PubMed]

63. Uygur, R.; Aktas, C.; Tulubas, F.; Uygur, E.; Kanter, M.; Erboga, M.; Caglar, V.; Topcu, B.; Ozen, O.A. Protective effects of fish omega-3 fatty acids on doxorubicin-induced testicular apoptosis and oxidative damage in rats. *Andrologia* **2014**, *46*, 917–926. [CrossRef] [PubMed]

64. Luchtman, D.W.; Meng, Q.; Wang, X.; Shao, D.; Song, C. Omega-3 fatty acid eicosapentaenoic acid attenuates MPP+-induced neurodegeneration in fully differentiated human SH-SY5Y and primary mesencephalic cells. *J. Neurochem.* **2013**, *124*, 855–868. [CrossRef] [PubMed]

65. Kou, W.; Luchtman, D.; Song, C. Eicosapentaenoic acid (EPA) increases cell viability and expression of neurotrophin receptors in retinoic acid and brain-derived neurotrophic factor differentiated SH-SY5Y cells. *Eur. J. Nutr.* **2008**, *47*, 104–113. [CrossRef] [PubMed]

66. Song, C.; Horrobin, D. Omega-3 fatty acid ethyl-eicosapentaenoate, but not soybean oil, attenuates memory impairment induced by central IL-1beta administration. *J. Lipid Res.* **2004**, *45*, 1112–1121. [CrossRef] [PubMed]

67. Wu, Y.Q.; Dang, R.L.; Tang, M.M.; Cai, H.L.; Li, H.D.; Liao, D.H.; He, X.; Cao, L.J.; Xue, Y.; Jiang, P. Long Chain Omega-3 Polyunsaturated Fatty Acid Supplementation Alleviates Doxorubicin-Induced Depressive-Like Behaviors and Neurotoxicity in Rats: Involvement of Oxidative Stress and Neuroinflammation. *Nutrients* **2016**, *8*, 243. [CrossRef] [PubMed]

68. Green, K.N.; Martinez-Coria, H.; Khashwji, H.; Hall, E.B.; Yurko-Mauro, K.A.; Ellis, L.; LaFerla, F.M. Dietary docosahexaenoic acid and docosapentaenoic acid ameliorate amyloid-beta and tau pathology via a mechanism involving presenilin 1 levels. *J. Neurosci.* **2007**, *27*, 4385–4395. [CrossRef] [PubMed]

69. Begum, G.; Yan, H.Q.; Li, L.; Singh, A.; Dixon, C.E.; Sun, D. Docosahexaenoic acid reduces ER stress and abnormal protein accumulation and improves neuronal function following traumatic brain injury. *J. Neurosci.* **2014**, *34*, 3743–3755. [CrossRef] [PubMed]

70. Plante, M.; Bailey, B.; Acworth, I. *Analysis of Lipids by RP-HPLC Using the Corona Analysis of Lipids by RP-HPLC Using the Corona Ultra*; ESA-A Dionex Company: Chelmsford, MA, USA, 2009.

71. Delobette, S.; Privat, A.; Maurice, T. In vitro aggregation facilities beta-amyloid peptide-(25-35)-induced amnesia in the rat. *Eur. J. Pharmacol.* **1997**, *319*, 1–4. [CrossRef]

marine drugs

MDPI

Article

Variation Quality and Kinetic Parameter of Commercial *n*-3 PUFA-Rich Oil during Oxidation via Rancimat

Kai-Min Yang and Po-Yuan Chiang *

Department of Food Science and Biotechnology, National Chung Hsing University, 250 Kuokuang Road, Taichung 40227, Taiwan; a9241128@gamil.com
* Correspondence: pychiang@nchu.edu.tw; Tel.: +886-4-2285-1665

Academic Editors: Rosário Domingues, Ricardo Calado and Pedro Domingues
Received: 13 January 2017; Accepted: 20 March 2017; Published: 28 March 2017

Abstract: Different biological sources of *n*-3 polyunsaturated fatty acids (*n*-3 PUFA) in mainstream commercial products include algae and fish. Lipid oxidation in *n*-3 PUFA-rich oil is the most important cause of its deterioration. We investigated the kinetic parameters of *n*-3 PUFA-rich oil during oxidation via Rancimat (at a temperature range of 70~100 °C). This was done on the basis of the Arrhenius equation, which indicates that the activation energies (Ea) for oxidative stability are 82.84–96.98 KJ/mol. The chemical substrates of different oxidative levels resulting from oxidation via Rancimat at 80 °C were evaluated. At the initiation of oxidation, the tocopherols in the oil degraded very quickly, resulting in diminished protection against further oxidation. Then, the degradation of the fatty acids with *n*-3 PUFA-rich oil was evident because of decreased levels of PUFA along with increased levels of saturated fatty acids (SFA). The quality deterioration from *n*-3 PUFA-rich oil at the various oxidative levels was analyzed chemometrically. The anisidine value (p-AV, r: 0.92) and total oxidation value (TOTOX, r: 0.91) exhibited a good linear relationship in a principal component analysis (PCA), while oxidative change and a significant quality change to the induction period (IP) were detected through an agglomerative hierarchical cluster (AHC) analysis.

Keywords: *n*-3 PUFA; oxidative stability index; Rancimat test; kinetic parameter

---

1. Introduction

Fish oil accounts for less than 1% of all the global edible oil produced, but it is the main source of *n*-3 PUFA, specifically, eicosapentaenoic acid (EPA) and docosahexaenoic acid (DHA) [1]. These oils are becoming increasingly popular with consumers in view of the clear evidence of the health benefits of *n*-3 PUFA for individuals with cardiovascular conditions, including their beneficial role as antithrombotic, anti-inflammatory, and hypolipidemic fatty acids [2,3]. The *n*-6/*n*-3 ratio in a person's dietary intake is also an important consideration due to its influences on cardiovascular health and inflammation, with high intake of *n*-6 PUFA potentially attenuating the known profitable effects of *n*-3 PUFA. Due to the nutritional changes described above in the Western diet, the *n*-6/*n*-3 ratio has now increased to falls between 10 and 20. The dietary recommendations regarding *n*-3 PUFA intake are, thus, of increasing importance [4].

In Europe, the recommended daily intake of EPA and DHA is 450 mg per day, that is, around 3 g per week, while the WHO/FAO recommends daily consumption of 250 mg (for primary prevention) to 2 g (for secondary prevention) of EPA and DHA to prevent cardiovascular conditions [5]. In contrast, the FDA and American Dietetic Association suggest a minimum intake of close to 500 mg/day to prevent coronary health diseases [6]. In recent years, consumers have identified *n*-3 PUFA supplements as options for reducing the probability of illness and avoiding expensive medical bills, so that the rate of sales growth for such supplements is currently around 15% per annum. The calculated market value of packaged products containing *n*-3 PUFA, which primarily consist of infant formula, has been estimated to reach \$34.7 billion by 2016, and to grow at a CAGR of 9.1% from 2015 to 2022 [7].

Worldwide fish stocks peaked some years ago, but in recent years there has been a shortage in the supply of fish oils. In the future, as the global population continues to grow, in turn increasing the need for products allowing consumers to meet the suggested intake of EPA and DHA, sustainable sources of *n*-3 PUFA-containing products will be needed to meet the growing demand [8,9]. With that in mind, the use of other forms of marine life, including Antarctic krill and algae, to provide *n*-3 PUFA continues to be developed, with these sources already having been commercialized. These sources may provide some high value and highly concentrated products for human consumption, including contaminant-free products with good sensory qualities that are also safe and environmentally friendly. In particular, the fermentation of algae can be used to shorten the growth process of algae and produce highly concentrated oil [10,11].

The drawback of using *n*-3 PUFA-rich oils for functional foods is that they are readily oxidized in the presence of oxygen, heat, light, and metal ions, and the secondary products of lipid oxidation can impair the sensory qualities and acceptability of products among consumers [12]. In addition, previous studies involving animal research have confirmed that oxidative products contain genotoxic and cytotoxic compounds [13]. Moreover, these oxidative compounds, when present in diets, have been considered as the possible causative agents of several diseases, such as chronic inflammation, neurodegenerative diseases, atherogenesis, diabetes, and certain types of cancer. Among these oxidative products, the oxygenated aldehydes are the most broadly studied, and their adsorption capacity and functional group profiles are most closely related to toxicity [13,14].

The Global Organization for EPA and DHA (GOED) voluntary monograph is the quality standard for EPA- and DHA-rich oils, and is used to help ensure that consumers have access to high-quality products; it is applicable to the EPA and DHA fatty acids obtained from fish, plant, or microbial sources [15]. Numerous analytical methods of lipid oxidation are used to measure food quality. However, there is no common and standard method for detecting all oxidative variations in multiple food systems. Therefore, it is necessary to select a proper and adequate method for the analysis of any fatty acid composition and its substrates.

Lipid oxidation occurs very slowly at room temperature and, hence, accelerated methods should be applied in order to estimate the oxidative stability of a product or the induction time of the autoxidation reaction in a more rapid manner, especially as the temperature and the rate of said reaction are exponentially related [16]. Well-established accelerated aging testing methods include the active oxygen method, Schaal Oven test, and Rancimat test. For the determination of IP, which is the time needed for oil deterioration to commence, the Rancimat test observes the changes to the conductivity of samples while the other two methods look at peroxide value (POV). So the Rancimat test is easy to use and has good reproducibility [17]. The Rancimat method has been widely used to evaluate the shelf lives of various products, including the kinetic parameters of antioxidants in oil samples, as well as the inhibition of lipid peroxidation in such antioxidants [16–18].

The Rancimat test promotes the oxidation process by exposing oil samples to a high temperature or temperatures and a sufficient amount of oxygen. In the current study, we collected oils with different oxidation levels during the induction period. This study discusses the evolution of substrate changes and the formation of primary and secondary oxidation products, which were characterized in terms of oxidative stability through the use of agglomerative hierarchical cluster AHC analysis and PCA. These parameters were used to evaluate the oxidative properties of *n*-3 PUFA-rich oil, and the resulting information can further be used to control the stability and shelf lives of PUFA-containing products.

2. Results and Discussion

2.1. Kinetic Analysis

Given the uncertainty regarding the best temperature conditions for accelerated methods, the oxidative stability of *n*-3 PUFA-rich oils was studied using the Rancimat test with temperatures ranging from 70 °C to 100 °C. The IP for the lipid oxidation of *n*-3 PUFA-rich oils at different temperatures are presented in Figure 1. For use of the Rancimat test at temperatures of 70 °C, 80 °C, 90 °C, and 100 °C, the induction times were 15.4, 6.8, 3.4, and 0.92 h, respectively, for the VA; 15.9, 7.6, 3.6, and 0.97 h, respectively, for the SuF; and 4.8, 2.3, 1.01, and 0.47 h, respectively, for the SiF. The temperature used affects the degree of oxygen solubility in a given oil sample, with the oxygen solubility decreasing by almost 25% for each 10 °C rise in temperature [19]. Generally, the induction time was halved with each 10 °C increase in temperature. Previous research has shown that the PIs (at 50~80 °C) of fish oils without antioxidants ranged from 24.3~0.6 h, and that the IP of the same oils ranged from 52.3~2.4 h when 400 ppm of α-tocopherol were added [20].

Many studies have shown that the Rancimat test can be used to determine and evaluate the kinetic parameters of oils. The determination of such kinetic parameters is valuable for the purpose of distinguishing the origins of various oils, for characterizing the differences or similarities in the oils, and for predicting the oxidative stability of oils under various storage conditions [17,19]. In this study, there was semi-logarithmic relationship with Equation (1) for all the oil samples, including a linear dependency with good correlation of determination, with R^2 being 0.979 for the VA, 0.977 for the SuF, and 0.998 for the SiF (Figure 1).

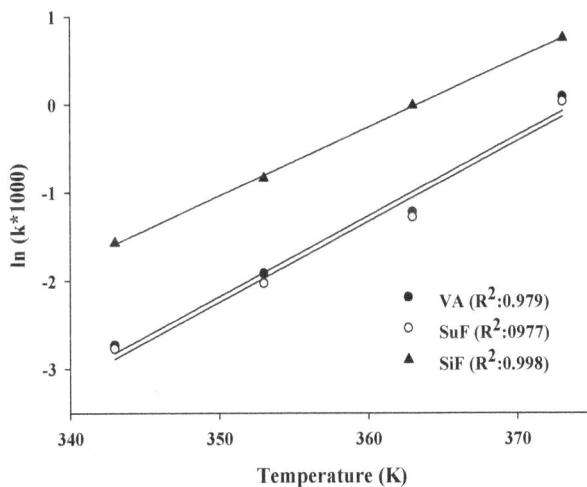

Figure 1. Semi-logarithmic relationship between k and temperature values for lipid oxidation of the *n*-3 PUFA-rich oils.

The *Ea* value is of interest for the properties of oils, which demonstrates the delay of the initial oxidation reaction due to the bond scission that takes place to form primary oxidation products [21]. Table 1 shows *Ea* values of the assayed oils were 96.98 kJ/mol for the VA, 96.97 kJ/mol for the SuF, and 82.84 kJ/mol for the SiF (Table 1). The *Ea* is influenced by unsaturated number of oil samples, as the *Ea* seems to be lower for oils with higher PUFA levels. According to the reference, the *Ea* values of DHA and EPA ethyl esters (of 95–97% purity), which were in the range of 52.1~62.4 kJ/mol [22]. However, this is contradicted by the example of the VA assayed in this study. Specifically, while the

PUFA content of the VA was higher than those of the other tested oils, because the process of molecular distillation can increase the IP of VA, it had a higher *Ea* value than the other oils [23].

Table 1. Regression parameters for Arrhenius relationships between the reaction rate constant and the temperature for the *n*-3 PUFA-rich oils.

Groups	VA	SuF	SiF
	$\ln(k) = a(1/T) + b$		
a	−11.66	−11.66	−9.96
b	31.16	31.1	27.44
R^2	0.973	0.971	0.998
Ea (kJ/mol)	96.98	96.97	82.84

2.2. Monitoring Substrate Variants

As shown in Table 2, we identified eight types of fatty acids as presented under the VA column, 12 types of fatty acids as presented under the SuF column, and 14 types of fatty acids as presented under the SiF column. We observed that the constituents of 100 g of VA consisted of 34.4 g of SFA, 34.5 g of MUFA, and 30.3 g of PUFA, while 100 g of SuF contained 44.5 g of SFA, 20.2 g of MUFA, and 17.6 g of PUFA. One hundred grams of SiF consisted of 43.9 g of SFA, 28.0 g of MUFA, and 24.0 g of PUFA. The concentration (g/100 g) of *n*-3 PUFA was 26.8 for VA, 16.4 for SuF, and 16.3 for SiF. The literature on this topic indicated that DHA could be efficiently synthesized in microalgae via an anaerobic pathway involving polyketide synthases [11].

Table 2. The fatty acid composition and tocopherol levels of the *n*-3 PUFA-rich oils.

Groups	VA	SuF	SiF
	Fatty Acid (g/100 g)		
SFA [a]	**34.4**	**44.5**	**43.9**
C14:0	2.6	8.6	15.6
C16:0	30.2	29.8	25.6
C18:0	1.6	6.1	2.7
MUFA [b]	**34.5**	**20.2**	**24.0**
C14:1	1.3	1.9	1.1
C16:1	N.D [d]	4.8	18.0
C18:1	33.2	11.7	7.1
C20:1	N.D	1.8	1.8
PUFA [c]	**30.3**	**17.6**	**24.0**
C18:2	3.5	1.2	3.5
C20:2	N.D	N.D	3.3
C20:3	N.D	1.9	0.6
AA	N.D	N.D	0.9
EPA	N.D	4.6	9.9
DPA	3.5	1.0	1.1
DHA	23.3	8.9	4.7
	Tocopherol (mg/kg)		
δ-	219.3	252.9	167.3
γ-	529.4	445.3	N.D
α-	106.6	124.4	N.D

[a] SFA, Saturated fatty acid; [b] MUFA, monounsaturated fatty acid; [c] PUFA, polyunsaturated fatty acid; [d] N.D, not detected.

The combination of fatty acids in edible oil is the most important factor in determining the oil's oxidation stability. Processing sophistication and antioxidants can improve the oxidation stability of commercial products. The levels of fatty acids change at 80 °C under the Rancimat test (Figure 2), with the difference in the PUFA amounts of the VA, SuF, and SiF blends being −16.60%, −13.11%, and −2.76% at the 100% oxidation level, and −29.43%, −24.08%, and −16.61% at the 125% oxidation level, mainly as a result of DHA degradation. The difference in the SFA amounts of the VA, SuF, and SiF blends were 1.01%, 4.11%, and 2.29% at the 100% oxidation level, and −0.67%, 4.34%, and 12.36% at the 125% oxidation level. It is known that the thermal treatment of oils and fats generates hydroperoxide breakdown of any fatty acids with a chain shorter than ten carbon atoms, such that such breakdown can be considered a chemical indicator of the fat degradation grade [24]. This study was similar to such results, as SFA formation appeared to be correlated with PUFA loss, and with monounsaturated fatty acid (MUFA) increases and decreases.

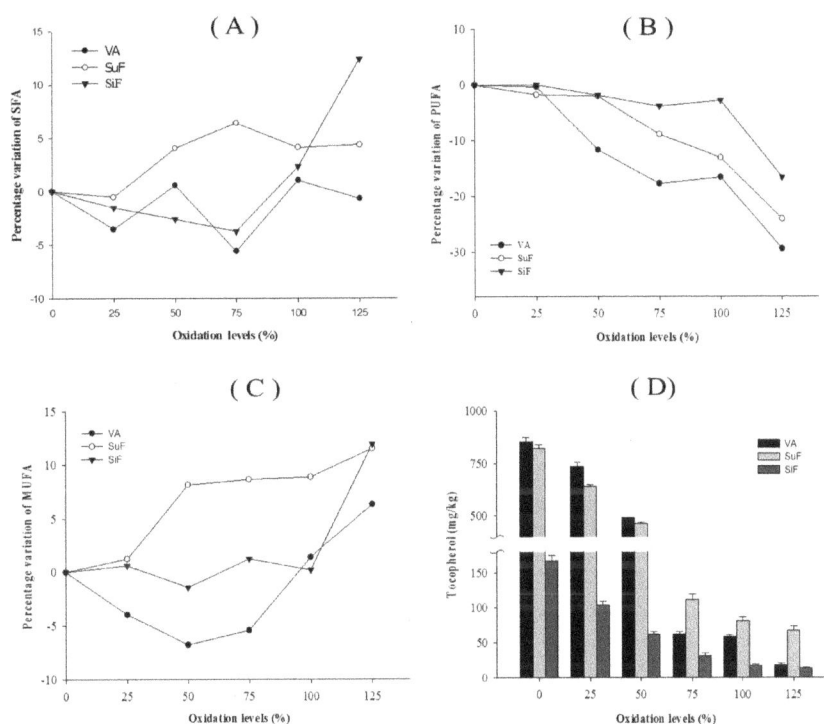

Figure 2. Percentage variations (g/100 g oil) of the (**A**) SFA, (**B**) MUFA, (**C**) PUFAs, and (**D**) total tocopherol measured in the *n*-3 PUFA-rich oils.

Tocopherols are the most abundant antioxidants in *n*-3 PUFA concentrates, because of their capacity to inhibit hydroperoxides and C-3 aldehydes [25]. In commercial products, tocopherol can be used independently or in combination with other compounds, such as ascorbic acid palmitate, lecithin, and catechin, and has shown significantly in fish oil stabilization. On the other hand, the structure conformation also affects the physical property of tocopherol, as γ- and δ-tocopherol have thermal resistance. Hydroperoxyl radical-scavenging activity occurs in the order of α- > β- > γ- > δ-tocopherol [26]. As shown in Figure 2, half of the total tocopherols of the tested *n*-3 PUFA-rich oils was lost at the 50% oxidation level under the Rancimat test, while 90% of the total was lost at the 100% oxidation level under the Rancimat test. The fact that the tocopherol content dropped very quickly under the Rancimat test conditions could be due mainly to the very high susceptibility of this molecule to oxidation to tocopherol quinones at high temperatures, which diminishes the protection of unsaturated fatty acids against oxidation [27].

2.3. Monitoring Oxidation Products

Lipid oxidation products have negative impacts on the flavor and odor of sensory parameters, which can, in turn, have harmful effects on human health. The progress of lipid oxidation can be evaluated by the monitoring of a diverse series of primary, secondary, and tertiary oxidation products over time. The quality standards of GOED require specific levels of acid value (AV, ≤3 mg KOH/g), POV (≤5 meq/kg), p-AV (≤20), and TOTOX (≤26) throughout the stated lifetime of a product [15]. We found that the VA and SuF tested in this study met their stated label claims for GOED (Table 3). POV and AV are used in the industry's on-line quality control index. POV has a significant correlation to the off-flavour compounds created during the initial oxidation. AV represents the content of free fatty acids which are easily oxidized to hydro-peroxides [28].

The EC regulations suggest that absorbances at 234 nm (K234), 270 nm (K270), and 280 nm (K280) be used to measure oils according to their oxidated products such as ethylenic diketones, conjugated ketodienes, and the dienal formation of conjugated dienes and trienes. K234 is a primary oxidation index that has been found to be closely related to hydroperoxide content [29]. In addition, the measurement of UV absorbance at 270 and 280 nm has been used previously in analyzing edible oils. Absorption at these wavelengths is mainly due to secondary oxidation. In a spectral analysis conducted for this study (Table 3), we found that the absorbances of 10 mg/mL of VA, SuF, and SiF were 3.33, 14.79, and 7.58, respectively, at K234; 1.12, 0.66, and 1.78, respectively, at K270; and 1.08, 0.51 and 1.48, respectively, at K280. Most of them increased as the oxidation levels increased. The K234 of SuF had high values at the beginning, a finding which can be attributed to the residual oxidated products or fat-soluble compounds that were produced from the simplified refining process.

Table 3. Initial quality characteristics of the *n*-3 PUFA-rich oils.

Groups	VA	SuF	SiF
Quality Indicators			
AV (mg KOH/g)	0.48 ± 0.02	0.49 ± 0.01	0.65 ± 0.01
CVD (%)	0.22 ± 0.05	1.18 ± 0.17	0.58 ± 0.03
POV (meq/kg)	1.98 ± 0.27	4.12 ± 0.34	13.62 ± 0.42
p-AV (meq/kg)	6.33 ± 0.71	15.12 ± 0.64	29.23 ± 1.84
TOTOX (meq/kg)	10.30 ± 0.92	23.26 ± 1.24	56.46 ± 2.11
Visible Spectra (10 mg/mL)			
K234 [a]	3.33 ± 0.07	14.79 ± 0.14	7.58 ± 0.17
K270	1.12 ± 0.01	0.66 ± 0.02	1.78 ± 0.04
K280	1.08 ± 0.02	0.51 ± 0.01	1.48 ± 0.07

[a] K234, K270, and K280, specific absorption at 234, 270, and 280 nm.

In the literatures, hierarchical cluster was showed that could apply to grouping basis on quality, sensory attributes and refined level [30,31]. The AHC analysis was applied to identify clusters of samples with similar oxidation properties. Three main clusters were extracted (Figure 3). Cluster 1 contained the 0–100% oxidation levels of the VA, the 0–75% oxidation levels of the SuF, and the 0–75% oxidation levels of the SiF. Cluster 2 contained the 100% oxidation levels of the SuF and SiF. Cluster 3 contained the 125% oxidation levels of the VA, SuF, and SiF. These results showed that there are significant changes of quality in the IP of a given oil when it is subjected to the Rancimat test. These results were similar to those of a previous study reported by our team, which oxidized volatiles of *n*-3 PUFA [32].

With regard to the quality standard change at 80 °C under the Rancimat test, according to the PCA analysis, two dimensions were extracted and together account for approximately 91.51% of the variability from the original data (Figure 3). In our results, we observed p-AV (r: 0.92), TOTOX (r: 0.91), K270 (r: 0.88), and POV (r: 0.87) with PAC1; and K234 (r: 0.83) and conjugated dienes (CVD, r: 0.81) with PCA2. The POV and p-AV are measures of primary oxidation and secondary oxidation, respectively. The TOTOX value gives an overall indication of the complete oxidation status of oil, and consists of the combination of the POV and the p-AV values to determine the oxidation level of the oil. Correlations between the quality parameters and spectral analysis is that K234, with AV, POV, p-AV, and TOTOX is low ($R < 0.5$), while K270 and K280 with acid value (AV), POV, p-AV, and TOTOX is high ($R > 0.7$). These results were based on measuring the formation of secondary oxidation products; for example, p-AV and K270 are available to support the AV evaluation of frying oil quality, and aldehyde molecules were a common marker for K270, K280, and p-AV [28,33].

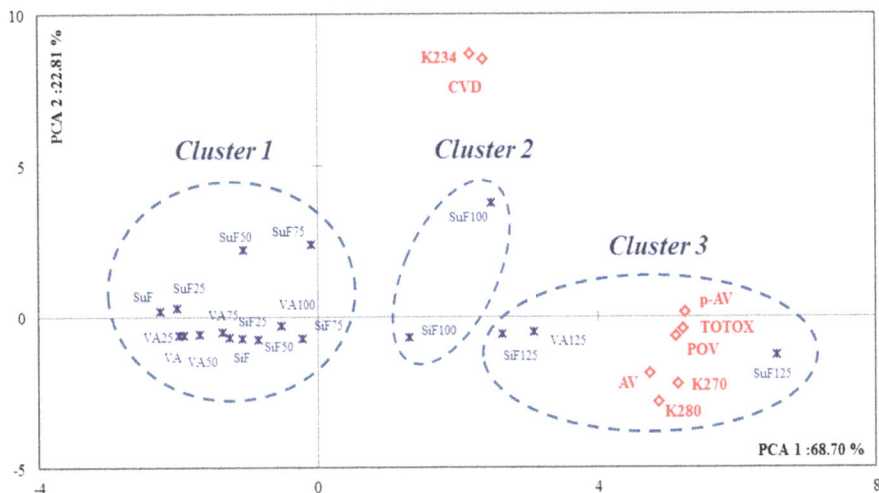

Figure 3. PCA plots of quality changes for the different oxidation levels of the *n*-3 PUFA-rich oils under the Rancimat method; the circles represent the clusters detected with AHC analysis; the solid fill type highlights the values with PCA.

3. Materials and Methods

3.1. Materials

Conventional algae oils (VA) were provided by VEDAN Enterprise Corporation (Taichung, Taiwan), and conventional fish oils from mackerel (SuF) were provided by SUN AGRICULTURE (Ilan, Taiwan). Additional fish oils (SiF) were provided by Sigma (Sigma-Aldrich, Taufkirchen, Germany). A fatty acid methyl ester standard (FAME) mixture, Supelco 37 Component FAME Mix, was purchased from Supelco (Sigma-Aldrich, Taufkirchen, Germany). Standards of α-, γ-, and δ- tocopherol were purchased from Merck (Darmstadt, Germany).

3.2. Rancimat Test

The oxidative stability index of the *n*-3 PUFA-rich oil was previously determined at four different temperatures (70 °C, 80 °C, 90 °C, and 100 °C) as the induction period (hours) that was recorded using a Rancimat 743 apparatus and a 5 ± 0.05 g sample of oil with an air flow of 10 L/h. Then, oil samples were oxidized at 80 °C for periods of time that corresponded to 25%, 50%, 75%, 100%, and 125% of their respective induction periods. The oil samples that were used to determine the oxidative stability were also analyzed for their volatile oxidation compounds. The IP of the *n*-3 PUFA-rich oils were automatically recorded and taken as the break point of the plotted curves (the intersection point of the two extrapolated parts of the curve).

3.3. Kinetic Data Analysis

The kinetic parameters were determined according to the method previously utilized as reported in [17]. The IP of the oil samples were automatically recorded and taken as the break point of the plotted curves (the intersection point of the two extrapolated parts of the curve). A kinetic rate constant was taken as the inverse of the IP (k, h^{-1}).

Temperature coefficients (T Coeff, K^{-1}) were determined from the slopes of the lines generated by regressing $\ln(k)$ vs. the absolute temperature (T, K):

$$\ln(k) = a(T) + b \qquad (1)$$

where a and b are the equation parameters.

Activation energies (Ea, kJ/mol) and pre-exponential or frequency factors (A, h^{-1}) were determined from the slopes and intercepts, respectively, of the lines generated by regressing $\ln(k)$ vs. $1/T$ using the Arrhenius equation:

$$\ln(k) = \ln(A) - (Ea/RT) \qquad (2)$$

where k is the reaction rate constant or reciprocal IP (h^{-1}), and R is the molar gas constant (8.3143 J/mol K).

3.4. Analysis of Tocopherol

Each oil sample (0.1 g) was diluted with 2-propanol to a volume of 10 mL and filtered through an MS nylon syringe filter with a 0.45 μm pore size directly to vials and then immediately analyzed using an HPLC system. Aliquots of 10 μL of the filtrate were injected into the injection port and analyzed with HPLC (Hitachi L-2130 pump, Hitachi, Tokyo, Japan). The remaining procedures were carried out as previously reported [34] using an HPLC attached to a detector L-2400 UV and a Hitachi L-2130 pump. An RP-18GP250 Mightysil column (l = 250 mm; i.d. = 4.6 mm; thickness = 0.32 μm; Kanto Chemical Co., Inc., Tokyo, Japan) was used for separation. The same mobile phase and elution conditions were adopted. The calibration curves were, respectively, established for tocopherol by plotting the peak area vs. each corresponding concentration, from which quantitations of the standards were achieved.

3.5. Fatty Acid Analysis

The *n*-3 PUFA-rich oil was analyzed for its fatty acid composition via GC/FID. The triacylglycerols were converted to methyl esters using the AOCS Official Method Ce 2–66 [35]. The methyl esters were separated using a column that was coated with DB-23 (30 m × 0.25 mm × 0.25 μm, Agilent, Palo Alto, CA, USA), and helium was used as the carrier gas at a flow rate of 1.0 mL/min. The oven temperature was initially held for 8 min at 200 °C, and then increased at 10 °C/min to 220 °C and then held there for 40 min. The FID was maintained at 270 °C, and the injector (split mode 1:40, 4 mm liner) was maintained at 250 °C. The contents of the fatty acids were determined using the normalization method, with heneicosanoic methyl ester used as an internal standard to quantitation.

3.6. Quality Analytical Determination

The POV and UV spectrophotometric were measured using the analytical methods described in European Regulation EEC 2568/91 [36]. The CDA measures the formed from PUFA during lipid oxidation according to the AOCS method Ti 1a-64 [35]. The p-AV measures secondary oxidation products, such as 2-alkenal and 2,4-alkadienal. The p-AV of each sample was determined according to the AOCS method Cd 18–90 [35]. The AV was determined using a titration with 0.1 N potassium hydroxide alcoholic solution.

3.7. Statistical Analysis

The data reported were obtained from triplicate measurements of each sample and were expressed as means. The data were subjected to an AHC analysis with squared Euclidean distances. Subsequently, the data were analyzed using PCA combined with VARIMAX rotation. For the AHC and PCA analysis, XLSTAT software (version 2010.2.01, Addinsoft Deutschland, Andernach, Germany) was used.

4. Conclusions

In this work, the Rancimat test was applied to the analysis of the oxidation properties of *n*-3 PUFA-rich oil. The results of the present study indicated: (1) the degree of unsaturation in fatty acids is not the only parameter for assessing oil quality, which is mainly influenced by commercialization. The oxidation stability of *n*-3 PUFA-rich oils is improved by commercialization, especially with respect to antioxidant protection; and (2) chemometric applications have also clarified the nature of the differences in correlation among oxidative and chemical parameters. The IP of primary and secondary oxidation product formations represented the quality changes as detected via AHC analysis. Not all quality parameters increased linearly through PCA analysis, which can allow for the selection of a proper and adequate method for a particular application. It is recommended for p-AV, TOTOX, K270, and POV to be used when carrying out quality control for *n*-3 PUFA- rich oils.

Acknowledgments: This work was supported by the Ministry of Science and Technology (NSC 105-2221-E-005-077), Taiwan.

Author Contributions: Kai-Min Yang carried out all the experiments Kai-Min Yang and Po-Yuan Chiang designed all of the experiments and analyzed the data; Kai-Min Yang and Po-Yuan Chiang wrote the manuscript.

Conflicts of Interest: The authors declare no conflict of interest.

Abbreviations

The following abbreviations are used in this manuscript:

SFA	Saturated fatty acids
MUFA	Monounsaturated fatty acid
PUFA	Polyunsaturated fatty acid
AA	Arachidonic acid
EPA	Eicosapentaenoic acid
DPA	Docosapentaenoic acid
DHA	Docosahexaenoic acid
GOED	Global Organization for EPA and DHA
IP	Induction period
Ea	Activation energies
POV	Peroxide value
p-AV	Anisidine value
AV	Acid value
CD	Conjugated dienes
TOTOX	Total oxidation value
PCA	Principal component analysis
AHC	Agglomerative hierarchical cluster

References

1. Pike, I.H. Fish oil: Supply and demand as a source of long-chain *n*-3 polyunsaturated fatty acids in the human diet. *Eur. J. Lipid Sci. Technol.* **2015**, *117*, 747–750. [CrossRef]
2. Nichols, P.D.; McManus, A.; Krail, K.; Sinclair, A.J.; Miller, M. Recent advances in omega-3: Health benefits, Sources, Products and bioavailability. *Nutrients* **2014**, *9*, 3727–3733. [CrossRef] [PubMed]
3. Kris-Etherton, P.M.; Harris, W.S.; Appel, L.J.; Committee, N. Fish consumption, fish oil, omega-3 fatty acids, and cardiovascular disease. *Circulation* **2002**, *106*, 2747–2757. [CrossRef] [PubMed]

4. Watanabe, N.; Onuma, K.; Fujimoto, K.; Miyake, S.; Nakamura, T. Long-term effect of an enteral diet with a different *n*-6/*n*-3 ratio on fatty acid composition and blood parameters in rats. *J. Oleo Sci.* **2011**, *60*, 109–115. [CrossRef] [PubMed]
5. Lim, D.K.; Garg, S.; Timmins, M.; Zhang, E.S.; Thomas-Hall, S.R.; Schuhmann, H.; Li, Y.; Schenk, P.M. Isolation and evaluation of oil-producing microalgae from subtropical coastal and brackish waters. *PLoS ONE* **2012**, *7*, e40751. [CrossRef] [PubMed]
6. Opperman, M.; Marais, D.W.; Benadé, A.S. Analysis of omega-3 fatty acid content of South African fish oil supplements. *Cardiovasc. J. Afr.* **2011**, *22*, 324–329. [CrossRef] [PubMed]
7. Facts, P. *Global Market for EPA/DHA Omega-3 Products*; Packaged Facts: Rockville, MD, USA, 2012.
8. Adarme-Vega, T.C.; Thomas-Hall, S.R.; Schenk, P.M. Towards sustainable sources for omega-3 fatty acids production. *Curr. Opin. Biotechnol.* **2014**, *26*, 14–18. [CrossRef] [PubMed]
9. Adarme-Vega, T.C.; Lim, D.K.; Timmins, M.; Vernen, F.; Li, Y.; Schenk, P.M. Microalgal biofactories: A promising approach towards sustainable omega-3 fatty acid production. *Microb. Cell Fact.* **2012**, *11*, 96. [CrossRef] [PubMed]
10. Kuratko, C.N.; Norman, S. Docosahexaenoic acid from algal oil. *Eur. J. Lipid Sci. Technol.* **2013**, *115*, 965–976. [CrossRef]
11. Martins, D.A.; Custódio, L.; Barreira, L.; Pereira, H.; Ben-Hamadou, R.; Varela, J.; Abu-Salah, K.M. Alternative sources of *n*-3 long-chain polyunsaturated fatty acids in marine microalgae. *Mar. Drugs* **2013**, *11*, 2259–2281. [CrossRef] [PubMed]
12. Drusch, S.; Groß, N.; Schwarz, K. Efficient stabilization of bulk fish oil rich in long-chain polyunsaturated fatty acids. *Eur. J. Lipid Sci. Technol.* **2008**, *110*, 351–359. [CrossRef]
13. Guillén, M.D.; Goicoechea, E. Toxic oxygenated α, β-unsaturated aldehydes and their study in foods: A review. *Crit. Rev. Food Sci. Nutr.* **2008**, *48*, 119–136. [CrossRef] [PubMed]
14. Guillén, M.D.; Cabo, N.; Ibargoitia, M.L.; Ruiz, A. Study of both sunflower oil and its headspace throughout the oxidation process. Occurrence in the headspace of toxic oxygenated aldehydes. *J. Agric. Food Chem.* **2005**, *53*, 1093–1101. [CrossRef] [PubMed]
15. GOED. Available online: http://www.goedomega3.com/healthcare (accessed on 10 January 2016).
16. Tan, C.; Man, Y.C.; Selamat, J.; Yusoff, M. Application of Arrhenius kinetics to evaluate oxidative stability in vegetable oils by isothermal differential scanning calorimetry. *J. Am. Oil Chem. Soc.* **2001**, *78*, 1133–1138. [CrossRef]
17. Farhoosh, R.; Niazmand, R.; Rezaei, M.; Sarabi, M. Kinetic parameter determination of vegetable oil oxidation under Rancimat test conditions. *Eur. J. Lipid Sci. Technol.* **2008**, *110*, 587–592. [CrossRef]
18. Chen, M.H.; Huang, T.C. Volatile and Nonvolatile Constituents and Antioxidant Capacity of Oleoresins in Three Taiwan Citrus Varieties as Determined by Supercritical Fluid Extraction. *Molecules* **2016**, *21*, 1735. [CrossRef] [PubMed]
19. Robertson, G.L. Shelf life of packaged foods, its measurements and prediction. In *Developing New Food Products for a Changing Marketplace*; CRC Press: Boca Raton, FL, USA, 2000; pp. 329–353.
20. Rupasinghe, H.V.; Erkan, N.; Yasmin, A. Antioxidant protection of eicosapentaenoic acid and fish oil oxidation by polyphenolic-enriched apple skin extract. *J. Agric. Food Chem.* **2009**, *58*, 1233–1239. [CrossRef] [PubMed]
21. Adhvaryu, A.; Erhan, S.; Liu, Z.; Perez, J. Oxidation kinetic studies of oils derived from unmodified and genetically modified vegetables using pressurized differential scanning calorimetry and nuclear magnetic resonance spectroscopy. *Thermochim. Acta* **2000**, *364*, 87–97. [CrossRef]
22. Yoshii, H.; Furuta, T.; Siga, H.; Moriyama, S.; Baba, T.; Maruyama, K.; Misawa, Y.; Hata, N.; Linko, P. Autoxidation kinetic analysis of docosahexaenoic acid ethyl ester and docosahexaenoic triglyceride with oxygen sensor. *Biosci. Biotechnol. Biochem.* **2002**, *66*, 749–753. [CrossRef] [PubMed]
23. Wang, W.; Li, T.; Ning, Z.; Wang, Y.; Yang, B.; Ma, Y.; Yang, X. A process for the synthesis of PUFA-enriched triglycerides from high-acid crude fish oil. *J. Food Eng.* **2012**, *109*, 366–371. [CrossRef]
24. De Leonardis, A.; Macciola, V. Heat-oxidation stability of palm oil blended with extra virgin olive oil. *Food Chem.* **2012**, *135*, 1769–1776. [CrossRef] [PubMed]
25. Kulås, E.; Ackman, R.G. Properties of α-, γ-, and δ-tocopherol in purified fish oil triacylglycerols. *J. Am. Oil Chem. Soc.* **2001**, *78*, 361–367. [CrossRef]

26. Fujisawa, S.; Kadoma, Y. Kinetic study of the radical-scavenging activity of vitamin E and ubiquinone. *In Vivo* **2005**, *19*, 1005–1011. [PubMed]

27. Verleyen, T.; Kamal-Eldin, A.; Dobarganes, C.; Verhé, R.; Dewettinck, K.; Huyghebaert, A. Modeling of α-tocopherol loss and oxidation products formed during thermoxidation in triolein and tripalmitin mixtures. *Lipids* **2001**, *36*, 719–726. [CrossRef] [PubMed]

28. Navas, J.A.; Tres, A.; Codony, R.; Guardiola, F. Optimization of analytical methods for the assessment of the quality of fats and oils used in continuous deep fat frying. *Grasas Aceites* **2007**, *58*, 154–162.

29. Shahid Chatha, S.A.; Anwar, F.; Manzoor, M.; Rehman Bajwa, J.U. Evaluation of the antioxidant activity of rice bran extracts using different antioxidant assays. *Grasas Aceites* **2006**, *57*, 328–335.

30. Navarro, T.; de Lorenzo, C.; Pérez, R. SPME analysis of volatile compounds from unfermented olives subjected to thermal treatment. *Anal. Bioanal. Chem.* **2004**, *379*, 812–817. [CrossRef] [PubMed]

31. Tu, D.; Li, H.; Wu, Z.; Zhao, B.; Li, Y. Application of headspace solid-phase microextraction and multivariate analysis for the differentiation between edible oils and waste cooking oil. *Anal. Bioanal. Chem.* **2014**, *7*, 1263–1270. [CrossRef]

32. Yang, K.M.; Cheng, M.C.; Chen, C.W.; Tseng, C.Y.; Lin, L.Y.; Chiang, P.Y. Characterization of Volatile Compounds with HS-SPME from Oxidized *n*-3 PUFA Rich Oils via Rancimat tests. *J. Oleo Sci.* **2017**, *66*, 113–122. [CrossRef] [PubMed]

33. Grau, A.; Guardiola, F.; Boatella, J.; Baucells, M.D.; Codony, R. Evaluation of lipid ultraviolet absorption as a parameter to measure lipid oxidation in dark chicken meat. *J. Agric. Food Chem.* **2000**, *48*, 4128–4135. [CrossRef] [PubMed]

34. Shehata, A.B.; Rizk, M.S.; Farag, A.M.; Tahoun, I.F. Development of two reference materials for all trans-retinol, retinyl palmitate, α- and γ-tocopherol in milk powder and infant formula. *J. Food Drug Anal.* **2015**, *23*, 82–92. [CrossRef]

35. American Oil Chemists' Society (AOCS). *Official Methods and Recommended Practices of the American Oil Chemists' Society*; Firestone, D., Ed.; AOCS Press: Champaign, IL, USA, 1998.

36. Regulation, H. Commission Regulation (EEC) No. 2568/91 of 11 July 1991 on the characteristics of olive oil and olive-residue oil and on the relevant methods of analysis Official Journal L 248, 5 September 1991. *Off. J. L* **1991**, *248*, 1–83.

marine drugs

MDPI

Review

Marine Lipids on Cardiovascular Diseases and Other Chronic Diseases Induced by Diet: An Insight Provided by Proteomics and Lipidomics

Lucía Méndez [1],*, Gabriel Dasilva [1], Nùria Taltavull [2], Marta Romeu [2] and Isabel Medina [1]

[1] Instituto de Investigaciones Marinas (IIM-CSIC), Eduardo Cabello 6, E-36208 Vigo, Spain; gabrielsilva@iim.csic.es (G.D.); medina@iim.csic.es (I.M.)

[2] Unitat de Farmacologia, Facultat de Medicina i Ciències de la Salut, Universitat Rovira i Virgili, Sant Llorenç 21, E-43201 Reus, Spain; nuria.taltavull@urv.cat (N.T.); marta.romeu@urv.cat (M.R.)

* Correspondence: luciamendez@iim.csic.es; Tel.: +34-986-231930; Fax: +34-986-292762

Received: 16 May 2017; Accepted: 15 August 2017; Published: 18 August 2017

Abstract: Marine lipids, especially ω-3 polyunsaturated fatty acids (PUFAs) eicosapentaenoic acid (EPA) and docosahexaenoic acid (DHA), have largely been linked to prevention of diet-induced diseases. The anti-inflammatory and hypolipidemic properties of EPA and DHA supplementation have been well-described. However, there is still a significant lack of information about their particular mechanism of action. Furthermore, repeated meta-analyses have not shown conclusive results in support of their beneficial health effects. Modern "omics" approaches, namely proteomics and lipidomics, have made it possible to identify some of the mechanisms behind the benefits of marine lipids in the metabolic syndrome and related diseases, i.e., cardiovascular diseases and type 2 diabetes. Although until now their use has been scarce, these "omics" have brought new insights in this area of nutrition research. The purpose of the present review is to comprehensively show the research articles currently available in the literature which have specifically applied proteomics, lipidomics or both approaches to investigate the role of marine lipids intake in the prevention or palliation of these chronic pathologies related to diet. The methodology adopted, the class of marine lipids examined, the diet-related disease studied, and the main findings obtained in each investigation will be reviewed.

Keywords: marine lipids; EPA; DHA; proteomics; lipidomics; metabolic syndrome; cardiovascular disease; type 2 diabetes

1. Introduction

1.1. Chronic Diseases Induced by Diet: A World Health Problem

The intake of westernized diets, which are rich in refined carbohydrates, cholesterol, saturated and trans fats, and have an increased ratio of ω-6/ω-3 PUFAs, along with a sedentary lifestyle can quickly lead to the development of different pathologies and metabolic disorders [1]. These chronic diseases related to diet, such as cardiovascular diseases (CVD), obesity, overweight, hypertension, type 2 diabetes, hyperlipidemia, hyperinsulinemia, osteoporosis, osteopenia or cancer, are considered the epidemic of modern societies. It is estimated that 20–30% of the adult population in these countries suffers Metabolic Syndrome (MetS) [2]. The MetS is a compilation of risk factors associated with CVD and type 2 diabetes. These factors include dyslipidemia (decreased high-density lipoprotein (HDL) cholesterol levels and increased low-density lipoprotein (LDL) cholesterol and triglycerides levels), hyperglycemia, hypertension, insulin resistance and obesity. Inflammation and oxidative stress have also been closely related to MetS and derived diseases [3].

According to the Eurostat report of May 2016, the first two leading causes of death in Europe were CVD and cancer, well above respiratory diseases, which were the third most common cause of death [4]. The average in Europe was 132 deaths/100,000 inhabitants due to CVD in 2013, being 350/100,000 in countries like Hungary or Slovakia, and less than 100/100,000 in Spain, Portugal or Greece. Data from the World Health Organization (WHO) in 2015 also showed that the leading causes of death in high-income countries were CVD and cancer, while infectious diseases were the leading cause of death in low-income countries [5].

Therefore, the study of diet components and their relation to the progression of chronic diseases and metabolic disorders has arisen as a field of great interest for the scientific community. Accordingly, the number of publications found in the Scopus database (www.scopus.com) which are focused on this topic has exponentially grown, especially since 1990. Considering the search criteria: TITLE-ABS-KEY (nutrition OR diet) AND TITLE-ABS-KEY ("metabolic disorder" OR "metabolic alteration" OR obesity OR "metabolic syndrome" OR diabetes OR "cardiovascular disease" OR cholesterol OR insulin OR atherosclerosis OR inflammation OR "oxidative stress"), around 272,000 were found. From these, 75% are research articles and 14% are reviews. The rest of the published work consists of book chapters, notes, conferences, etc. In relation to the field of study, 76% of the publications belong to the field of medicine, 27% to biochemistry, and the rest of them are framed in the fields of agricultural and biological sciences, nursing, pharmacology, toxicology and pharmaceutics, neuroscience, chemistry, immunology and microbiology, etc.

As a consequence of these investigations, huge quantities of bioactive compounds have been found to exert beneficial effects on human health. Among that, ω-3 PUFAs from marine origin have quickly gained more attention. Currently, there is considerable evidence that the intake of marine-origin polyunsaturated fatty acids (PUFAs), especially eicosapentaenoic acid (EPA) and docosahexaenoic acid (DHA), can help in the prevention/palliation of inflammatory processes and metabolic diseases, although their mechanisms of action are not entirely understood yet.

1.2. Marine Lipids as Bioactive Compounds against MetS and Chronic Diseases Induced by Diet

Marine lipids, especially EPA and DHA, have largely demonstrated their bioactivity in human health. The interest in their intake arose from a series of pioneering studies in the Inuit Eskimo population during the 1960s and 1970s. These studies reported a lower incidence of cardiovascular pathologies in that population associated with a high intake of fish in their diet. Later, similar relationships were discovered in human populations from other regions, particularly in Iceland and Alaska Natives [6].

In addition to these investigations, marine ω-3 PUFAs have been shown to alleviate metabolic disorder symptoms, such as heart disease, diabetes, obesity and insulin resistance. Since the late 1950s, the number of publications found in the Scopus database (TITLE-ABS-KEY ("omega 3" OR "PUFA" OR "eicosapentaenoic" OR "docosahexaenoic" OR "marine fatty acids" OR "marine lipid" OR "marine oil" OR "fish oil" OR "fish lipid") AND TITE-ABS-KEY (nutrition OR diet) AND TITLE-ABS-KEY ("metabolic disorder" OR "metabolic alteration" OR "metabolic disease" OR obesity OR "metabolic syndrome" OR diabetes OR "cardiovascular disease" OR cholesterol OR insulin OR atherosclerosis OR inflammation OR "oxidative stress")), which have related the consumption of marine fatty acids or lipids to these metabolic disorders, has increased exponentially, mainly since the late 1970s. On the whole, more than 13,100 publications can currently be found in the Scopus database. Sixty-five percent of these publications are research articles, 23% are reviews and the rest of them are book chapters, notes, conferences, etc.

These research articles have used a plethora of different approaches (from chemistry or biochemistry to phycology) and a huge variety of methodologies to address this topic. As a consequence, the publications are categorized in very diverse fields of study, medicine (74%), biochemistry (30%), agricultural and biological sciences (20%), nursing (19%) and the rest of the articles are framed in the fields of pharmacology, toxicology and pharmaceutics, chemistry, neuroscience, immunology and microbiology, phycology, social sciences, mathematics, computer science, etc.

The aim of the present review is to comprehensively show those research articles currently available in the literature which have specifically used proteomics, lipidomics or both approaches to investigate the role of marine lipids intake in the prevention/palliation of chronic pathologies induced by the consumption of westernized diets, namely MetS, CVD and type 2 diabetes. Our ultimate objective is to highlight the usefulness of the application of these approaches to nutrition research, in order to discover the underlying mechanisms of the beneficial effects of marine lipids in human metabolic health, since its translation to practice via nutritional interventions is still in its infancy.

2. Omics for Unrevealing Mechanisms: Proteomics and Lipidomics

Almost all cellular processes, from gene expression to synthesis, degradation and protein activity, can be affected by diet and lifestyle. Therefore, nutrients and other food components (including marine lipids) may alter metabolic functions in cells in a quite complex way. This complex relationship between nutrition and health makes nutrition research an ideal field for the application of systems biology approaches [7], as being holistic and integrated. Systems biology encompasses a broad range of functional areas called "omics" These areas include genomics, transcriptomics, proteomics, metabolomics, lipidomics and bioinformatics [8]. This new outlook started as a result of the human genome project in the early 2000s [9].

The number of publications found in the literature which incorporates the most novel omics approaches (proteomics and lipidomics) to the study of the beneficial effects of marine lipids against MetS and their associated chronic diseases is still scarce. Proteomics and lipidomics studies applied to nutrition research have to cope with a variety of difficulties ranging from experimental design, sample processing and optimization strategies to data analysis and identification. The need for high-sensitivity modern mass spectrometers combined with bioinformatics resources increases costs and requires highly qualified staff. In spite of these difficulties, the interest of the scientific community in these approaches has been progressively growing, in parallel with analytical advancements in liquid chromatography, mass spectrometry and bioinformatics.

In the following subsections, we review the research articles currently available in the literature which have used proteomics (Section 2.1), lipidomics (Section 2.2) or both (Section 2.3) to analyze the effects of the consumption of marine lipids on the development and progression of human MetS as well as their associated chronic diseases (CVD and type 2 diabetes).

2.1. Beneficial Effects of Marine Lipids Intake Assayed by Proteomics

Proteins are important mediators of biological activities in all living cellular units. Proteomics is the large-scale study of proteins and therefore comprises the methodologies used for the study of a proteome [10]. The "proteome" a term firstly coined by Wilkins et al. in the early 2000s, is comprised of all expressed proteins encoded by the genome of a cellular system, including all cellular proteins and all protein species (isoforms and protein modifications) [11,12].

The study of how ω-3 PUFAs regulate proteins and metabolic pathways may significantly contribute to understanding the putative mechanisms by which marine lipids elicit their beneficial health effects. In fact, nutrition research has lately adopted the proteomics tools to measure changes in the protein complement of a biological system. This adoption has enabled modeling of biological processes in response to dietary marine lipids, as well as the elucidation of novel biomarkers for health or disease which are sensitive to such lipids [13]. There are limited studies which have addressed the influence of marine lipids on proteome with the aim to investigate their effects against human metabolic disorders. These studies have mainly used classical two-dimensional gel electrophoresis (2-DE) combined with mass spectrometry (MS) to elucidate changes in metabolic pathways and target proteins which may explain a certain beneficial effect on human health. This traditional methodology has made it possible to identify changes in several metabolic pathways, such as glucose and fatty acid metabolism, oxidative stress, antioxidant defense mechanisms and redox status. The study of minor abundant proteins such as those participating in inflammatory pathways requires the introduction in nutrition research of more quantitative and sensitive methods, like multiple reaction monitoring (MRM) and multiplexed immunoassays. Both strategies might be useful for the evaluation and validation of newly discovered candidate biomarkers in human biofluids.

By using Scopus database, 53 publications have been found following the search criteria: TITLE-ABS-KEY ("omega 3" OR "PUFA" OR "eicosapentaenoic" OR "docosahexaenoic" OR "marine fatty acids" OR "marine lipid" OR "marine oil" OR "fish oil" OR "fish lipid") AND TITE-ABS-KEY (nutrition OR diet) AND TITLE-ABS-KEY ("metabolic disorder" OR "metabolic alteration" OR "metabolic disease" OR obesity OR "metabolic syndrome" OR diabetes OR "cardiovascular disease" OR cholesterol OR insulin OR atherosclerosis OR inflammation OR "oxidative stress") AND TITE-ABS-KEY (proteomic OR proteomics OR proteome). These publications, mainly research articles (56%) and reviews (30%), belong, in descending order of abundance, to the fields of biochemistry, genetics and molecular biology (63%) and medicine (58%), and to a lesser extent nursing, agricultural and biological sciences, pharmacology, toxicology and pharmaceutics, immunology and microbiology, neuroscience, chemistry, etc.

In this review, only research articles focusing on proteomics are presented. The studies which have used ω-3 PUFAs from a vegetal origin, i.e., α-linolenic acid (ALA), have been excluded. Only articles which have assayed the effect of marine lipids on human metabolic disorders related to diet (i.e., features of MeS, cardiovascular diseases and type 2 diabetes) and those that have used animal or cellular models to mimic these alterations are presented. Moreover, articles which have combined both lipidomics and proteomics tools in the same research are shown jointly in Section 2.3.

According to these filters, 14 research articles have been found. All of them investigated the effect of EPA or DHA or both on metabolic health. Most of these studies used bottom-up proteomics approaches, based on 2-DE coupled with MS protein identification. Novel gel-free proteomics techniques based on MS, such as shotgun proteomics, have scarcely been used yet, as noted in Table 1.

2.1.1. Proteomics in Clinical Trials

Two studies analyzed the effect of the supplementation of a mix of EPA and DHA on the modulation of the peripheral blood mononuclear cells (PBMCs) proteome after acute intake [14] or after 12 weeks of dietary intervention [15] in patients with MetS. In the first case, proteomics analysis identified five proteins related to cell signaling and interaction, DNA repair, cellular assembly and organization, and cell morphology which were regulated by acute consumption of marine PUFAs.

In the second study, the prolonged intake of EPA and DHA regulated 17 proteins of PBMCs proteome involved in immunological diseases and inflammatory response, down-regulating proteins directly related to oxidative stress, inflammation, endoplasmic reticulum stress and DNA repair. A third study, which was carried out with the same dietary intervention and proteomics tools, evaluated the changes induced in subcutaneous white adipose tissue (WAT) proteome after 12 weeks of EPA and DHA intake [16]. Three proteins of glucose metabolism were found down-regulated by marine PUFAs supplementation. These proteins were correlated with lower systemic insulin resistance and improved insulin signaling in subcutaneous WAT of the MetS patients.

Also in humans, De Roos et al. [17] studied mechanisms involved in preventing the early onset of coronary heart disease (CHD) through fish oil intake. Authors identified ten serum proteins that were down-regulated in healthy volunteers after the consumption of a daily dose of 3.5 g of fish oil for 6 weeks. These altered serum proteins, metabolically related to lipoprotein metabolism and inflammation, led to a significant shift towards the larger, more cholesterol-rich HDL2 particle which might imply that fish oil activated anti-inflammatory and lipid modulating mechanisms believed to impede the early onset of CHD. This specific effect of fish oil on HDL metabolism has later been investigated by Burillo et al. [18]. They compared the proteome of HDL before and after ω-3 PUFAs intake for 5 weeks. Healthy smoker volunteers ingested a commercial mixture of marine and non-marine ω-3 fatty acids. The consumption of marine ω-3 PUFAs up-regulated seven proteins related to the antioxidant, anti-inflammatory and anti-atherosclerotic properties of HDL. Likewise, marine lipids down-regulated six proteins involved in the regulation of complement activation and acute phase response. Moreover, the modification of lipoprotein containing apoAI (LpAI) proteome suggested that the protein changes found might have improved the functionality of the particle.

2.1.2. Proteomics in Animal Models and Cell Cultures

Besides clinical trials, several studies have used animal or cell models to investigate the role of marine ω-3 PUFAs in regulating cellular proteomes which are rather difficult to analyze in humans, such as the liver proteome. In a study carried out in C57BL/6 mice, Ahmed et al. [19] investigated the regulation of the liver proteome by a dietary intervention with 10% ω-3 PUFAs from menhaden oil for 4 months. Proteomics data showed that fish oil up-regulated eight proteins related to lipid, carbohydrate, protein and one-carbon metabolism and the citric acid cycle, which involved an integrated regulation of metabolic pathways by fish oil. This proteome modulation was correlated with a significant reduction of plasma triglycerides and free fatty acid (FFA) levels.

Likewise, other studies investigated the effect of fish oil when it was added to high fat, cholesterol and sucrose diets. The first study [20] evaluated changes in the liver mitochondrial subproteome, an organelle which plays a critical role in cell metabolism and the development of metabolic alterations, induced in rats by the supplementation with 10% fish oil of a high fat diet for 50 weeks as compared to rats fed low fat diet. As a result, Wrzesinski et al. identified 54 mitochondrial proteins regulated by fish oil. These proteins were involved in fatty acid oxidation and amino acid metabolism as well as in the increase of oxidative phosphorylation.

In a second research, De Roos et al. [21] studied the modulation of the liver proteome by fish oil supplementation of a diet high in saturated fat and cholesterol (HFC) for 3 weeks in APOE*3 Leiden transgenic mice, a model for lipid metabolism and atherosclerosis. These authors identified up to 44 proteins altered by fish oil as compared to HFC, which were mainly involved in glucose and lipid metabolism, as well as oxidation and aging processes. These changes were correlated with lower plasma and liver cholesterol and triglycerides levels as well as minor plasma FFA and glucose but higher plasma insulin levels, revealing new insights into mechanisms by which these fish oils can regulate lipid metabolism and related pathways.

A third research [22] reported the physiological modulation of rat liver proteome induced by 24-week supplementation with fish oil considering two different background diets: standard or high in fat and sucrose (HFHS). Méndez et al. demonstrated a different capacity of fish oil for regulating liver proteins depending on the background diet (6 proteins into standard diet and 31 into HFHS diet). Proteome changes induced by fish oil, especially under HFHS diet, consisted of decreasing the level of enzymes from lipogenesis and glycolysis, enhancing fatty acid beta-oxidation and insulin signaling and ameliorating endoplasmic reticulum stress. Moreover, the consumption of fish oil decreased protein oxidation and improved several biochemical parameters.

Other authors have used cell culture experiments to evaluate EPA and DHA effects on the proteome. Kalupahana et al. [23] investigated the effects of EPA on proteins from 3T3-L1 adipocytes treated with either EPA or arachidonic acid (ARA). EPA-treated cells presented higher levels of 19 proteins involved in carbohydrate and fatty acid metabolism and several other proteins related to cellular metabolism including response to stress. Likewise, EPA treatment resulted in lower levels of eight proteins belong to lipogenesis and other cellular metabolic processes such as cytoskeleton organization and biogenesis.

Proteomics has also been employed to evaluate the differential effect of the main marine ω-3 PUFAs. Mavrommantis et al. [24] analyzed the potentially different effects on the liver proteome regulation exerted by the dietary supplementation either with fish oil (EPA and DHA) or DHA alone in apoE knockout mice (model of atherosclerosis) fed HFC diet for 2 weeks. Both DHA and fish oil regulated 35 liver proteins, mainly involved in the metabolism of lipoproteins and oxidative stress. However, their effects on the proteome were not the same, since four of these proteins were differentially modulated by DHA or fish oil. As a consequence, although both fish oil and DHA could beneficially affect lipoprotein metabolism and oxidative stress, intervention with fish oil but not with DHA resulted in significantly lower levels of hepatic soluble epoxide hydrolase, an enzyme closely related to cardiovascular disease, as compared to control oil. This different behavior between EPA and DHA was also highlighted by Johnson et al. [25]. These authors evaluated the potentially independent protective/reversal effect of dietary EPA or DHA against mitochondrial dysfunction in aging skeletal muscle, after diet supplementation of young or older C57BL/6 mice. Authors found that 10-weeks of dietary supplementation with EPA but not with DHA partially attenuated the age-related decay in mitochondrial function. Thirty-nine mitochondrial proteins changed between old control and old EPA-treated mice. Thirty-two mitochondrial proteins changed between old control and old DHA-treated mice. However, only three proteins for EPA supplementation and seven for DHA were coincident with those proteins which were found to differ in young and old control mice. Authors concluded that neither EPA nor DHA attenuated the age-related drop in mitochondrial protein content, although these changes demonstrated that both EPA and DHA exerted some common biological effects (anticoagulation, anti-inflammatory, reduced FXR/RXR activation). Additionally, proteomics data showed that EPA improved muscle protein quality, specifically by decreasing mitochondrial protein carbamylation, a post-translational protein modification (PTM) that is driven by inflammation.

2.1.3. Proteomics for Studying Post-Translational Protein Modifications (PTMs)

Abundant evidence has shown that ω-3 PUFAs influence redox homeostasis [26] and several researchers have described both antioxidant and pro-oxidant properties for these fatty acids. Since diet-induced metabolic diseases are often associated with increased oxidative stress, the characterization of oxidative PTMs seems to be critical for understanding the mechanisms underlying the action of marine ω-3 PUFAs in theses pathologies. However, studies focused on the effects of EPA and DHA in modulating protein quality, besides protein quantity, are scarce likely due to methodological limitations.

Among the oxidative PTMs, protein carbonyl moieties formed in proteins (protein carbonylation), which are the more common oxidative PTMs, are considered as a major hallmark of oxidative protein damage. Metabolic alterations have strongly been correlated with high levels of protein carbonylation. However, only a very few studies have used proteomics tools to identify protein carbonylation targets in metabolic diseases and evaluate the potential effects that marine lipids exerted on them. A study addressed the effect of various dietary EPA and DHA ratios (1:1, 2:1, and 1:2, respectively) on protein carbonylation from plasma, kidney, skeletal muscle, and liver [27] in rats after 13 weeks of dietary intervention. Rats fed soybean (rich in ω-6 linoleic acid (LA)) or linseed oil (rich in ω-3 ALA) were used as controls. Authors identified targets of protein carbonylation in all the analyzed tissues. The three fish oil ratios, especially the 1:1 EPA:DHA, exerted a selective-protective effect against carbonylation of six proteins from plasma and liver, which was correlated with the improvement of biochemical features.

Jourmard-Cubizolles et al. [28] studied a specific oxidative PTM, namely 4-hydroxynonenal (4-HNE) protein adducts derived from PUFA peroxidation. These authors investigated the modulation of the aortic proteome in atherosclerotic prone (LDLR$^{-/-}$) mice fed an atherogenic diet supplemented with DHA for 20 weeks. Nineteen proteins were differentially regulated in the aorta of DHA-supplemented group. Most of them were related to glucose or lipid metabolism, including the up-regulation of superoxide dismutase by DHA which suggested an impact on vascular antioxidant defenses. This up-regulation was in agreement with data from the quantification of proteins with 4-HNE adducts. None of the twelve different identified proteins with 4-HNE adducts enhanced their oxidation in response to DHA supplementation. The articles cited in this section of the review are summarized in Table 1.

Table 1. Research articles found in literature which used proteomics to assay health marine lipid effects.

Reference	Marine Lipids Intervention	Experimental Model	Proteomics Tools	Target Proteome	Main Effects
Camargo et al., 2013 [14]	Acute intake of EPA/DHA (1.4:1)	Human suffering MetS	Quantitative 2-DE-MS/MS	PBMCs	5 proteins regulated from cell signaling and interaction, DNA repair, cellular assembly and organization and cell morphology
Rangel-Zúñiga et al., 2015 [15]	EPA/DHA (1.4:1) for 12 weeks	Human suffering MetS	Quantitative 2-DE-MS/MS	PBMCs	17 proteins regulated from immunological diseases and inflammatory response, oxidative stress, inflammation, endoplasmic reticulum stress and DNA repair
Jiménez-Gómez et al., 2014 [16]	EPA/DHA (1.4:1) for 12 weeks	Human suffering MetS	Quantitative 2-DE-MS/MS	White adipose tissue	3 proteins regulated from glucose metabolism
De Roos et al., 2008 [17]	EPA/DHA (2:1) for 6 weeks	Healthy humans	Quantitative 2-DE-MS/MS	Serum	10 proteins regulated from lipoprotein metabolism and inflammation
Burillo et al., 2012 [18]	0.6 g/d EPA and DHA for 5 weeks	Healthy smokers humans	2-DIGE-MS/MS	HDL	12 proteins regulated related to antioxidant, anti-inflammatory and anti-atherosclerotic properties, regulation of complement activation and acute phase response
Ahmed et al., 2014 [19]	EPA/DHA (1:1) for 4 months	Healthy C57BL/6 mice	Quantitative 2-DE-MS/MS	Liver	11 proteins regulated from lipid, carbohydrate, one-carbon, citric acid cycle and protein metabolisms
Wrzesinski et al., 2013 [20]	EPA/DHA (2:1) for 50 weeks	Wistar rats fed HFHS diet	Quantitative 2-DE-MS/MS	Liver mitochondria	54 proteins regulated from fatty acid and amino acid metabolisms, fatty acid oxidation and oxidative phosphorylation
De Roos et al., 2005 [21]	EPA/DHA (2:1) for 3 weeks	APOE*3 Leiden transgenic mice fed HFC diet	Quantitative 2-DE-MS/MS	Liver	44 proteins regulated from glucose and lipid metabolism, oxidation and aging processes
Méndez et al., 2017 [22]	EPA/DHA (1:1) for 28 weeks	Wistar Kyoto rats fed HFHS diet or STD diet	2-DIGE-MS/MS iTRAQ-nanoLC-MS/MS	Liver	6 proteins regulated in STD diet 31 proteins regulated in HFHS diet from lipogenesis and glycolysis, fatty acid beta-oxidation, insulin signaling, oxidative stress and ameliorating endoplasmic reticulum stress
Kalupahana et al., 2010 [23]	EPA	Cell culture	2-DIGE-MS/MS	3T3-L1 adipocytes	27 proteins regulated from carbohydrate and fatty acid and cell metabolism, response to stress, lipogenesis, cytoskeleton organization and biogenesis
Mavrommatis et al., 2010 [24]	EPA/DHA (1.4:1) or DHA for 2 weeks	apoE knockout mice fed HFC diet	Quantitative 2DE-MS/MS	Liver	35 proteins regulated from of lipoproteins metabolism and oxidative stress; 4 of them different between DHA and fish oil
Johnson et al., 2015 [25]	0.5% EPA or 0.5% DHA for 10 weeks	6- or 24-months C57BL/6 mice	Quantitative untargeted nanoLC-MS/MS	Quadriceps muscle	39 proteins regulated by EPA-treated and 32 proteins regulated by DHA-treated old mice related to anticoagulation, anti-inflammatory, reduced FXR/RXR activation EPA decrease protein carbamylation
Méndez et al., 2013 [27]	EPA:DHA 1:1 or 2:1 or 1:2 for 13 weeks	Wistar Kyoto rats	FTSC-carbonyl protein labeling Quantitative 1DE- and 2DE-MS/MS	Plasma, kidney, skeletal muscle, and liver	6 carbonylated protein targets regulated by 1:1 EPA:DHA in plasma and liver
Jourmard-Cubizolles et al., 2013 [28]	2% DHA for 20 weeks	LDLR-/- mice fed atherosclerotic diet	Quantitative 2DE-MS/MS	Aorta	19 proteins regulated from glucose and lipid metabolisms and oxidative stress 12 identified 4-HNF-proteins

2.2. Beneficial Effects of Marine Lipids Intake Assayed by Lipidomics

In the era of genomics, transcriptomics and proteomics, metabolomics is becoming a critical component of the omics revolution and systems biology [9]. Among the four types of biological molecules that compose the human body, i.e., nucleic acids, amino acids, carbohydrates and lipids, the study of lipid homeostasis has been attracting increasing interest, especially during the last decade. As a subfield of metabolomics, lipidomics addresses the large-scale study of lipids, including the detailed characterization of lipid metabolites, their interactions and influence on biological systems. Lipidomics comprises the methodologies used for the study of a lipidome, which can be defined as the comprehensive and quantitative description of a set of lipid species present in an organism [29].

In spite of a lack of a commonly accepted definition, lipids are hydrophobic or amphipathic compounds of relatively small molecular weight. They include a vast plethora of different structures which play a critical role in cell physiology. Based on their chemical structure, lipids are divided into eight main classes: (a) fatty acyls, such as polyunsaturated fatty acids (PUFAs); (b) glycerolipids (GLs), such as triglycerides (TGs); (c) glycerophospholipids (GPs), also referred to as phospholipids (PLs), such as phosphatidylcholine; (d) sphingolipids (SPs), such as ceramides; (e) sterol lipids (STs) such as cholesterol; (f) prenol lipids (PRs); (g) saccharolipids (SLs); (h) polyketides (PKs). Some of the biological functions of these lipids are energy storage and structural components of cellular membranes, cell signaling, endocrine actions and essential role in signal transduction, membrane trafficking and morphogenesis [30].

Lipids are potent signaling molecules which can act as biosynthetic precursors of lipid mediators. In this review, the lipid mediators derived from ω-6 and ω-3 PUFAs, such as ARA, EPA, and DHA, are particularly noteworthy. These lipid mediators, which are generally known as eicosanoids (derivatives from the oxidation of C20 PUFAs, i.e., ARA, EPA, and dihomo-γ-linolenic acid (DGLA)) and docosanoids (derivatives from the oxidation of C22 PUFAs, i.e., DHA and docosapentaenoic acid (DPA)), are bioactive compounds related to inflammatory processes [31]. While most of them are formed by the action of lipoxygenases (LOX), cyclooxygenases (COX), cytochrome P450 (CyP450), and several subsequent enzymes on PUFAs, several compounds, for instance, isoprostanes, are the consequence of non-enzymatic process [32]. The main lipid mediators originated from ARA, EPA, and DHA are shown in Table 2.

The enormous complexity of lipidomes implies that lipidomics must consider the characterization of thousands of pathways and networks which involve cellular lipids species and their interactions with other molecules. Lipidomics is a valuable tool to investigate the influence of marine lipids on health. However, its use is still limited due to its novelty but also to several difficulties associated with data interpretation, among other limitations [33]. In consequence, only a few studies have addressed the effect of marine lipids on the lipidome in the context of human metabolic disorders. Some of them used lipidomics to characterize total lipid classes, mainly by using gas chromatography (GC)-MS or LC-MS, and others were focused on lipid mediators, mainly by using solid phase extraction (SPE) coupled to identification and quantification by LC-MS/MS.

By using Scopus database, 53 publications were found following these search criteria: TITLE-ABS-KEY ("omega 3" OR "PUFA" OR "eicosapentaenoic" OR "docosahexaenoic" OR "marine fatty acids" OR "marine lipid" OR "marine oil" OR "fish oil" OR "fish lipid") AND TITLE-ABS-KEY (nutrition OR diet) AND TITLE-ABS-KEY ("metabolic disorder" OR "metabolic alteration" OR "metabolic disease" OR obesity OR "metabolic syndrome" OR diabetes OR "cardiovascular disease" OR cholesterol OR insulin OR atherosclerosis OR inflammation OR "oxidative stress") AND TITLE-ABS-KEY (lipidomic OR lipidomics OR lipidome). These publications, mainly research articles (69%) and reviews (19%), belong, in descending order of abundance, to the fields of medicine (65%) and biochemistry, genetics and molecular biology (61%), and to a lesser extent nursing, agricultural and biological sciences, pharmacology, toxicology and pharmaceutics, chemistry, immunology and microbiology, neuroscience, etc.

Table 2. Main classes of lipid mediators derived from arachidonic acid (ARA), eicosapentaenoic acid (EPA) and docosahexaenoic acid (DHA).

Family	Lipid Mediators from ARA		Lipid Mediators from EPA		Lipid Mediators from DHA	
	Nomenclature	Isomers	Nomenclature	Isomers	Nomenclature	Isomers
Monohydroxys	HETE	3-, 5-, 8-, 9-, 11-, 12-, 15, 18-, 19- and 20HETE	HEPE	5-, 8-, 9-, 11-, 12-, 15 and 18HEPE	HDoHE	4-, 7-, 8-, 10-, 11-, 13-, 14-, 16-, 17- and 20HDoHE
Dihydroxys	DiHET (DiHETrE)	5,6-,8,9-, 11,12- and 14,15 DiHETrE	DiHETE	5,6-, 5,12-, 5,15-, 8,15-, 14,15- and 17,18DiHETE	DiHDPA	10,11-, 14,21- and 19,20DiHDPA
Leukotrienes	LT_4	LTA_4, $-B_4$, $-C_4$, $-D_4$ and $-E_4$	LT_5	LTA_5, $-B_5$, $-C_5$, $-D_5$ and $-E_5$		
Trihydroxys (lipoxins)	LX_4	LXA_4 and $-B_4$	LX_5	LXA_5		
Hydroperoxides	HpETE	5-, 8-, 9-, 11-, 12-, 15-, 19- and 20HpETE	HpEPE	5-, 8-, 9-, 11-, 12-, 15 and 18HpEPE	HpDoHE	4-, 7-, 8-, 10-, 11-, 13-, 14-, 16-, 17- and 20HpDoHE
Epoxides	EET (EpETrE)	5,6-,8,9-, 11,12- and 14,1EET	EEQ (EpETE)	8,9-, 11,12-, 14,15- and 17,18EEQ	EDP (EpDPA)	7,8-,10,11-, 13,14, 16,17- and 19,20EDP
Thromboxanes	TX_2	TXA_2 and $-B_2$	TX_3	TXA_3 and $-B_3$		
Prostaglandins	PG_2	PGA_2, $-B_2$, $-D_2$, $-E_2$, $-G_2$, $-H_2$, $-I_2$, $-J_2$ and $-F_{2\alpha}$	PG_3	PGA_3, $-B_3$, $-C_3$, $-D3$, $-E_3$, $-I_3$, $-H_3$ and $-F_{3\alpha}$		
Isoprostanes	$IsoP_2$	$8isoPGJ_2$, $-A_2$, $-E_2$ and-D_2	$IsoP_3$	8-, 5-, 11-, 12-, 15- and $18isoPGF_{3\alpha}$		
Resolvins	8-, 5-, 12 and $15isoPGF_{2\alpha}$		RvE	RvE_1, $-E_2$ and $-E_3$	RvD	RvD_1, $-_2$, $-_3$ and $-_4$
Neuroprotectins					PD	PD_1
Maresins					MaR	MaR_2 (13,14DiHDPA) 7-MaR_1
Keto-derivatives						
Keto-PG	oxoETE	5-, 8-, 9-, 11-, 12-, 15, 19- and 20 oxoETE				

In this section, the same inclusion criteria exposed above (Section 2.1) for proteomics articles is followed (i.e., only research articles focusing on lipidomics, which have assayed the effect of marine lipids on features of MetS, cardiovascular diseases and type 2 diabetes, excluding the studies which have used ω-3 PUFAs from vegetal origin, i.e., ALA). Likewise, works which have combined both proteomics and lipidomics approaches will be presented in Section 2.3.

According to these filters, 35 research articles have been found and are summarized below.

2.2.1. Lipidomics in Clinical Trials

Several researches have used lipidomics tools in human clinical trials to gain insights of physiological mechanisms behind the evidence of the multiple beneficial health effects of fish consumption.

Some of them have investigated the modulation of plasma/serum lipidome by fish oil in healthy subjects. Ottestad et al. [34] studied the plasma lipidomic profile in healthy subjects which received either 8 g/day of fish oil from cod liver or high oleic sunflower oil for 7 weeks. Authors identified and quantified 260 different lipids in plasma, being 23 lipids significantly decreased and 51 significantly increased by fish oils. Data analysis demonstrated that fish oil supplementation altered lipid metabolism and increased the plasma proportion of PLs and TGs containing long-chain PUFAs. Therefore, the beneficial effects of fish oil supplementation could be explained in part by a remodeling of the plasma lipids. Rudkowska et al. [35] combined transcriptomics and metabolomics technologies to investigate molecular and metabolic changes in healthy subjects underwent 6-week supplementation with EPA and DHA. Authors measured 107 lipids in plasma, which were further subdivided into three different classes: 15 sphingomyelins (SMs) and SM derivatives, 15 lysophosphatidylcholines (lysoPCs) and 77 glycerophosphatidylcholines (glyPCs). Results showed some gender differences in lipidomic profiles between pre- and post- ω-3 supplementation, although the main differences were found as a result of fish oil supplementation. These differences were principally due to changes in glyPCs, and overall, results demonstrated that there was an increase in unsaturated fatty acids after the ω-3 PUFAs supplementation period. These and others data given by authors supported the cardioprotective effects of the ω-3 PUFAs supplementation, although some of their mechanisms can be dependent on gender. The high variability in lipid profiles and lipidome responses to PUFAs supplementation was addressed in a third study performed by Nording et al. in healthy subjects. These authors used a multi-platform lipidomics approach to investigate both the consistent and inconsistent responses to a defined ω-3 intervention for 6 weeks [36]. Thus, Nording et al. evaluated the changes induced by ω-3 intervention in total lipidomic plasma profile (including fatty acids, lipid classes and lipoprotein distribution) but also the effects of ω-3 on lipid mediators. Authors measured 7 lipid classes, and a total of 87 lipid mediators. Results showed significant changes in both total lipidomic and lipid mediator profiles after ω-3 supplementation, as well as a strong correlation between lipid mediator profiles and EPA and DHA incorporated into different lipid classes. However, authors found that both ω-3 and ω-6 fatty acid metabolites displayed a large degree of variation among the subjects; for instance, only the 50% of the subjects presented significantly decreased levels of PGE_2, TXB_2 and 12-HETE whereas the other 50% did not show any change or even increased levels. Specifically, 12-HEPE showed high heterogeneity, decreasing up to 82% in some subjects and increasing up to 5% in others. This work pointed out the highly variable response to ω-3 fatty acids supplementation and the need for an in-depth lipidomic phenotype characterization in order to properly assess their effectiveness against diseases.

Since lipidomic characterization requires an accurate determination of the huge range of lipid mediators and fatty acid derivatives, some authors have tried to shed light on this matter. Mas et al. [37] published the development of a SPE-LC-MS/MS assay to measure resolvins and protectins families generated from the ω-3 EPA and DHA in human blood after fish oil supplementation (4 g fish oil containing 35% EPA and 25% DHA/day for 3 weeks). It was the first time that 17R/SHDHA RvD_1 and RvD_2 were detected in plasma/serum after oral ω-3 fatty acid supplementation. Authors found that those RvD_1 and RvD_2 were within the biological range of anti-inflammatory and pro-resolving activities detected in isolated human leukocytes and in vivo studies in mice. This methodology was further employed to examine the effect of short-term (5 days) ω-3 fatty acid supplementation. As compared to baseline, ω-3 intake significantly increased plasma levels of RvE_1, 18R/S-HEPE, 17R/S-HDHA, and 14R/S-HDHA up to concentrations biologically active in healthy humans. Therefore, ω-3 PUFAs were able to exhibit their anti-inflammatory action even after short interventions [38].

Furthermore, Keelan et al. [39] determined if the supplementation with EPA and DHA during pregnancy could modify placental PUFAs composition and the accumulation of lipid mediators. In this case, only resolvins (RvD_1, $17R-RvD_1$ and RvD_2) and protectins from the D-series (PD_1 and 10S, 17SdiHDHA) and upstream precursors (18-HEPE and 17-HDHA) were measured. Authors found that the ω-3 PUFAs supplementation increased placental DHA levels, as well as the levels of precursors 18-HEPE and 17-HDHA, but the concentration of EPA was not significantly increased neither concentrations of RvD_1, $17R-RvD_1$, RvD_2 and PD_1. Placental pro-resolving lipid mediator levels seemed to be modulated by maternal dietary PUFAs, although their biological significance in the placenta remains unknown.

Besides healthy subjects, other authors have used lipidomics tools to evaluate changes in lipid homeostasis induced by ω-3 in patients suffering metabolic disorders. The modulation of lipid mediators by EPA and DHA was analyzed in plasma of humans with MetS [40]. Dietary intervention consisted in a daily supplementation with EPA and DHA in the form of TGs for 3 weeks. Plasma lipid mediators (i.e., 18-HEPE, E-series resolvins, 17-HDHA, D-series resolvins, 14-HDHA, and maresin-1) from MetS volunteers and their healthy controls (at baseline and after dietary intervention) were measured. Results showed that ω-3 PUFAs supplementation increased E-series resolvins to a similar extent in MetS subjects and controls. However, only the healthy controls presented increased concentrations of E- and D-series resolvin precursors and 14-HDHA in response to ω-3 PUFAs supplementation. The action of EPA and DHA supplementation was also investigated in hyperlipidemic men (cholesterol >200 mg/dL; triglyceride >150 mg/mL) after 12-weeks daily intake [41]. Schuchardt et al. measured serum levels of 44 free hydroxy, epoxy and dihydroxy fatty acids and found that after supplementation, all subjects (including healthy controls) showed considerably elevated levels of EPA-derived lipid mediators and a less pronounced increment of DHA-derived ones. However, the supplementation with higher amounts of DHA than EPA (DHA:EPA 5:1) for 3 months induced a significant increase of pro-resolving DHA derivatives in plasma of obese women [42]. All these different results in the level of EPA and DHA derivatives seemed to be correlated with the EPA and DHA supplement content.

In another study, Lankinen et al. [43] investigated how fatty fish or lean fish in a diet affect serum lipidomic profiles in subjects with coronary heart disease after their consumption for 8 weeks. Lipidomic changes among groups were detected in at less 59 bioactive lipid plasma species, including ceramides, lysoPCs and diacylglycerols (DGs), which were found significantly diminished in the fatty fish group, whereas in the lean fish group cholesterol esters and specific long-chain TGs increased significantly. Therefore, fatty fish intake reduced lipid species which are potential mediators of lipid-induced insulin resistance and inflammation, and these results might be associated with the protective effects of fatty fish on the progression of atherosclerotic vascular diseases or insulin resistance. However, Midtbø et al. [44] demonstrated in mice fed western diets that the consumption of farmed salmon, which had previously fed fish feed with a reduced ratio of ω-3/ω-6 PUFAs, led to a selectively increased abundance of ARA in the liver PLs pool of the mice. This increment was accompanied by higher levels of hepatic ceramides and ARA-derived pro-inflammatory mediators and a reduced abundance of lipid mediators derived from EPA and DHA. Therefore, the studies made after fish consumption rather than fish oil have to consider the PUFA composition of fish to get proper conclusions.

Two articles have used enriched ω-3 PUFAs dairy products to test their effects on lipidome. In the first one, mildly hypertriacylglycerolemic subjects consumed yogurt supplemented with 3 g of EPA and DHA per day for 10 weeks [45]. Results showed that a daily intake of supplemented yogurt significantly increased plasma EPA-derived mediators (PGE$_3$, 12-, 15-, 18-HEPE), as well as EPA and DHA levels in plasma and red blood cells, and improved some cardiovascular risk factors. In the second one, overweight and moderately hypercholesterolemic subjects consumed 250 mL of enriched milk with EPA and DHA for 28 days [46]. In this case, the changes induced on LDL-lipidome composition were primarily addressed. Enriched milk significantly reduced TGs and very low-density lipoprotein (VLDL) cholesterol and caused significant changes in the LDL lipid metabolite pattern, increasing the long-chain polyunsaturated cholesteryl esters and the ratio PC36:5/lysoPC16:0. All these modifications were associated with its reduced inflammatory activity.

2.2.2. Lipidomics in Animal Models

In animal models, several authors have investigated the influence of dietary EPA and DHA on lipid homeostasis in the context of metabolic alterations through lipidomics approaches. Some of them have used healthy models to look into the potentially different effects of EPA and DHA to find their optimal proportions in the diet. Dasilva et al. [47] tested whether the intake of three proportions of EPA/DHA (1:1, 2:1 or 1:2) for 22 weeks provoked a different modulation of the formation of lipid mediators. Then, authors examined their influence on various indexes of inflammation and oxidative stress in Wistar Kyoto rats. A total of nine compounds derived from PUFA oxidative metabolism (namely five EPA eicosanoids -12HEPE, 15HEPE, 12HpHEPE, 15HpHEPE, TXB$_3$-, two DHA docosanoids -17HDoHE, 17HpDoHE-, and two ARA eicosanoids -11HETE, PGE$_2$) were identified and quantified in plasma. Results evidenced that the ratios 1:1 and 2:1 EPA:DHA exerted a remarkable healthy effect generating a less oxidative environment and modulating LOX and COX activities towards a decrease in the production of pro-inflammatory ARA eicosanoids and oxidative stress biomarkers from EPA and DHA. On the other hand, the higher DHA amount in the diet (i.e., 1:2 ratio) reduced the health benefits described in terms of inflammation and oxidative stress. The beneficial effect of the ratios with higher EPA amount was further evaluated in a rat model of MetS (SHROB rats), in which the 1:1 and 2:1 ratios exerted the highest health benefits. In this other work, EPA and DHA supplementation also decreased the level of pro-inflammatory ARA eicosanoids produced, in agreement with the results found in the healthy model [48]. Other authors [49] reported that the benefit of ω-3 PUFA-rich diets could be attributed to the generation of electrophilic oxygenated metabolites that transduce anti-inflammatory actions rather than the suppression of pro-inflammatory ARA metabolites. This work was focused on the endogenous production of ω-3 PUFAs electrophilic ketone derivatives and their hydroxy precursors in human neutrophils. Authors evaluated in vitro endogenous generation of these lipid mediators from DHA and DPA in neutrophils isolated from healthy subjects, both at baseline and upon stimulation with calcium ionophore. Additionally, their potential modulation by diet was assessed through a randomized clinical trial carried out with healthy adults receiving daily oil capsule supplements, which contained either 1.4 g of EPA and DHA or soybean oil, for 4 months. Results reported the 5-LOX-dependent endogenous generation of 7-oxo-DHA, 7-oxo-DPA and 5-oxo-EPA and their hydroxy precursors stimulated in human neutrophils, whereas the dietary supplementation with EPA and DHA increased the formation of 7-oxo-DHA and 5-oxo-EPA, without significant modulation of ARA metabolite levels.

In C57BL/6 mice fed a diet containing 10% ω-3 PUFAs from menhaden oil for 4 months, Balogun et al. [50] analyzed the effect of fish oil on the fatty acid composition of various bioactive lipids in plasma and liver by using lipidomics. Results demonstrated a significantly higher concentration of EPA containing phosphatidylcholine (PCs), lysophosphatidylcholine (LPCs), and cholesteryl esters (CEs) after fish oil intake in plasma and liver, as well as a higher concentration of free ω-3 PUFAs.

In healthy rats, lipidomics was also used to determine if LOX-generated lipid mediators were presented in bone marrow and if so, their modulation by dietary EPA and DHA supplementation [51]. Data analysis revealed the presence of LOX-pathway lipid mediators derived from ARA, EPA and DHA, including lipoxins, resolving D$_1$, resolvin E$_1$, and protectin D$_1$ in bone marrow. Moreover, the daily supplementation with DHA or with EPA ethyl ester for 4 months increased the percentage of

DHA and EPA in bone marrow, and the proportion of LOX mediators biosynthesized from DHA or EPA, respectively. Given the potent bioactivities of the lipoxins, resolvins and protectins, their presence and changes in their profile found after EPA and DHA ethyl ester supplementation may be of interest in bone marrow function and as a potential source of these mediators in vivo.

Several articles have used lipidomics to assay the protective effects of EPA and DHA against metabolic alterations induced by unhealthy diets. In rats, Taltavull et al. [52] investigated how supplementation of a high fat and sucrose diet with both EPA and DHA modified the hepatic ceramide profile triggered by the unhealthy diet in a dietary intervention of 24 weeks. Authors found that ω-3 PUFAs reduced total liver ceramide content and altered ceramide profiles in pre-diabetic rats. They also observed a significant positive linear correlation between long chain ceramide 18:1/18:0 and the HOMA index, and negative between very long chain ceramides 18:1/24:0 and 18:1/20:0 and plasma insulin levels and the HOMA index. Overall, these data may help explain the protective action of ω-3 PUFAs against liver insulin resistance induced by diet. Caesar et al. [53] used lipidomics tools to evaluate the regulation of lipid composition in mice liver and serum by dietary fish oil as compared to lard oil, but considering their interaction with gut microbiota. After 11 weeks, menhaden fish oil supplementation induced significant changes in abundance of most lipid classes. The gut microbiota affected lipid composition by increasing hepatic levels of cholesterol and cholesteryl esters in mice fed high-fat lard diet but not in mice fed high-fat fish oil diet. These results highlighted that the regulation of hepatic cholesterol metabolism induced by gut microbiota was dependent on dietary lipid composition.

Animal models of obesity have also been used to investigate the modulation of lipid metabolism in the adipose tissue. Kuda et al. [54] identified cells producing lipid mediators in epididymal WAT of mice fed for 5 weeks obesogenic high-fat diet, which was supplemented or not with EPA and DHA. Results demonstrated selectively increased levels of anti-inflammatory lipid mediators in WAT in response to ω-3, reflecting either their association with adipocytes (endocannabinoid-related Ndocosahexaenoylethanolamine) or with stromal vascular cells (pro-resolving lipid mediator protectin D_1). In parallel, tissue levels of obesity-associated pro-inflammatory endocannabinoids were suppressed. Moreover, they found that adipose tissue macrophages (ATMs) were not the main producers of protectin D_1 and that ω-3 PUFAs lowered lipid load in ATMs while promoting their less-inflammatory phenotype. Besides these specific roles of various cell types in WAT, the kind of fat depots seemed to be also critical. In the abdominal (epididymal) fat but not in other fat depots, Flachs et al. [55] found a synergistic induction of the mitochondrial oxidative capacity and lipid catabolism after combining ω-3 PUFAs intake and caloric restriction in high-fat fed mice. This combination resulted in an increased oxidation of metabolic fuels in the absence of mitochondrial uncoupling, while low-grade inflammation was suppressed, reflecting changes in tissue levels of anti-inflammatory lipid mediators, namely 15-deoxy-Δ(12,15)-prostaglandin J_2 and protectin D_1.

Lipidomics analysis [56] also revealed that ω-3 PUFAs supplementation could alleviate hepatic steatosis in *ob/ob* mice, an obesity model of insulin resistance and fatty liver disease. ω-3 PUFAs inhibited the formation of ω-6 PUFAs derived eicosanoids while triggering the formation of ω-3 PUFAs derived resolvins and protectins. Moreover, representative members of these lipid mediators, namely resolvin E_1 and protectin D_1, mimicked the insulin sensitizing and anti-steatotic effects of ω-3 PUFAs and induced adiponectin expression to a similar extent that the antidiabetic drug rosiglitazone. These findings uncovered beneficial actions of ω-3 PUFAs and their bioactive lipid derivatives in preventing obesity-induced insulin resistance and hepatic steatosis. Similar conclusions were obtained by Kalish et al. [57] who demonstrated in a mouse model of steatosis that parental nutrition with fish oil-based lipid emulsions was associated with the production of anti-inflammatory and pro-resolving lipid mediators. The preventive effect of EPA and DHA against necroinflammatory injury in liver was also investigated in mice fed high saturated fat diets containing either DHA or both EPA and DHA for 5 weeks [58]. Both marine ω-3-rich diets induced an increased hepatic formation of DHA-derived lipid mediators (i.e., 17S-hydroxy-DHA (17S-HDHA) and protectin D_1), which was correlated with significant protection of liver injury. This work reported a potential role for DHA-derived products, specifically 17SHDHA and protectin D$_1$, in mediating the protective effects of dietary DHA against necroinflammatory liver injury.

The potential role of fish oil to prevent glomerulosclerosis in a rat model of MetS (JCR:LA-cp rats) via renal eicosanoid metabolism and lipidomics analysis was addressed by Aukema et al. [59]. MetS rats were supplemented with 5% or 10% fish oil for 16 weeks. Dietary fish oil reduced glomerulosclerosis and albuminuria and the 11- and 12-HETE levels, as well as other (5-, 9- and 15-) HETE. Also, fish oil reduced endogenous renal levels of 6-keto PGF$_{1\alpha}$ (PGI$_2$ metabolite), thromboxane B$_2$ (TXB$_2$), PGF$_{2\alpha}$, and PGD$_2$ by approximately 60% in rats fed 10% fish oil as compared to untreated MetS rats. Whereas in rats fed 5% fish oil, TXB$_2$ decreased in 250% and PGF2a in 241%. These results suggested that dietary fish oil might improve dysfunctional renal eicosanoid metabolism associated with kidney damage during conditions of the MetS.

The impact of DHA supplementation on the profiles of PUFA oxygenated metabolites and their contribution to atherosclerosis prevention were investigated by Gladine et al. [60]. The study was conducted with atherosclerosis prone mice which received increasing doses of DHA (0%, 0.1%, 1% or 2% of energy) during 20 weeks. Targeted lipidomics analysis determined a significant modulation of EPA and DHA and their respective oxygenated metabolites in plasma and liver. Remarkably, hepatic F4-neuroprostanes were strongly correlated with the hepatic DHA level. The hepatic level of F4-neuroprostanes was the variable most negatively correlated with the plaque extent and plasma EPA-derived diols. Thus, oxygenated ω-3 PUFAs derivatives, particularly F4-neuroprostanes, were revealed as potential biomarkers of DHA-associated_atherosclerosis prevention which might contribute to the anti-atherogenic effects of DHA.

It is well known that oils from marine organisms have a different fatty acid composition and differ in their molecular composition. Fish oil has a high content of EPA and DHA mostly esterified to TGs, while in krill oil these fatty acids are mainly esterified to PLs. Considering that, Skorve et al. [61] studied the effects of these oils on the lipid content and fatty acid distribution in the various lipid classes in liver and brain of mice. After 6 weeks of feeding a high-fat diet supplemented with fish oil or with krill oil, shotgun lipidomics showed that in both fish and krill oil fed mice, the TGs content in the liver was more than doubled compared to control mice. The fatty acid distribution was affected by the oils in both liver and brain with a decrease in the abundance of LA and ARA, and an increase in EPA and DHA in both study groups. LA decreased in all lipid classes in the fish oil group but with only minor changes in the krill oil one. Differences were especially evident in some of the minor lipid classes associated with inflammation and insulin resistance. Ceramides and DGs were decreased, and cholesteryl esters increased in the liver of the krill oil group, while plasmalogens were diminished in the fish oil group. In the brain, DGs were decreased, more by krill than fish oil, while ceramides and lactosylceramides were increased, more by fish than krill oil. Changes in hepatic sphingolipids and ARA fatty acid levels were higher in the krill oil group than in the fish oil one. These changes were consistent with a hypothesis that krill oil may have a stronger anti-inflammatory action and enhance insulin sensitivity more potently than fish oil.

2.2.3. Lipidomics in Cell Cultures

Finally, some authors have applied lipidomics approaches in cell cultures and in vitro assays to deeply analyze the metabolic effect of dietary marine ω-3 PUFAs. Polus et al. demonstrated that the addition of EPA during differentiation of human subcutaneous adipose tissue stromal vascular fraction cells induced the formation of small lipid droplets and reduced the production of pro-inflammatory mediators in adipose tissue in comparison to ARA addition. These changes were the consequence of the production of anti-inflammatory eicosanoids derived from EPA [62].

The preventive effect of DHA at physiological doses against insulin resistance was investigated in C2C12 myotubes exposed to palmitate. DHA decreased protein kinase C activation, restored cellular acylcarnitine profile, insulin-dependent AKT phosphorylation and glucose uptake. Results showed that DHA participated in the regulation of muscle lipid and glucose metabolism by preventing lipotoxicity, inflammation and insulin resistance in skeletal muscle [63]. Likewise, in in vitro experiments, Ting et al. [64] investigated structural changes in cardiolipins after DHA or EPA supplementation and compared them to ARA treatment, using H9c2 cardiac myoblast as a cell model. Among the 116 cardiolipin species with 36 distinct mass identified, the three PUFAs treatments differentially perturbed the fatty acyl chain compositions in the mitochondrial of the H9c2 cardiac myoblast, suggesting that both mitochondrial membrane composition and function were susceptible to exogenous lipids. Additionally, DHA supplementation correlated with an elevation of less unsaturated and ω-3 cardiolipin species, which appeared to be a minor effect on EPA but not on ARA.

2.2.4. Marine Lipids and Other Bioactive Compounds Assayed by Lipidomics

Several articles have also used lipidomics to study the combined action of ω-3 PUFAs with other bioactive compounds which have shown potential benefits in subjects susceptible to cardiovascular diseases.

The protective effects of fatty fish consumption on the progression of insulin resistance were tested in combination with other products with recognized effect on glucose metabolism (whole grain and low postprandial insulin response grain products, and bilberries) in a clinical trial [65]. Plasma lipidomic profiles of people with impaired glucose metabolism and with at least two other features of the MetS were evaluated after 12 weeks of dietary intervention. Among the 364 characterized lipids in plasma, 25 changed significantly in the treated group, including multiple TGs incorporating the long chain ω-3 PUFAs. These results were supported by biochemical data and suggested an improvement of glucose metabolism and a beneficial effect in preventing type 2 diabetes in population groups at considerably higher risk of suffering it.

Other three articles found in literature have employed lipidomics tools to study the combined action of ω-3 PUFAs with other bioactive compounds which have shown potential benefits in subjects suffering CVD or at considerably high risk. The first one [66] was designed to assay the combined effect of L-alanyl-L-glutamine and fish oil supplementation for 3 months on skeletal muscle function and metabolism in patients with chronic heart failure. Patients were randomized to either L-alanyl-L-glutamine and PUFAs or placebo (safflower oil and milk powder). Regular uptake of the bioactive compounds led to the expected increase in unsaturated fatty acids. Moreover, the lipidomic analysis revealed a decrease in circulating levels in total ceramides and two ceramide subspecies (C22:1 and C20:1) induced by supplements at 4 weeks, which was not detectable in samples at 3 months. In another study, the combined effect of ω-3 fatty acids with Coenzyme Q10 was evaluated on plasma lipid mediators profile in patients with chronic kidney disease (CKD), which are highly predisposed to suffer CVD, partially due to their chronic inflammation [67]. Patients received a daily dose of ω-3 PUFAs, Coenzyme Q10 (CoQ), or both supplements for 8 weeks. Compounds as 18-HEPE, 17-HDHA, RvD$_1$, 17R-RvD$_1$, and RvD$_2$ were measured in plasma before and after the intervention. Results showed that ω-3 PUFAs but not CoQ significantly increased plasma levels of the upstream precursors of the E and D-series resolvins (18-HEPE and 17-HDHA, respectively) as well as RvD$_1$. This finding may have important implications for limiting ongoing low-grade inflammation in CKD. Finally, Bondía-Pons et al. [68] investigated the effects of ω-3 PUFAs and polyphenol rich diets on plasma and HDL fraction lipidomic profiles in MetS patients. Authors compared the effects of diets contained low or high ω-3 PUFAs (EPA and DHA) in combination with low or high of polyphenols, resulting in 4 isoenergetic diets, differing in their natural ω-3 PUFAs and polyphenols amount. Authors successfully identified 350 and 293 lipid species in total plasma and HDL fraction samples respectively. Results showed that the two diets high in ω-3 PUFAs highly increased unsaturated long-chain TGs and EPA and DHA-containing PLs levels and decreased levels in total plasma of low unsaturated PLs, and PCes, LysoPCs, and PCps with ARA in their structure. With regards to HDL, PCs and TGs with DHA or EPA in their structure increased after the consumption of high ω-3 PUFAs diets, while PCes and PCps with ARA in their structure, and medium-chain PCs decreased. The diet high in both ω-3 and polyphenols significantly reduced PCs and PEs levels, especially of those alkyl and alkenyl ether lipids with 16:0 in their structure as well as saturated and low-unsaturated PCs and PEs. The study found a relevant association among lipidomics data, dietary and clinical/anthropometric variables. The most remarkable and complex association among variables was observed after the intervention with the diet high in both ω-3 PUFAs and polyphenols. Different types of TGs were positively or negatively associated with waist circumference, which was positively associated with insulin levels. Glucose was positively associated with body weight, which was negatively associated with EPA. This latter association was only detected after feeding the combined diet, which may mean that dietary polyphenols interacted with ω-3 PUFAs in the regulation of body weight in MetS subjects. Overall, data reflected different lipid rearrangements after a nutritional intervention with diets rich in ω-3 PUFAs and polyphenols in these patients at a high CVD risk.

The promising cooperative effect between fish oil and polyphenols was further analyzed considering the ability of polyphenols as antioxidants to potentially ameliorate oxidative damage of ω-3 PUFAs when they are consumed together and to enhance their individual potential effects on metabolic health through the modulation of fatty acids profiling and the formation of lipid mediators. Dasilva et al. [69] evaluated the effect of diet supplementation with EPA and DHA, grape polyphenols or both in rats fed either standard or high fat and sucrose diets (a total of eight diets), on the inflammatory response and redox unbalance triggered by these unhealthy diets. Authors analyzed total fatty acid composition in the liver, plasma, adipose tissue, erythrocytes as well as circulating FFA in plasma across experimental groups and calculated fatty acid desaturases (FADs) indexes (stearoyl-CoA desaturases SCD-16 and SCD-18 as well as desaturases Δ4, Δ5, and Δ6). Likewise, a total of nine compounds derived from PUFA oxidative metabolism (namely five EPA eicosanoids, two DHA docosanoids and two ARA eicosanoids) were identified and quantified in plasma. Data analysis reflected that the supplementation with fish oil led to an anti-inflammatory situation associated with a lower ω-6/ω-3 index in plasma and membranes, a lower production of ARA pro-inflammatory lipid mediators, an up-regulation of desaturases related to EPA and DHA synthesis and a down-regulation of these desaturases to synthesize ARA. However, polyphenols bioactivity was influenced by the background diet. In a standard diet, they seemed to modulate enzymes towards an anti-inflammatory and antioxidant response, and the combination with fish oil down-regulated Δ5D related with ARA synthesis, decreased COX activity on ARA, enhanced the antioxidant enzymes and decreased total FFA in plasma. Similarly, the combination of both supplements also produced a significant improvement in the antioxidant balance and oxidative stress in unhealthy diets. However, the efficacy of polyphenols to reduce inflammation was lower when were added to the unhealthy diet, and some pro-inflammatory pathways were found even up-regulated. Therefore, fish oil seemed to be the main responsible for the anti-inflammatory effects observed in the combined group in the unhealthy diet. The combination of both bioactive supplements may improve the metabolic health in both background diets by acting on inflammation and oxidative stress pathways. A summary of the articles cited in this section of the review is shown in Table 3.

Table 3. Research articles found in literature which used lipidomics to assay health marine lipid effects.

Reference	Marine Lipids Intervention	Experimental Model	Lipidomics Tools	Target Lipidome	Main Effects
Ottestad et al., 2012 [34]	0.7 g/day EPA and 0.9 g/day DHA for 7 weeks	Healthy humans	UPLC-MS	Plasma	Decreased 23 lipids / Increased PLs and TGs containing EPA and DHA
Rudkowska et al., 2013 [35]	1.9 g/day EPA and 1.1 g/day DHA for 6 weeks	Healthy humans	MS assay kit	Plasma	Increased glyPCs in unsaturated FA
Nording et al., 2013 [36]	1.9 g/day EPA and 1.5 g/day DHA for 6 weeks	Healthy humans	HPLC-GS-MS / SPE-LC-MS/MS	Plasma	Increased incorporation of EPA and DHA into 7 lipid classes / High variability in 87 lipid mediators measured
Mas et al., 2012 [37]	4 g fish oil/day (35% EPA and 25% DHA) for 3 weeks	Healthy humans	SPE-LC-MS/MS	Plasma/serum	Measured for first time 17R/S-HDHA, RvD_1, and RvD_2 concentrations / RvD_1 and RvD_2 into anti-inflammatory and pro-resolving concentration range
Barden et al., 2014 [38]	4 g fish oil/day (35% EPA and 25% DHA) for 5 days	Healthy humans	SPE-LC-MS/MS	Plasma	Increased RvE1, 18R/S-HEPE, 17R/S-HDHA and 14R/S-HDHA
Keelan et al., 2015 [39]	3.7 g/day (27.7% EPA and 56.% DHA) from 20 pregnancy-week	Healthy pregnant women	GC SPE-LC-MS/MS	Placenta	Increased DHA / Increased 18-HEPE and 17-HDHA
Barden et al., 2015 [40]	1.4 g EPA/day and 1 g DHA/day in the form of triglycerides for 3 weeks.	Human suffering metabolic syndrome	SPE-LC-MS/MS	Plasma	Increased E-series resolvins in MetS patients and controls, in which also increased D-series resolvin precursors and 14-HDHA
Schuchardt et al., 2014 [41]	1.14 g/day DHA and 1.56 g/day EPA for 12 weeks	Hyperlipidemic men	SPE-LC-MS/MS	Plasma	Increased EPA-derived lipid mediators / Less increased DHA-derived lipid mediators
Polus et al., 2016 [42]	3× (430 mg of DHA and 90–150 mg of EPA)/day for 3 months	Obese women	GC-MS / LC-MS/MS	Plasma	Increased pro-resolving DHA derivatives
Lankinen et al., 2009 [43]	Fatty or lean fish for 8 weeks	Coronary heart disease patients	GC-MS / UPLC-ESI-MS	Plasma	Decreased 59 bioactive lipid species (ceramides, lysoPCs and DCs) by fatty fish / Increased cholesterol esters and specific long-chain TGs by lean fish
Midtbø et al., 2015 [44]	Farmed salmon fed with a reduced ratio of ω-3/ω-6 for 10 weeks	C57BL/6J mice fed western diets	LC-MS/MS	Liver	Increased ARA in PLs / Increased ceramides / Increased ARA-derived pro-inflammatory mediators / Decreased lipid mediators derived from EPA and DHA
Dawczynski et al., 2013 [45]	3 g of EPA and DHA (in 1:1 ratio)/day for 10 weeks	Mildly hypertriacylglycerolemic subjects	LC-MS/MS	Plasma Red blood cells	Increased EPA and DHA levels in plasma and red blood cells / Increased plasma EPA-derived mediators (PGE_3, and 12-, 15- and 18-HEPE)
Padro et al., 2015 [46]	0.375 EPA and DHA g/day for 28 days	Overweight and moderately hypercholesterolemic subjects	LC-MS/MS	LDL	Increased long-chain polyunsaturated CEs / Increased ratio PC36:5/lysoPC16:0

Table 3. Cont.

Reference	Marine Lipids Intervention	Experimental Model	Lipidomics Tools	Target Lipidome	Main Effects
Dasilva et al., 2015 [47]	EPA:DHA 1:1 or 2:1 or 1:2 for 13 weeks	Wistar Kyoto rats	SPE-LC-MS/MS	Plasma	Decreased pro-inflammatory ARA eicosanoids by 1:1 and 2:1 ratios
Dasilva et al., 2016 [48]	EPA:DHA 1:1 or 2:1 or 1:2 for weeks	SHROB rats	SPE-LC-MS/MS	Plasma	Decreased pro-inflammatory ARA eicosanoids by 1:1 and 2:1 ratios
Cipollina et al., 2014 [49]	1 g/day EPA and 0.4 g/day DHA for 4 months	Healthy humans	BME reaction	Blood neutrophils	Increased 7-oxo-DHA and 5-oxo-EPA
Balogun et al., 2013 [50]	EPA:DHA 1:1 for 4 months	C57BL/6 mice	LC-MS	Plasma Liver	Increased EPA containing PCs, LPCs, and CEs / Increased free ω-3 PUFAs
Poulsen et al., 2008 [51]	0.5 g DHA or EPA ethyl ester/kg body weight/day 4 months	Sprague-Dawley rats	LC-MS/MS	Bone marrow	Increased EPA and DHA / Increased LOX mediators biosynthesized from DHA and EPA (lipoxins, resolving D_1, resolvin E_1 and protectin D_1)
Taltavull et al., 2016 [52]	EPA/DHA (1:1) for 24 weeks	Wistar Kyoto rats fed HFHS diet	GS-MS SPE-LC-MS/MS	Liver	Decreased total ceramides / Decreased long chain ceramide 18:1/18:0 / Increased very long chain ceramides 18:1/24:0 and 18:1/26:0
Caesar et al., 2016 [53]	Menhaden fish oil (25.2g EPA and 18.2 g DHA/100 g) for 11 weeks	C57BL/6 mice fed HF diet	UPLC-MS	Serum Liver	Interaction with gut microbiota increased hepatic levels of cholesteryl and cholesteryl esters by land but not by fish oil
Kuda et al., 2016 [54]	4.3 mg EPA and 14.7 mg DHA/g diet for 5 weeks	C57BL/6J mice fed obesogenic HF diet	SPE-LC-MS/MS	White adipose tissue	Increased anti-inflammatory lipid mediators (endocannabinoid-related N-docosahexaenoylethanolamine) and pro-resolving lipid mediator protectin D_1
Flachs et al., 2011 [55]	46% DHA and 14% EPA for 5 weeks	Mice fed obesogenic MF diet	LC-MS/MS	White adipose tissue	Increased anti-inflammatory lipid mediators (15-deoxy-Δ(12,15)-prostaglandin J_2 and protectin D_1) in epididymal fat
González-Périz et al., 2009 [56]	6 g/100 g ω-3 PUFAs for 5 weeks	ob/ob mice (B6.V-Lep/J)	SPE-LC-MS/MS	Liver	Inhibited formation of ω-6 PUFAs derived eicosanoids Induced formation of ω-3 PUFAs derived resolvins and protectins
Kalish et al., 2013 [57]	Parental nutrition with fish oil-based lipid emulsions	C57BL6/J mice high-carbohydrate diet	LC-MS/MS	Liver	Induced production of anti-inflammatory and pro-resolving lipid mediators
González-Périz et al., 2006 [58]	1.37% DHA or 1.37% EPA and DHA for 5 weeks	129S2/SvPasCrl mice fed high saturated fat diets	HPLC-GC/MS	Liver	Increased DHA-derived lipid mediators (17S-hydroxy-DHA (17S-HDHA) and protectin D_1 by both supplementations
Aukema et al., 2013 [59]	5% or 10% fish oil for 16 weeks	JCR:LA-cp rats	LC-MS/MS	Kidney	Decreased 5, 9, 11, 12- and 15-HETE / Decreased endogenous renal levels of 6-keto $PGF_{1\alpha}$, TXB_2, $PGF_{2\alpha}$ and PGD_2
Gladine et al., 2014 [60]	DHA (0%, 0.1%, 1% or 2% of energy) for 20 weeks	LDLR$^{-/-}$ mice	GC-MS SPE-LC-MS/MS	Plasma Liver	Increased DHA / Increased F4-neuroprostanes (DHA peroxidized metabolites)
Skorve et al., 2015 [61]	Fish oil or krill oil for 6 weeks	C57BL/6J mice fed HF diet	GC-MS UPLC-MS/MS	Liver Brain	Decreased unsaturated fatty acids by fish and krill oils / Decreased ceramides and DGs in liver and brain by krill oil / Increased CEs by krill oil in liver / Decreased plasmalogens by fish oil in liver / Increased hepatic sphingolipids and ARA fatty acid levels more by krill than fish oil in liver / Increased ceramides and lactosylceramides more by fish than krill oil in brain

Table 3. *Cont.*

Reference	Marine Lipids Intervention	Experimental Model	Lipidomics Tools	Target Lipidome	Main Effects
Polus et al., 2015 [62]	EPA	Cell culture	GS-MS LC-MS/MS	Human subcutaneous adipose tissue stromal vascular fraction cells	Decreased pro-inflammatory mediators from ARA Increased anti-inflammatory eicosanoid from EPA
Capel et al., 2015 [63]	DHA	Cell culture	GC-FID LC-MS/MS	C2C12 myotubes	Restoring cellular acylcarnitine profile
Ting et al., 2015 [64]	EPA or DHA	Cell culture	LC-MS/MS	H9c2 cardiac myoblast	Elevation of less unsaturated and ω-3 cardiolipin species mainly by DHA
Lankinen et al., 2011 [65]	Fatty fish and other bioactive compounds for 12 weeks	Metabolic syndrome patients	UPLC-ESI-MS	Plasma	25 altered lipids, including multiple TGs incorporating the long chain ω-3 PUFAs
Wu et al., 2015 [66]	ω-3 PUFA (6.5 g/day) and L-alanyl-l-glutamine (8 g/day) for 3 months	Patients with chronic heart failure	LC-MS	Plasma Skeletal muscle	Increased uptake EPA and DHA Decreased total ceramides and ceramides 22:1 and 20:1
Mas et al., 2016 [67]	ω-3 fatty acids (4 g), Coenzyme Q10 (CoQ) (200 mg) or both for 8 weeks	Patients with chronic kidney disease	LC-MS/MS	Plasma	Increased 8-HEPE, 17-HDHA and RvD1 by ω-3 PUFAs
Bondia-Pons et al., 2014 [68]	0.5% or 1.5% total energy intake EPA and DHA 365 mg or 2900 mg of polyphenols	Patients with metabolic syndrome	UPLC-QTOF-MS	Plasma and HDL fraction	Increased plasma highly unsaturated long-chain TGs and EPA and DHA-containing PLs by ω-3 diets Decreased plasma low unsaturated PLs, PCs, LysoPCs and PCps containing ARA by ω-3 diets Increased PCs and TGs containing DHA or EPA by ω-3 diets in HDL fraction Decreased PCes and PCps containing ARA and medium-chain PCs by ω-3 diets in HDL fraction Decreased PCs and Pes, several alkyl and alkenyl etherlipids containing 16:0 and saturated and low-unsaturated PCs and PEs by both ω-3 and polyphenols diet
Dasilva et al., 2017 [69]	EPA/DHA (1:1) Grape polyphenols for 24 weeks	Wistar Kyoto rats fed HFHS diet or STD diet	GS-MS SPE-LC-MS/MS	Plasma Liver Adipose tissue	Decreased ω-6/ω-3 index in plasma and membranes by ω-3 diets Decreased ARA pro-inflammatory lipid mediators by ω-3 diets Increased desaturases related to EPA and DHA synthesis by ω-3 diets Decreased desaturases related to ARA synthesis by ω-3 diets Combination ω-3&polyphenols cooperative down-regulated Δ5D related with ARA synthesis, decreased COX activity on ARA and total FFA in plasma into STD and HFHS diets

2.3. Beneficial Effects of Marine Lipids Intake Assayed by Both Proteomics and Lipidomics

The number of research articles which have combined both proteomics and lipidomics approaches to address the beneficial effects of marine lipids in metabolic alterations induced by unhealthy diet intake is scant yet. More articles have used genomics or transcriptomics tools, but gene expression or transcription is not always correlated with protein levels or activities. The application of proteomics and lipidomics approaches can help overcome this drawback and identify molecular pathways actually affected by marine lipids. The integration of different omics seems essential to have a complete picture of how marine lipids affect health. In spite of this observation, only three research articles which have combined both proteomics and lipidomics are currently available.

The first study was carried out in healthy overweight men with mildly elevated plasma C-reactive protein concentrations. Bakker et al. [70] performed a dietary intervention with a mix of several products selected for their evidence-based anti-inflammatory properties. That "anti-inflammatory dietary mix" consisted of fish oil, green tea extract, resveratrol, vitamin E, vitamin C, and tomato extract. Regarding fish oil, subjects consumed 1200 mg cold water fish oil/day, composed of 380 mg EPA and 260 mg DHA, and 60 mg other ω-3 PUFAs. Authors measured inflammatory and oxidative stress defense markers in plasma and urine. Furthermore, 120 plasma proteins, 274 plasma metabolites (lipids, FFA, and polar compounds) and the transcriptomes of peripheral blood mononuclear cells and adipose tissue were also quantified. Therefore, this study combined proteomics, metabolipidomics and transcriptomics tools. Plasma adiponectin concentrations increased by 7%, whereas C-reactive protein (principal inflammation marker) was unchanged. However, a multitude of subtle changes was detected by an integrated analysis of the omics data, which indicated modulated inflammation of adipose tissue, improved endothelial function and oxidative stress, and increased liver fatty acid oxidation. Using the same subjects, Pellis et al. [71] determined their postprandial response to the consumption of a standardized 500 mL high-fat dairy shake by using metabolomics and proteomics tools. During a 6 h time course after Postprandial Challenge Test (PCT), authors quantified several plasma metabolites, 79 plasma proteins and 7 clinical biochemistry parameters (glucose, insulin, total FFA, total TGs, hsCRP, IL6, and TNFa). Among these, 31 had different responses over time between treated and control groups, revealing differences in amino acid metabolism, oxidative stress, inflammation, and endocrine metabolism. Results showed different short-term metabolic responses to the PCT in subjects previously supplemented with the anti-inflammatory mix compared to the controls. Additional metabolic changes related to the dietary intervention were also detected as compared to non-perturbed conditions.

In the third research [72] the obesogenic effect of diets heavily enriched in ω-6 PUFAs and poor in ω-3 PUFAs was assayed by combined proteomics and lipidomics. The study was carried out in aging mice, which had previously suffered a myocardial infarction, after 5 months of feeding a ω-6 enriched diet (ω-6:ω-3 442:1 ratio). Plasma proteomic profiling revealed higher VCAM-1, macrophage inflammatory protein-1 and D40 and myeloperoxidase in the ω-6 PUFAs group. Lipidomic analysis showed higher levels of ARA and 12(S)-HETE and altered levels of inflammation-resolving enzymes 5-LOX, COX-2, and heme oxygenase-1, which reflected that excess of ω-6 stimulate prolonged neutrophil trafficking and pro-inflammatory lipid mediators after myocardial infarction. Table 4 summarizes the articles cited in this section of the review.

Table 4. Research articles found in literature which used both proteomics and lipidomics to assay health marine lipid effects.

Reference	Marine Lipids Intervention	Experimental Model	Proteomics and Lipidomics Tools	Target Proteome and Lipidome	Main Effects
Bakker et al., 2010 [70]	380 mg EPA and 260 mg DHA and other anti-inflammatory compounds for 5 weeks	Healthy overweight men	HumanMAP GS-MS LC-MS/MS	Plasma	Regulated plasma proteins and plasma metabolites (lipids, free fatty acids, and polar compounds) related to modulation of inflammation, improved endothelial function, oxidative stress and increased fatty acid oxidation.
Pellis et al., 2012 [71]	Postprandial response in anti-inflammatory mix-supplemented men Acute intake	Healthy overweight men	HumanMAP GS-MS	Plasma	31 regulated proteins and lipids involved in amino acid metabolism, oxidative stress, inflammation and endocrine metabolism.
López et al., 2015 [72]	ω-6:ω-3 in 442:1 ratio for 5 months	Aging C57BL/6J mice previously suffered myocardial infarction	Protein immunoblot analysis LC-MS/MS	Plasma	Increased VCAM-1, macrophage inflammatory protein-1, D40 and myeloperoxidase Increased ARA and 12(S)-HETE and altered levels of inflammation-resolving enzymes 5-LOX, COX-2, and heme oxygenase-1

3. Mechanisms behind the Beneficial Effects of Marine Lipids Assayed by Proteomics and Lipidomics

The information obtained from both proteomics and lipidomics approaches has confirmed mechanisms proposed by genomics and transcriptomics data on the role of marine lipids in diet-induced metabolic diseases. But interestingly, these techniques have also provided new insights suggesting the modulation of new molecular pathways and proteins.

Lipidomics tools have revealed the potential mechanisms related to the influence exerted on the regulation of lipid profiles in plasma/blood, tissues and membranes and lipid mediator synthesis by marine lipids in human, animal and cell-model experiments. These lipid profiles and their derivative metabolites from fatty acids have closely been associated with inflammation, oxidative stress and the endogenous antioxidant system [73,74].

Besides their known effects in decreasing plasma TGs and cholesterol levels, some of the lipidomics studies reported in this review have found a remodeling of the plasma/membrane/tissue lipids into PLs, TGs, lipoproteins and other lipid species of long chain PUFAs, including plasma FFA profiles [34–36,39,43–48,50–53,60,61,63–66,68,69]. In general, the consumption of marine ω-3 led to a replacement of ARA with EPA and DHA in cell membranes and lipid species presented in plasma, erythrocytes and liver and adipose tissue, but also kidney and muscle. Such replacement provoked the consequent enrichment on ω-3 long chain PUFAs accompanied by modulation of LOX and COX activities, which was attributed to a competence mechanism, especially between ARA and EPA. As a result, the production of lipid mediators was affected. Additionally, the uptake of ω-3 PUFAs modulated the synthesis de novo of ARA through elongases and desaturases. In fact, lipidomics data from the liver, which is in charge of the de novo fatty acids synthesis, confirmed the preferential substrate competition of $\Delta 5D$, which controls de novo synthesis of EPA and ARA, for ω-3 PUFAs over ω-6 PUFAs [44,50,53,60,61,69]. Marine lipids also regulated the formation of bioactive lipids such as ceramides. The level of long chain ceramides associated with insulin resistance was found to be reduced in plasma [43,66] and liver [52,61] together to an increment of the concentration of very long chain ceramides which seemed to be IR protective.

The fatty acids modulation in the liver due to the consumption of marine lipids was reflected in the total fatty acids profile of plasma and adipose tissue [34–36,43,45–48,50,53,65,68,69]. Such modulation was also observed in circulating plasma FFA and the incorporation of fatty acids into erythrocyte membranes [47,48,69]. Fatty acids released from the adipose tissue, which become circulating FFA in plasma revealed the influence of the diet and synthesis de novo in the accumulation of fat in adipocytes. In consequence, the $\omega 6/\omega 3$ ratio in plasma, circulating FFA, membranes and tissues was lower after feeding marine lipids. The $\omega 6/\omega 3$ ratio is an excellent clinical marker for cellular inflammation [75] and higher values are correlated with increased prevalence of chronic inflammatory diseases [76].

In agreement with these effects, lipidomics studies focused on the formation of lipid mediators from marine PUFAs demonstrated that the intake of these lipids promoted the generation of anti-inflammatory and pro-resolving lipid mediators derived from EPA and DHA while decreasing pro-inflammatory mediators derived from ARA [37–42,44,45,47–49,51,54–58,60,62,64,67,69]. This anti-inflammatory response can be further due to the fact that ω-3 PUFAs, especially EPA, and ω-6 PUFAs can compete for the same enzymes, including phospholipases, desaturases, lipoxygenases and cyclooxygenases, resulting in a higher production of derived metabolites from EPA and DHA than derived from ARA.

By using genomics and transcriptomics tools, some authors had previously reported that PUFAs inhibited the nuclear factor kb (NF-κB) and reduced cytokine production because EPA and DHA can inhibit the binding between saturated fats and toll-like receptors of membranes by direct competition. The binding to saturated fats would activate NF-κB gene transcription factor, which induces inflammatory responses through COX and cytokine synthesis [77]. Moreover, although it is not fully known, marine ω-3 PUFAs may also control gene expression by direct interaction with at least another 4 metabolic nuclear receptors: PPAR (peroxisome proliferator activated receptor),

LXR (liver X receptor), HNF-4α (hepatic nuclear factor 4) and farnesol X receptor (FXR). Likewise, marine PUFAs can reduce the levels of sterol regulatory element binding proteins (SREBPs) and the carbohydrate response element binding protein (ChREBP) [78]. Therefore, the modulation of EPA and DHA levels and their outcomes revealed by lipidomics approaches may explain their actions in regulating gene expression. In addition to these findings, the use of proteomics approaches has confirmed the influence of EPA and DHA on several pathways modulated by these transcriptional factors. Interestingly, proteomics data identified specific proteins with a pivotal role in these pathways which were altered by marine lipids. In this regard, it is necessary to highlight that a direct correlation between the level of gene expression and the cellular content of proteins cannot always be found [79]. Consequently, proteomics becomes a critical tool for understanding marine lipid actions on cellular metabolism and for identifying biomarkers of modulation. Proteomics has specifically revealed that the influence of marine lipids intake is not limited to regulating cellular protein quantity but also protein quality by controlling the formation of oxidative PTMs on proteins, mainly carbonyl moieties. The control of these oxidative PTMs plays a key role in understanding the mechanism behind the beneficial effect of marine lipids in decreasing the risk of CVD and type 2 diabetes [80].

In the adipose tissue, aorta and especially liver, proteomics analysis revealed that ω-3 PUFAs from fish oil produced an improvement of lipid profiles and lower accumulation of fat through a mechanism that involved the down-regulation of proteins participating in lipogenesis and glycolysis while causing the up-regulation of proteins involved in fatty acid beta-oxidation. Fish oil also showed an important action on proteins implicated in the urea cycle and protein metabolism. Additionally, EPA and DHA demonstrated to act on insulin signaling by modulating proteins such as proteasome system [16,19–24,27,28].

Proteomics also found a substantial up-regulation of the antioxidant system in blood/plasma and the rest of tissue analyzed. In fact, fish oil induced higher levels of antioxidant enzymes, ameliorating of endoplasmic oxidative stress and stimulating protein and DNA cellular system repair [14–23,25,27,28]. Moreover, EPA and DHA altered protein carbonylation levels of specific liver proteins, demonstrating an additional mechanism of protein regulation by marine lipids, which is particularly interesting in the investigation of diseases induced by diet. It is important to point out that the impairment of normal redox homeostasis, and the consequent accumulation of oxidized biomolecules, has been linked to the onset and/or development of a great variety of diet-induced diseases [81,82]. Fish oil reduced carbonylation of proteins related to the antioxidant system such as albumin or 3-α-hydroxysteroid dehydrogenase in plasma and liver, respectively. Proteomics findings in liver demonstrated that EPA and DHA improved ammonia detoxification by decreasing carbonylation level of argininosuccinate synthetase while increasing oxidation of aspartate aminotransferase. Finally, in skeletal muscle, fish oil intake exerted a protection from cellular dysfunction by ameliorating actin carbonylation level [27].

Therefore, proteomics and lipidomics can largely help understand some of the mechanisms behind the beneficial effect of marine lipids against chronic disease induced by diet. These mechanisms are mainly related to competence from enzymes involved in lipid de novo synthesis and oxidation, modulation of anti-inflammatory and antioxidant pathways as well as protein homeostasis. Data also reflected that the effect of dietary marine lipids is closely dependent on their doses, the EPA and DHA ratio, diet components or health status of patients, among others. A schematic representation of mechanisms and beneficial effects of EPA and DHA intake found by proteomics and lipidomics tools is shown in Figure 1.

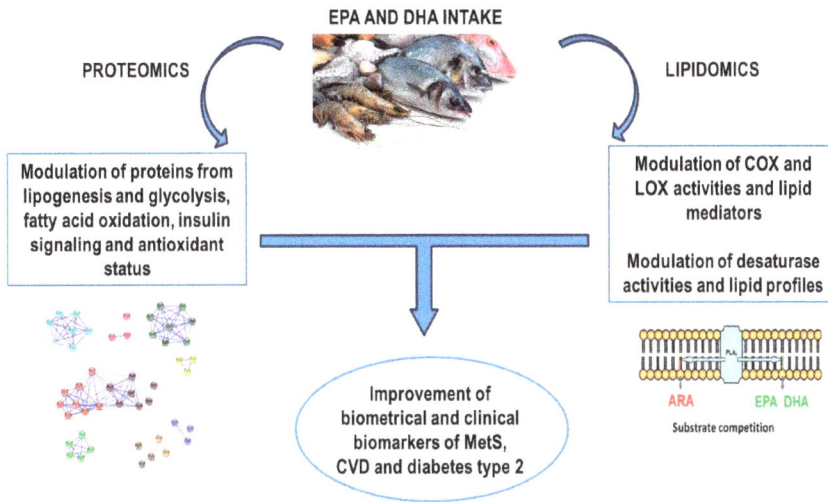

Figure 1. Schematic representation of mechanisms and beneficial effects of EPA and DHA intake found by proteomics and lipidomics tools.

4. Concluding Remarks and Final Considerations

Proteomics and lipidomics approaches constitute valuable tools for studying the effects of marine lipids on metabolic health and disease. In spite of their recent application to nutrition research, these omics have already allowed the identification of numerous metabolic molecules and pathways, tissues and physiological processes which are modulated by the consumption of marine lipids. These findings confirm some of the mechanisms previously suggested by genomics and transcriptomics tools, but they also provide new insights revealing the existence of novel mechanisms and target molecules of great interest for the growing field of food bioactives and personalized nutrition. Proteomics and lipidomics contribute to the discovering of new biomarkers of disease and support the optimal design of both preventive and palliative nutritional strategies against the pathologies in which marine lipids have previously demonstrated their beneficial effects. The progress on the characterization of mechanisms of action of marine lipids at proteomics and lipidomics levels might further identify other diseases in which marine lipids can exert a positive influence. Therefore, although the combination of proteomics and lipidomics approaches is still scarce, it could be the key to understanding the mechanisms involved in the beneficial effects of marine lipids. Such combination will offer a complete overview of cellular process contributing to clarify several controversial facts regarding the in vivo role of marine lipids.

Finally, it should be noted that the enormous complexity of proteomes and lipidomes together with the scant knowledge available have limited the use of omics in the field of marine lipids. This inconvenience is especially significant for lipidomics due to the lack of information for predicting the number of individual lipid molecules present in an organism. Additionally, it is necessary to mention the high variability of results derived from in vivo experiments as well as other caveats related to high instrumental costs or the need for highly qualified staff. The achievements on higher sensitivity and specificity of the modern MS developments, such as imaging MS or top-down approaches, constitute promising tools which can help solve these problems.

Acknowledgments: Spanish Ministry of Science and Innovation is acknowledged for grant AGL2013-49079-C2-1-R. The Consejo Superior de Investigaciones Científicas (CSIC) and the University of Santiago de Compostela (USC) are gratefully acknowledged for the doctoral fellowship to G.D. Language revision by Hannelore Lott is appreciated.

Author Contributions: L.M. performed search of the literature and wrote the review. I.M. conceived, revised and corrected the paper. G.D., N.T., M.R. revised and corrected the paper. All authors agreed with the final submitted version.

Conflicts of Interest: The authors declare no conflict of interest.

References

1. Cordain, L.; Eaton, S.B.; Sebastian, A.; Mann, N.; Lindeberg, S.; Watkins, B.A.; O'Keefe, J.H.; Brand-Miller, J. Origins and evolution of the Western diet: Health implications for the 21st century. *Am. J. Clin. Nutr.* **2005**, *81*, 341–354. [PubMed]
2. Hutcheson, R.; Rocic, P. The metabolic syndrome, oxidative stress, environment, and cardiovascular disease: The great exploration. *Exp. Diabetes Res.* **2012**, *2012*, 271028. [CrossRef] [PubMed]
3. Grundy, S.M. Metabolic syndrome pandemic. *Arterioscler. Thromb. Vasc. Biol.* **2008**, *28*, 629–636. [CrossRef] [PubMed]
4. Eurostat. News Releases, Product Code: 3-24052016-AP, Published on 24 May 2016. Available online: http://ec.europa.eu/eurostat/documents/2995521/7335847/3-24052016-AP-EN.pdf/4dd0a8ad-5950-4425-9364-197a492d3648 (accessed on 17 August 2017).
5. World Health Organization. The Top 10 Causes of Death. Available online: http://www.who.int/mediacentre/factsheets/fs310/en/index1.html (accessed on 17 August 2017).
6. Simopoulos, A.P. Essential fatty acids in health and chronic disease. *Am. J. Clin. Nutr.* **1999**, *70*, 560S–569S. [PubMed]
7. Moore, J.B.; Weeks, M.E. Proteomics and systems biology: Current and future applications in the nutritional sciences. *Adv. Nutr.* **2011**, *2*, 355–364. [CrossRef] [PubMed]
8. Ganesh, V.; Hettiarachchy, N.S. Nutriproteomics: A promising tool to link diet and diseases in nutritional research. *Biochim. Biophys. Acta BBA Proteins Proteom.* **2012**, *1824*, 1107–1117. [CrossRef] [PubMed]
9. Kussmann, M.; Raymond, F.; Affolter, M. OMICS-driven biomarker discovery in nutrition and health. *J. Biotechnol.* **2006**, *124*, 758–787. [CrossRef] [PubMed]
10. Tyers, M.; Mann, M. From genomics to proteomics. *Nature* **2003**, *422*, 193–197. [CrossRef] [PubMed]
11. Wasinger, V.C.; Cordwell, S.J.; Cerpa-Poljak, A.; Yan, J.X.; Gooley, A.A.; Wilkins, M.R.; Duncan, M.W.; Harris, R.; Williams, K.L.; Humphery-Smith, I. Progress with gene-product mapping of the Mollicutes: Mycoplasma genitalium. *Electrophoresis* **1995**, *16*, 1090–1094. [CrossRef] [PubMed]
12. Wilkins, M.; Gooley, A. Protein Identification in Proteome Projects. In *Proteome Research: New Frontiers in Functional Genomics*; Wilkins, M., Williams, K., Appel, R., Hochstrasser, D., Eds.; Springer: Berlin/Heidelberg, Germany, 1997; pp. 35–64.
13. Sauer, S.; Luge, T. Nutriproteomics: Facts, concepts, and perspectives. *Proteomics* **2015**, *15*, 997–1013. [CrossRef] [PubMed]
14. Camargo, A.; Rangel-Zúñiga, O.A.; Peña-Orihuela, P.; Marín, C.; Pérez-Martínez, P.; Delgado-Lista, J.; Gutierrez-Mariscal, F.M.; Malagón, M.M.; Roche, H.M.; Tinahones, F.J.; et al. Postprandial changes in the proteome are modulated by dietary fat in patients with metabolic syndrome. *J. Nutr. Biochem.* **2013**, *24*, 318–324. [CrossRef] [PubMed]
15. Rangel-Zúñiga, O.A.; Camargo, A.; Marín, C.; Peña-Orihuela, P.; Pérez-Martínez, P.; Delgado-Lista, J.; González-Guardia, L.; Yubero-Serrano, E.M.; Tinahones, F.J.; Malagón, M.M.; et al. Proteome from patients with metabolic syndrome is regulated by quantity and quality of dietary lipids. *BMC Genom.* **2015**, *16*, 509. [CrossRef] [PubMed]
16. Jiménez-Gómez, Y.; Cruz-Teno, C.; Rangel-Zúñiga, O.A.; Peinado, J.R.; Pérez-Martínez, P.; Delgado-Lista, J.; García-Ríos, A.; Camargo, A.; Vázquez-Martínez, R.; Ortega-Bellido, M.; et al. Effect of dietary fat modification on subcutaneous white adipose tissue insulin sensitivity in patients with metabolic syndrome. *Mol. Nutr. Food Res.* **2014**, *58*, 2177–2188. [CrossRef] [PubMed]
17. De Roos, B.; Geelen, A.; Ross, K.; Rucklidge, G.; Reid, M.; Duncan, G.; Caslake, M.; Horgan, G.; Brouwer, I.A. Identification of potential serum biomarkers of inflammation and lipid modulation that are altered by fish oil supplementation in healthy volunteers. *Proteomics* **2008**, *8*, 1965–1974. [CrossRef] [PubMed]

18. Burillo, E.; Mateo-Gallego, R.; Cenarro, A.; Fiddyment, S.; Bea, A.M.; Jorge, I.; Vázquez, J.; Civeira, F. Beneficial effects of omega-3 fatty acids in the proteome of high-density lipoprotein proteome. *Lipids Health Dis.* **2012**, *11*, 116. [CrossRef] [PubMed]
19. Ahmed, A.A.; Balogun, K.A.; Bykova, N.V.; Cheema, S.K. Novel regulatory roles of omega-3 fatty acids in metabolic pathways: A proteomics approach. *Nutr. Metab.* **2014**, *11*, 6. [CrossRef] [PubMed]
20. Wrzesinski, K.; León, I.R.; Kulej, K.; Sprenger, R.R.; Bjørndal, B.; Christensen, B.J.; Berge, R.K.; Jensen, O.N.; Rogowska-Wrzesinska, A. Proteomics identifies molecular networks affected by tetradecylthioacetic acid and fish oil supplemented diets. *J. Proteom.* **2013**, *84*, 61–77. [CrossRef] [PubMed]
21. De Roos, B.; Duivenvoorden, I.; Rucklidge, G.; Reid, M.; Ross, K.; Lamers, R.J.; Voshol, P.J.; Havekes, L.M.; Teusink, B. Response of apolipoprotein E*3-Leiden transgenic mice to dietary fatty acids: Combining liver proteomics with physiological data. *FASEB J.* **2005**, *19*, 813–815. [CrossRef] [PubMed]
22. Méndez, L.; Ciordia, S.; Fernández, M.S.; Juárez, S.; Ramos, A.; Pazos, M.; Gallardo, J.M.; Torres, J.L.; Nogués, M.R.; Medina, I. Changes in liver proteins of rats fed standard and high-fat and sucrose diets induced by fish omega-3 PUFAs and their combination with grape polyphenols according to quantitative proteomics. *J. Nutr. Biochem.* **2017**, *41*, 84–97. [CrossRef] [PubMed]
23. Kalupahana, N.S.; Claycombe, K.; Newman, S.J.; Stewart, T.; Siriwardhana, N.; Matthan, N.; Lichtenstein, A.H.; Moustaid-Moussa, N. Eicosapentaenoic acid prevents and reverses insulin resistance in high-fat diet-induced obese mice via modulation of adipose tissue inflammation. *J. Nutr.* **2010**, *140*, 1915–1922. [CrossRef] [PubMed]
24. Mavrommatis, Y.; Ross, K.; Rucklidge, G.; Reid, M.; Duncan, G.; Gordon, M.J.; Thies, F.; Sneddon, A.; De Roos, B. Intervention with fish oil, but not with docosahexaenoic acid, results in lower levels of hepatic soluble epoxide hydrolase with time in apoE knockout mice. *Br. J. Nutr.* **2010**, *103*, 16–24. [CrossRef] [PubMed]
25. Johnson, M.L.; Lalia, A.Z.; Dasari, S.; Pallauf, M.; Fitch, M.; Hellerstein, M.K.; Lanza, I.R. Eicosapentaenoic acid but not docosahexaenoic acid restores skeletal muscle mitochondrial oxidative capacity in old mice. *Aging Cell* **2015**, *14*, 734–743. [CrossRef] [PubMed]
26. Richard, D.; Kefi, K.; Barbe, U.; Bausero, P.; Visioli, F. Polyunsaturated fatty acids as antioxidants. *Pharmacol. Res.* **2008**, *57*, 451–455. [CrossRef] [PubMed]
27. Méndez, L.; Pazos, M.; Gallardo, J.M.; Torres, J.L.; Pérez-Jiménez, J.; Nogués, R.; Romeu, M.; Medina, I. Reduced protein oxidation in Wistar rats supplemented with marine omega-3 PUFAs. *Free Radic. Biol. Med.* **2013**, *55*, 8–20. [CrossRef] [PubMed]
28. Joumard-Cubizolles, L.; Gladine, C.; Gérard, N.; Chambon, C.; Brachet, P.; Comte, B.; Mazur, A. Proteomic analysis of aorta of LDLR-/- mice given omega-3 fatty acids reveals modulation of energy metabolism and oxidative stress pathway. *Eur. J. Lipid Sci. Technol.* **2013**, *115*, 1492–1498. [CrossRef]
29. Wenk, M.R. The emerging field of lipidomics. *Nat. Rev. Drug Discov.* **2005**, *4*, 594–610. [CrossRef] [PubMed]
30. Wymann, M.P.; Schneiter, R. Lipid signalling in disease. *Nat. Rev. Mol. Cell Biol.* **2008**, *9*, 162–176. [CrossRef] [PubMed]
31. Shearer, G.C.; Harris, W.S.; Pedersen, T.L.; Newman, J.W. Detection of omega-3 oxylipins in human plasma and response to treatment with omega-3 acid ethyl esters. *J. Lipid Res.* **2010**, *51*, 2074–2081. [CrossRef] [PubMed]
32. Murakami, M. Lipid mediators in life science. *Exp. Anim.* **2011**, *60*, 7–20. [CrossRef] [PubMed]
33. Smith, P.K.; Krohn, R.I.; Hermanson, G.T.; Mallia, A.K.; Gartner, F.H.; Provenzano, M.D.; Fujimoto, E.K.; Goeke, N.M.; Olson, B.J.; Klenk, D.C. Measurement of protein using bicinchoninic acid. *Anal. Biochem.* **1985**, *150*, 76–85. [CrossRef]
34. Ottestad, I.; Hassani, S.; Borge, G.I.; Kohler, A.; Vogt, G.; Hyötyläinen, T.; Orešič, M.; Brønner, K.W.; Holven, K.B.; Ulven, S.M.; et al. Fish Oil Supplementation Alters the Plasma Lipidomic Profile and Increases Long-Chain PUFAs of Phospholipids and Triglycerides in Healthy Subjects. *PLoS ONE* **2012**, *7*, e42550. [CrossRef] [PubMed]
35. Rudkowska, I.; Paradis, A.M.; Thifault, E.; Julien, P.; Tchernof, A.; Couture, P.; Lemieux, S.; Barbier, O.; Vohl, M.C. Transcriptomic and metabolomic signatures of an n-3 polyunsaturated fatty acids supplementation in a normolipidemic/normocholesterolemic Caucasian population. *J. Nutr. Biochem.* **2013**, *24*, 54–61. [CrossRef] [PubMed]
36. Nording, M.L.; Yang, J.; Georgi, K.; Hegedus Karbowski, C.; German, J.B.; Weiss, R.H.; Hogg, R.J.; Trygg, J.; Hammock, B.D.; Zivkovic, A.M. Individual variation in lipidomic profiles of healthy subjects in response to omega-3 fatty acids. *PLoS ONE* **2013**, *8*, e76575. [CrossRef] [PubMed]

37. Mas, E.; Croft, K.D.; Zahra, P.; Barden, A.; Mori, T.A. Resolvins D1, D2, and other mediators of self-limited resolution of inflammation in human blood following n-3 fatty acid supplementation. *Clin. Chem.* **2012**, *58*, 1476. [CrossRef] [PubMed]

38. Barden, A.; Mas, E.; Croft, K.D.; Phillips, M.; Mori, T.A. Short-term n-3 fatty acid supplementation but not aspirin increases plasma proresolving mediators of inflammation. *J. Lipid Res.* **2014**, *55*, 2401–2407. [CrossRef] [PubMed]

39. Keelan, J.A.; Mas, E.; D'Vaz, N.; Dunstan, J.A.; Li, S.; Barden, A.E.; Mark, P.J.; Waddell, B.J.; Prescott, S.L.; Mori, T.A. Effects of maternal n-3 fatty acid supplementation on placental cytokines, pro-resolving lipid mediators and their precursors. *Reproduction* **2015**, *149*, 171–178. [CrossRef] [PubMed]

40. Barden, A.E.; Mas, E.; Croft, K.D.; Phillips, M.; Mori, T.A. Specialized proresolving lipid mediators in humans with the metabolic syndrome after n-3 fatty acids and aspirin. *Am. J. Clin. Nutr.* **2015**, *102*, 1357–1364. [CrossRef] [PubMed]

41. Schuchardt, J.P.; Schmidt, S.; Kressel, G.; Willenberg, I.; Hammock, B.D.; Hahn, A.; Schebb, N.H. Modulation of blood oxylipin levels by long-chain omega-3 fatty acid supplementation in hyper- and normolipidemic men. *Prostaglandins Leukot. Essent. Fat. Acids PLEFA* **2014**, *90*, 27–37. [CrossRef] [PubMed]

42. Polus, A.; Zapala, B.; Razny, U.; Gielicz, A.; Kiec-Wilk, B.; Malczewska-Malec, M.; Sanak, M.; Childs, C.E.; Calder, P.C.; Dembinska-Kiec, A. Omega-3 fatty acid supplementation influences the whole blood transcriptome in women with obesity, associated with pro-resolving lipid mediator production. *Biochim. Biophys. Acta BBA Mol. Cell Biol. Lipids* **2016**, *1861*, 1746–1755. [CrossRef] [PubMed]

43. Lankinen, M.; Schwab, U.; Erkkilä, A.; Seppänen-Laakso, T.; Hannila, M.L.; Mussalo, H.; Lehto, S.; Uusitupa, M.; Gylling, H.; Orešič, M. Fatty fish intake decreases lipids related to inflammation and insulin signaling—A lipidomics approach. *PLoS ONE* **2009**, *4*, e5258. [CrossRef] [PubMed]

44. Midtbø, L.K.; Borkowska, A.G.; Bernhard, A.; Rønnevik, A.K.; Lock, E.-J.; Fitzgerald, M.L.; Torstensen, B.E.; Liaset, B.; Brattelid, T.; Pedersen, T.L.; et al. Intake of farmed Atlantic salmon fed soybean oil increases hepatic levels of arachidonic acid-derived oxylipins and ceramides in mice. *J. Nutr. Biochem.* **2015**, *26*, 585–595. [CrossRef] [PubMed]

45. Dawczynski, C.; Massey, K.A.; Ness, C.; Kiehntopf, M.; Stepanow, S.; Platzer, M.; Grün, M.; Nicolaou, A.; Jahreis, G. Randomized placebo-controlled intervention with n-3 LC-PUFA-supplemented yoghurt: Effects on circulating eicosanoids and cardiovascular risk factors. *Clin. Nutr.* **2013**, *32*, 686–696. [CrossRef] [PubMed]

46. Padro, T.; Vilahur, G.; Sánchez-Hernández, J.; Hernández, M.; Antonijoan, R.M.; Pérez, A.; Badimon, L. Lipidomic changes of LDL in overweight and moderately hypercholesterolemic subjects taking phytosterol- and omega-3-supplemented milk. *J. Lipid Res.* **2015**, *56*, 1043–1056. [CrossRef] [PubMed]

47. Dasilva, G.; Pazos, M.; García-Egido, E.; Gallardo, J.M.; Rodríguez, I.; Cela, R.; Medina, I. Healthy effect of different proportions of marine ω-3 PUFAs EPA and DHA supplementation in Wistar rats: Lipidomic biomarkers of oxidative stress and inflammation. *J. Nutr. Biochem.* **2015**, *26*, 1385–1392. [CrossRef] [PubMed]

48. Dasilva, G.; Pazos, M.; García-Egido, E.; Pérez-Jiménez, J.; Torres, J.L.; Giralt, M.; Nogués, M.R.; Medina, I. Lipidomics to analyze the influence of diets with different EPA:DHA ratios in the progression of Metabolic Syndrome using SHROB rats as a model. *Food Chem.* **2016**, *205*, 196–203. [CrossRef] [PubMed]

49. Cipollina, C.; Salvatore, S.R.; Muldoon, M.F.; Freeman, B.A.; Schopfer, F.J. Generation and dietary modulation of anti-inflammatory electrophilic omega-3 fatty acid derivatives. *PLoS ONE* **2014**, *9*, e94836. [CrossRef] [PubMed]

50. Balogun, K.A.; Albert, C.J.; Ford, D.A.; Brown, R.J.; Cheema, S.K. Dietary omega-3 polyunsaturated fatty acids alter the fatty acid composition of hepatic and plasma bioactive lipids in C57BL/6 mice: A lipidomic approach. *PLoS ONE* **2013**, *8*, e82399. [CrossRef] [PubMed]

51. Poulsen, R.C.; Gotlinger, K.H.; Serhan, C.N.; Kruger, M.C. Identification of inflammatory and proresolving lipid mediators in bone marrow and their lipidomic profiles with ovariectomy and omega-3 intake. *Am. J. Hematol.* **2008**, *83*, 437–445. [CrossRef] [PubMed]

52. Taltavull, N.; Ras, R.; Mariné, S.; Romeu, M.; Giralt, M.; Méndez, L.; Medina, I.; Ramos-Romero, S.; Torres, J.L. Protective effects of fish oil on pre-diabetes: A lipidomic analysis of liver ceramides in rats. *Food Funct.* **2016**, *7*, 3981–3988. [CrossRef] [PubMed]

53. Caesar, R.; Nygren, H.; Orešic, M.; Bäckhed, F. Interaction between dietary lipids and gut microbiota regulates hepatic cholesterol metabolism. *J. Lipid Res.* **2016**, *57*, 474–781. [CrossRef] [PubMed]

54. Kuda, O.; Rombaldova, M.; Janovska, P.; Flachs, P.; Kopecky, J. Cell type-specific modulation of lipid mediator's formation in murine adipose tissue by omega-3 fatty acids. *Biochem. Biophys. Res. Commun.* **2016**, *469*, 731–736. [CrossRef] [PubMed]

55. Flachs, P.; Ruhl, R.; Hensler, M.; Janovska, P.; Zouhar, P.; Kus, V.; Macek Jilkova, Z.; Papp, E.; Kuda, O.; Svobodova, M.; et al. Synergistic induction of lipid catabolism and anti-inflammatory lipids in white fat of dietary obese mice in response to calorie restriction and n-3 fatty acids. *Diabetologia* **2011**, *54*, 2626–2638. [CrossRef] [PubMed]

56. González-Périz, A.; Horrillo, R.; Ferré, N.; Gronert, K.; Dong, B.; Morán-Salvador, E.; Titos, E.; Martínez-Clemente, M.; López-Parra, M.; Arroyo, V.; et al. Obesity-induced insulin resistance and hepatic steatosis are alleviated by ω-3 fatty acids: A role for resolvins and protectins. *FASEB J.* **2009**, *23*, 1946–1957. [CrossRef] [PubMed]

57. Kalish, B.T.; Le, H.D.; Fitzgerald, J.M.; Wang, S.; Seamon, K.; Gura, K.M.; Gronert, K.; Puder, M. Intravenous fish oil lipid emulsion promotes a shift toward anti-inflammatory proresolving lipid mediators. *Am. J. Physiol. Gastrointest. Liver Physiol.* **2013**, *305*, G818–G828. [CrossRef] [PubMed]

58. González-Périz, A.; Planagumà, A.; Gronert, K.; Miquel, R.; López-Parra, M.; Titos, E.; Horrillo, R.; Ferré, N.; Deulofeu, R.; Arroyo, V.; et al. Docosahexaenoic acid (DHA) blunts liver injury by conversion to protective lipid mediators: Protectin D1 and 17S-hydroxy-DHA. *FASEB J.* **2006**, *20*, 2537–2539. [CrossRef] [PubMed]

59. Aukema, H.M.; Lu, J.; Borthwick, F.; Proctor, S.D. Dietary fish oil reduces glomerular injury and elevated renal hydroxyeicosatetraenoic acid levels in the JCR:LA-*cp* rat, a model of the metabolic syndrome. *Br. J. Nutr.* **2013**, *110*, 11–19. [CrossRef] [PubMed]

60. Gladine, C.; Newman, J.W.; Durand, T.; Pedersen, T.L.; Galano, J.M.; Demougeot, C.; Berdeaux, O.; Pujos-Guillot, E.; Mazur, A.; Comte, B. Lipid profiling following intake of the omega 3 fatty acid DHA identifies the peroxidized metabolites F4-neuroprostanes as the best predictors of atherosclerosis prevention. *PLoS ONE* **2014**, *9*, e89393. [CrossRef] [PubMed]

61. Skorve, J.; Hilvo, M.; Vihervaara, T.; Burri, L.; Bohov, P.; Tillander, V.; Bjørndal, B.; Suoniemi, M.; Laaksonen, R.; Ekroos, K.; et al. Fish oil and krill oil differentially modify the liver and brain lipidome when fed to mice. *Lipids Health Dis.* **2015**, *14*. [CrossRef] [PubMed]

62. Polus, A.; Kiec-Wilk, B.; Razny, U.; Gielicz, A.; Schmitz, G.; Dembinska-Kiec, A. Influence of dietary fatty acids on differentiation of human stromal vascular fraction preadipocytes. *Biochim. Biophys. Acta Mol. Cell Biol. Lipids* **2015**, *1851*, 1146–1155. [CrossRef] [PubMed]

63. Capel, F.; Acquaviva, C.; Pitois, E.; Laillet, B.; Rigaudière, J.P.; Jouve, C.; Pouyet, C.; Gladine, C.; Comte, B.; Vianey Saban, C.; et al. DHA at nutritional doses restores insulin sensitivity in skeletal muscle by preventing lipotoxicity and inflammation. *J. Nutr. Biochem.* **2015**, *26*, 949–959. [CrossRef] [PubMed]

64. Ting, H.-C.; Chao, Y.-J.; Hsu, Y.-H.H. Polyunsaturated fatty acids incorporation into cardiolipin in H9c2 cardiac myoblast. *J. Nutr. Biochem.* **2015**, *26*, 769–775. [CrossRef] [PubMed]

65. Lankinen, M.; Schwab, U.; Kolehmainen, M.; Paananen, J.; Poutanen, K.; Mykkänen, H.; Seppänen-Laakso, T.; Gylling, H.; Uusitupa, M.; Orešič, M. Whole grain products, fish and bilberries alter glucose and lipid metabolism in a randomized, controlled trial: The Sysdimet study. *PLoS ONE* **2011**, *6*, e22646. [CrossRef] [PubMed]

66. Wu, C.; Kato, T.S.; Ji, R.; Zizola, C.; Brunjes, D.L.; Deng, Y.; Akashi, H.; Armstrong, H.F.; Kennel, P.J.; Thomas, T.; et al. Supplementation of L-Alanyl-L-Glutamine and fish oil improves body composition and quality of life in patients with chronic heart failure. *Circ. Heart Fail.* **2015**, *8*, 1077–1087. [CrossRef] [PubMed]

67. Mas, E.; Barden, A.; Burke, V.; Beilin, L.J.; Watts, G.F.; Huang, R.C.; Puddey, I.B.; Irish, A.B.; Mori, T.A. A randomized controlled trial of the effects of n-3 fatty acids on resolvins in chronic kidney disease. *Clin. Nutr.* **2016**, *35*, 331–336. [CrossRef] [PubMed]

68. Bondia-Pons, I.; Pöhö, P.; Bozzetto, L.; Vetrani, C.; Patti, L.; Aura, A.M.; Annuzzi, G.; Hyötyläinen, T.; Rivellese, A.A.; Orešič, M. Isoenergetic diets differing in their n-3 fatty acid and polyphenol content reflect different plasma and HDL-fraction lipidomic profiles in subjects at high cardiovascular risk. *Mol. Nutr. Food Res.* **2014**, *58*, 1873–1882. [CrossRef] [PubMed]

69. Dasilva, G.; Pazos, M.; García-Egido, E.; Gallardo, J.M.; Ramos-Romero, S.; Torres, J.L.; Romeu, M.; Nogués, M.-R.; Medina, I. A lipidomic study on the regulation of inflammation and oxidative stress targeted by marine ω-3 PUFA and polyphenols in high-fat high-sucrose diets. *J. Nutr. Biochem.* **2017**, *43*, 53–67. [CrossRef] [PubMed]

70. Bakker, G.C.; van Erk, M.J.; Pellis, L.; Wopereis, S.; Rubingh, C.M.; Cnubben, N.H.; Kooistra, T.; van Ommen, B.; Hendriks, H.F. An antiinflammatory dietary mix modulates inflammation and oxidative and metabolic stress in overweight men: A nutrigenomics approach. *Am. J. Clin. Nutr.* **2010**, *91*, 1044–1059. [CrossRef] [PubMed]

71. Pellis, L.; van Erk, M.J.; van Ommen, B.; Bakker, G.C.M.; Hendriks, H.F.J.; Cnubben, N.H.P.; Kleemann, R.; van Someren, E.P.; Bobeldijk, I.; Rubingh, C.M.; et al. Plasma metabolomics and proteomics profiling after a postprandial challenge reveal subtle diet effects on human metabolic status. *Metabolomics* **2012**, *8*, 347–359. [CrossRef] [PubMed]

72. López, E.F.; Kabarowski, J.H.; Ingle, K.A.; Kain, V.; Barnes, S.; Crossman, D.K.; Lindsey, M.L.; Halade, G.V. Obesity superimposed on aging magnifies inflammation and delays the resolving response after myocardial infarction. *Am. J. Physiol. Heart Circ. Physiol.* **2015**, *308*, H269–H280. [CrossRef] [PubMed]

73. Warensjö, E.; Riserus, U.; Vessby, B. Fatty acid composition of serum lipids predicts the development of the metabolic syndrome in men. *Diabetologia* **2005**, *48*, 1999–2005. [CrossRef] [PubMed]

74. Filep, J.G. Resolution pathways in inflammation: The devil in the adipose tissues and in the details. Focus on "Diversity of lipid mediators in human adipose tissue depots". *Am. J. Physiol. Cell Physiol.* **2013**, *304*, C1127–C1128. [CrossRef] [PubMed]

75. Sears, B. *The Anti-Inflammation Zone*; Regan Books: New York, NY, USA, 2005.

76. McDaniel, J.C.; Massey, K.; Nicolaou, A. Fish oil supplementation alters levels of lipid mediators of inflammation in microenvironment of acute human wounds. *Wound Repair Regen.* **2011**, *19*, 189–200. [CrossRef] [PubMed]

77. Bouwens, M.; van de Rest, O.; Dellschaft, N.; Bromhaar, M.G.; de Groot, L.C.; Geleijnse, J.M.; Müller, M.; Afman, L.A. Fish-oil supplementation induces antiinflammatory gene expression profiles in human blood mononuclear cells. *Am. J. Clin. Nutr.* **2009**, *90*, 415–424. [CrossRef] [PubMed]

78. Davidson, M.H. Mechanisms for the Hypotriglyceridemic Effect of Marine Omega-3 Fatty Acids. *Am. J. Cardiol.* **2006**, *98* (Suppl. 1), 27–33. [CrossRef] [PubMed]

79. Krammer, J.; Digel, M.; Ehehalt, F.; Stremmel, W.; Fullekrug, J.; Ehehalt, R. Overexpression of CD36 and acyl-CoA synthetases FATP2, FATP4 and ACSL1 increases fatty acid uptake in human hepatoma cells. *Int. J. Med. Sci.* **2011**, *8*, 599–614. [CrossRef] [PubMed]

80. Ahmed, N.; Dobler, D.; Dean, M.; Thornalley, P.J. Peptide mapping identifies hotspot site of modification in human serum albumin by methylglyoxal involved in ligand binding and esterase activity. *J. Biol. Chem.* **2005**, *280*, 5724–5732. [CrossRef] [PubMed]

81. Ruskovska, T.; Bernlohr, D.A. Oxidative stress and protein carbonylation in adipose tissue—Implications for insulin resistance and diabetes mellitus. *J. Proteom.* **2013**, *92*, 323–334. [CrossRef] [PubMed]

82. Keaney, J.F.; Larson, M.G.; Vasan, R.S.; Wilson, P.W.F.; Lipinska, I.; Corey, D.; Massaro, J.M.; Sutherland, P.; Vita, J.A.; Benjamin, E.J. Obesity and Systemic Oxidative Stress: Clinical Correlates of Oxidative Stress in The Framingham Study. *Arterioscler. Thromb. Vasc. Biol.* **2003**, *23*, 434–439. [CrossRef] [PubMed]

marine drugs

MDPI

Article

Antibacterial Effect of Eicosapentaenoic Acid against *Bacillus cereus* and *Staphylococcus aureus*: Killing Kinetics, Selection for Resistance, and Potential Cellular Target

Phuc Nguyen Thien Le[1] and **Andrew P. Desbois**[2,*]

[1] School of Biotechnology, International University—Vietnam National University HCMC, Block 6, Linh Trung Ward, Thu Duc District, Ho Chi Minh City 700000, Vietnam; lntphuc@hcmiu.edu.vn
[2] Institute of Aquaculture, University of Stirling, Stirling FK9 4LA, UK
* Correspondence: ad54@stir.ac.uk; Tel.: +44-1786-467894

Received: 18 August 2017; Accepted: 23 October 2017; Published: 1 November 2017

Abstract: Polyunsaturated fatty acids, such as eicosapentaenoic acid (EPA; C20:5n-3), are attracting interest as possible new topical antibacterial agents, particularly due to their potency and perceived safety. However, relatively little is known of the underlying mechanism of antibacterial action of EPA or whether bacteria can develop resistance quickly against this or similar compounds. Therefore, the aim of this present study was to determine the mechanism of antibacterial action of EPA and investigate whether bacteria could develop reduced susceptibility to this fatty acid upon repeated exposure. Against two common Gram-positive human pathogens, *Bacillus cereus* and *Staphylococcus aureus*, EPA inhibited bacterial growth with a minimum inhibitory concentration of 64 mg/L, while minimum bactericidal concentrations were 64 mg/L and 128 mg/L for *B. cereus* and *S. aureus*, respectively. Both species were killed completely in EPA at 128 mg/L within 15 min at 37 °C, while reduced bacterial viability was associated with increased release of 260-nm-absorbing material from the bacterial cells. Taken together, these observations suggest that EPA likely kills *B. cereus* and *S. aureus* by disrupting the cell membrane, ultimately leading to cell lysis. Serial passage of the strains in the presence of sub-inhibitory concentrations of EPA did not lead to the emergence or selection of strains with reduced susceptibility to EPA during 13 passages. This present study provides data that may support the development of EPA and other fatty acids as antibacterial agents for cosmetic and pharmaceutical applications.

Keywords: antibiotic resistance; antimicrobial; fish oil; free fatty acid; omega-3; wound infections

1. Introduction

The marine-derived polyunsaturated fatty acid (PUFA) eicosapentaenoic acid (EPA; C20:5 n-3) has antimicrobial properties and there is increasing interest in developing fatty acids as new antibacterial agents, especially given the rise of bacterial pathogens with resistance against existing antibiotics [1–4]. Similar to many other PUFAs, EPA exerts potent effects against Gram-positive species, including human pathogens *Bacillus cereus* and *Staphylococcus aureus* [3]. *S. aureus* causes a multitude of clinical problems from mild skin complaints, such as impetigo, to more serious soft tissue infections, osteomyelitis, and systemic bacteraemia [5]. Meanwhile, *B. cereus* is a well-known foodborne pathogen that causes infections of the gastrointestinal tract, but this bacterium is also responsible for severe infections of the eyes, lungs, cutaneous tissues, and central nervous system [6]. Importantly, both pathogens can cause serious infections of wounds and surgical sites [5,6] and new effective treatment options are highly desirable.

In clinical and cosmetic applications, free fatty acids such as EPA could be applied topically to bolster the free fatty acids present naturally on the skin and mucosal surfaces as part of innate immunity to protect against microbial infection [3,4,7–10]. In addition to antimicrobial activities, EPA exerts beneficial anti-inflammatory actions [11] and has other positive attributes that would support its development as a new topical antibacterial agent, including wound healing properties [12], potency and perceived safety [1,4,13], and a suspected lack of acquired bacterial resistance mechanisms against this and other fatty acids [14]. However, little is known of whether or not bacteria can develop resistance quickly against this compound, or the underlying mechanisms of antibacterial action of EPA [3]. Addressing these knowledge gaps may hasten the development of EPA and other fatty acids as new topical antibacterial agents [1].

The aim of the present study was to characterize the antibacterial activity of EPA against two Gram-positive pathogens, *B. cereus* and *S. aureus*, and investigate whether the bacteria could develop reduced susceptibility to this fatty acid upon repeated exposure and determine the possible mechanism of action.

2. Results

The susceptibility of *B. cereus* NCIMB 9373 and *S. aureus* Newman to EPA was assessed by broth micro-dilution according to Clinical and Laboratory Standards Institute protocols [15,16]. EPA demonstrated both growth inhibitory and bactericidal activities against *B. cereus* and *S. aureus*. The minimum inhibitory concentration (MIC) was 64 mg/L for EPA against both *B. cereus* and *S. aureus*, while the minimum bactericidal concentration (MBC) for *B. cereus* and *S. aureus* was 64 and 128 mg/L, respectively. In trials to determine kill kinetics at 128 mg/L EPA, no colonies formed by surviving cells were detected after plating 5×10^5 colony forming units (CFU)/mL suspensions of *B. cereus* and *S. aureus* in Mueller-Hinton (MH) broth at 15 min or at subsequent sample times (Figure 1). As expected, some cell division occurred in the control suspensions in MH broth during the 4-h incubation (Figure 1). This experiment was repeated for cell suspensions prepared in phosphate-buffered saline (PBS) to determine whether active bacterial growth was necessary for the killing activity of EPA, but again no colonies formed by surviving cells were detected within 15 min and there was little change in CFU/mL during the 4-h incubation in the control suspensions (Figure 1).

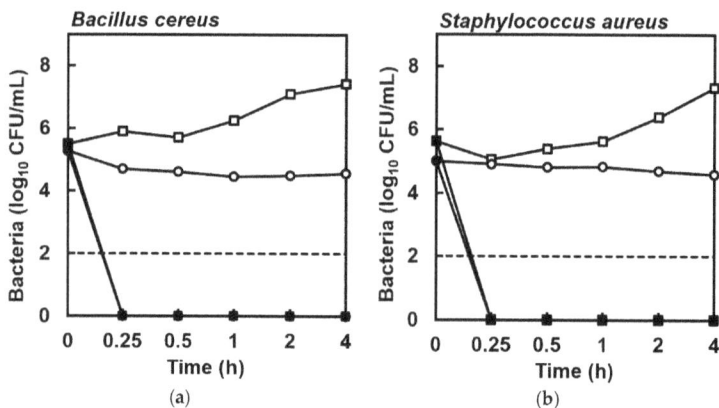

Figure 1. Enumeration of colonies from cell suspensions of (**a**) *Bacillus cereus* NCIMB 9373 and (**b**) *Staphylococcus aureus* Newman in MH broth (■) or PBS (●) exposed to eicosapentaenoic acid (EPA) at 128 mg/L compared to cells suspended in MH broth (□) or PBS (○) lacking EPA during 4 h at 37 °C, showing that no viable bacterial cells were detected in suspensions exposed to EPA within 15 min. The detection limit was 2 $\log_{10}$ CFU/mL (dashed line). Data are geometric mean ± standard error (not all error bars are visible); *n* = 4.

Next, to investigate the possibility to select experimentally for strains with reduced susceptibility to EPA quickly, *B. cereus* and *S. aureus* were serially passaged 13 times in the presence of sub-inhibitory concentrations of this fatty acid in the wells of a 96-well microtitre plate. At each sub-passage, the contents of the wells used to inoculate the subsequent cultures were stored at −70 °C in a cryogenic tube with 15% glycerol (v/v), so that the susceptibility of each passage isolate to EPA could be determined by MIC and compared to the parent strains. After passage of the *B. cereus* strain in sub-inhibitory concentrations of EPA, the isolates from each of 13 passages and the parent strain showed no change in susceptibility to EPA as all isolates had identical MIC and MBC values, indicating the lack of selection of *B. cereus* cells with reduced susceptibility to EPA (Figure 2). Similarly, the MIC values of the corresponding *S. aureus* passage isolates were the same as the parent strain, though the MBC value of each passaged isolate (except for the final passage isolate) was lower than the MBC of the parent strain (Figure 2). Still, there was no evidence for the selection of *S. aureus* cells with reduced susceptibility during repeated exposure to sub-inhibitory concentrations of EPA.

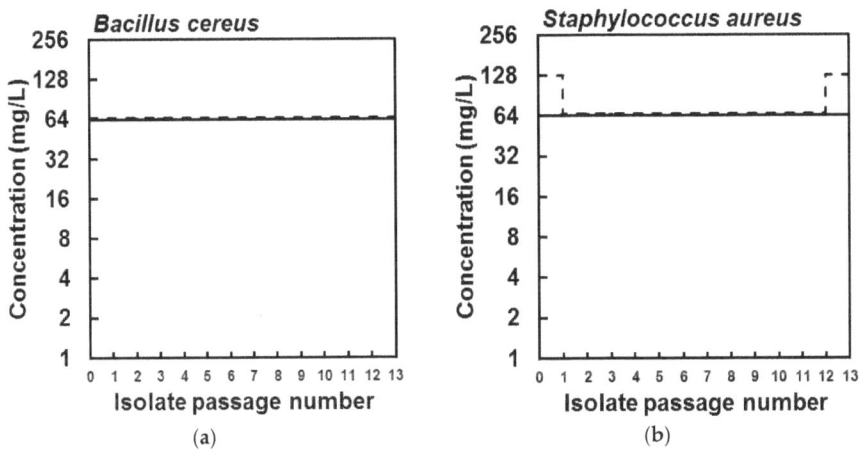

Figure 2. Minimum inhibitory (solid lines) and bactericidal concentration (dashed lines) values of the parent (**a**) *Bacillus cereus* NCIMB 9373 and (**b**) *Staphylococcus aureus* Newman isolates and isolates collected from each of 13 serial passages in sub-inhibitory concentrations of eicosapentaenoic acid (EPA), showing that susceptibility of the bacteria to EPA did not reduce during serial passage.

Finally, to determine the mechanism of antibacterial action of EPA, leakage of 260-nm (A260)-absorbing material from the bacterial cells in suspension was quantified after exposure to increasing concentrations of EPA for 30 min, according to a protocol modified from Carson et al. [17]. The detection of A260-absorbing material can indicate membrane perturbation and an increase in membrane permeability, and these measurements were taken concomitant with bacterial viability assessments by plating of the cell suspensions on agar. The bacterial inoculums at the start of incubation were $1.51 \times 10^9 \pm 0.41 \times 10^9$ CFU/mL (mean ± standard error) and $1.57 \times 10^9 \pm 0.15 \times 10^9$ CFU/mL for *B. cereus* and *S. aureus*, respectively, and thus were considerably greater than used in the killing kinetics experiment above. The carrier solvent (ethanol) had little effect on bacterial viability and at the greatest concentration of ethanol (2.56%, v/v) the bacteria recovered at 30 min was $1.31 \times 10^9 \pm 0.10 \times 10^9$ CFU/mL and $2.79 \times 10^9 \pm 0.05 \times 10^9$ for *B. cereus* and *S. aureus*, respectively (data not shown). Control incubations in the presence of carrier solvent (ethanol) showed that negligible quantities of A260-absorbing material were detected in cell-free filtrates (data not shown). However, leakage of A260-absorbing material was detected from *B. cereus* and *S. aureus* cell suspensions that had been incubated in the presence of ≥64 mg/L EPA for 30 min, and greater concentrations of EPA led to the detection of greater quantities of A260-absorbing material released from both species of bacteria (Figure 3). Importantly, the increasing quantities of A260-absorbing material coincided with reductions in viable CFU/mL in the suspensions (Figure 3). Taken together, these observations suggest membrane disruption and probable cell lysis of the bacterial cells by EPA.

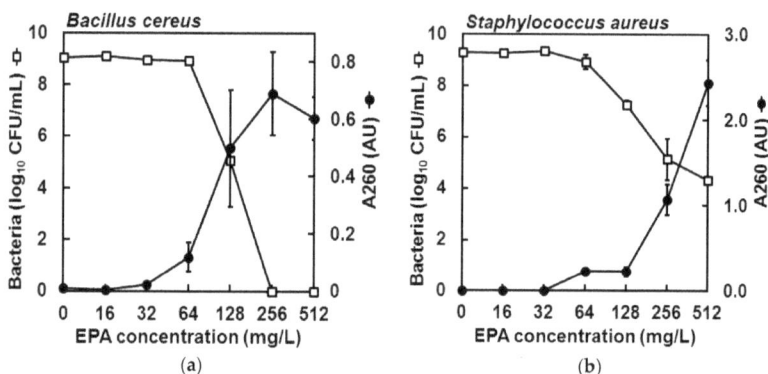

Figure 3. Enumeration of colonies from cell suspensions of (**a**) *Bacillus cereus* NCIMB 9373 and (**b**) *Staphylococcus aureus* Newman in PBS (□) and detection of 260-nm (A260)-absorbing material in cell-free filtrates (●) after exposure to increasing concentrations of eicosapentaenoic acid (EPA) during 30 min at 37 °C, showing that the number of viable cells in suspension reduced concomitant to the amount of A260-absorbing material in cell-free filtrates, indicating likely cell membrane perturbation and cell lysis. The bacterial inoculums at the start of incubation were $1.51 \times 10^9 \pm 0.41 \times 10^9$ CFU/mL (mean ± standard error) and $1.57 \times 10^9 \pm 0.15 \times 10^9$ CFU/mL for *B. cereus* and *S. aureus*, respectively. The carrier solvent (ethanol) had little effect on bacterial viability and at the greatest concentration of ethanol (2.56%, *v/v*) the bacteria recovered at 30 min was $1.31 \times 10^9 \pm 0.10 \times 10^9$ CFU/mL and $2.79 \times 10^9 \pm 0.05 \times 10^9$ for *B. cereus* and *S. aureus*, respectively (data not shown). Meanwhile, control incubations in the presence of carrier solvent (ethanol) showed that negligible quantities of A260-absorbing material were detected in cell-free filtrates (data not shown). Data are mean (geometric mean for CFU values) ± standard error (not all error bars are visible); $n = 3$ for A260 values, $n = 2$ for CFU/mL determinations (detection limit was 2 $\log_{10}$ CFU/mL); note that the secondary *y*-axis (A260) scales differ for (**a**,**b**).

3. Discussion

EPA is antimicrobial and this property is being exploited in the development of new topical cosmetics and pharmaceuticals [1–3]; however, relatively little is known for its antibacterial mechanisms or the ease with which it is possible to select for strains with reduced susceptibility. In this present study, EPA was observed to kill rapidly two species of Gram-positive pathogen, probably by causing cell lysis, and there was little evidence for the selection of strains with reduced susceptibility after 13 passages.

In this present study, EPA inhibited the growth and killed both *B. cereus* and *S. aureus* at concentrations similar to previous reports for these and other Gram-positive species [2,3,18,19]. For *B. cereus*, MIC and MBC values were the same (64 mg/L) and killing was observed in PBS and culture medium within 15 min, indicating that actively dividing cells were not essential for bactericidal action. In conjunction with evidence of leakage of A260-absorbing material from cells at growth inhibitory and bactericidal concentrations, these data support the likely catastrophic loss of bacterial cell membrane integrity once a concentration threshold of EPA is reached. Meanwhile, for *S. aureus*, there was a two-fold difference between MIC and MBC values of EPA (64 and 128 mg/L, respectively) and EPA killed cells within 15 min, which is consistent with previous reports [2,18]. Similar to *B. cereus*, there was no evidence of the need for actively dividing cells to exert antibacterial action, as EPA killed

S. aureus equally as effectively in PBS and MH broth. Notably, the leakage experiment showed that 2.10×10^4 CFU/mL of *S. aureus* survived even at the greatest concentration of EPA (i.e., 512 mg/L) despite the release of A260-absorbing material being four times greater at this concentration than observed for *B. cereus* when no surviving cells were detected. This observation may derive from differences in the physiology of the cell membranes of these species, and it could be that EPA causes differential effects on membrane permeability of the two bacteria. Moreover, *S. aureus* may tolerate greater membrane perturbation but remain viable (for longer at least), whereas similar disruption of the *B. cereus* membrane may be lethal. Indeed, the species-specific variations in the action of EPA observed in this present study are worthy of further investigation. In addition, the composition and quantity of A260-absorbing material (typically nucleic acids [17]) in the cytoplasmic components could differ between the species and this also would need to be determined. Taken together, these observations suggest possible concentration-dependent inhibitory and bactericidal mechanisms of action [4], and EPA probably affects cell membrane-associated metabolic systems or increases permeability, ultimately leading to cell lysis. This suggestion is consistent with other studies that have proposed the cell membrane to be the main site of action for antibacterial free fatty acids, with detrimental effects caused through disruption of vital metabolic processes including cellular respiration [8,20–23] and nutrient uptake [24], or physical disturbance leading to increased permeability, leakage of cellular components [8,23,25–28], and cell lysis (reviewed by Desbois and Smith [14]).

Certain bacteria intrinsically resist the actions of fatty acids and the cell wall of Gram-positive species can confer protection against free fatty acids [25]. Some bacteria increase cell wall synthesis or decrease cell surface hydrophobicity on exposure to free fatty acids [29–31]. The presence of cell-membrane-stabilizing carotenoids that decrease fluidity may also reduce susceptibility to free unsaturated fatty acids [32,33]. Still, there have been few studies on the selection of bacterial strains with reduced susceptibility to antibacterial free fatty acids, particularly for Gram-positive species, and, to our knowledge, this present study is the first to perform serial passage to select for any strains with reduced susceptibility to EPA. Previously, strains of *Escherichia coli* with reduced susceptibility to caprylic acid (C8:0) or capric acid (C10:0) were selected successfully after 10 serial transfers on agar containing the fatty acids at sub-inhibitory concentrations [34], though the mechanisms underlying this phenomenon were not investigated further. Moreover, Petschow et al. [35] reported the isolation of *Helicobacter pylori* mutants on agar containing $10\times$ MIC of the medium-chain length saturated fatty acid lauric acid (C12:0) at a rate of 10^{-8}, but the susceptibility of individual colonies was not subsequently confirmed. Additionally, Obonyo et al. [36] isolated *H. pylori* cells that resisted a previously bactericidal concentration of linolenic acid (C18:3 n-3) after just three sub-cultures in a sub-bactericidal concentration of this fatty acid. In contrast, Sun et al. [37] reported no change in susceptibility to lauric acid of *H. pylori* in response to serial passage six times in sub-inhibitory concentrations. Meanwhile, Lacey and Lord [38] were unable to select stable *S. aureus* strains with reduced susceptibility to linolenic acid including mutants generated by chemical mutagenesis, and, elsewhere, no *S. aureus* strains resistant to linoleic acid (C18:2 n-6) were detected in a 5000 clone transposon insertion library [30]. The opportunity for *B. cereus* to develop resistance to EPA may be reduced by MIC and MBC values being the same as the concentration inhibiting growth is close to the concentration having a lethal effect. However, for *S. aureus*, culturable cells were detected after 30 min exposure to EPA at 512 mg/L and there exists a difference between MIC and MBC values, meaning cells are inhibited but not killed at concentrations in this window, thus potentially permitting the opportunity to select for spontaneous resistant mutants in the surviving population; nevertheless, this was not borne out in practice and serial passage of both bacteria in the presence of sub-inhibitory concentrations of EPA did not lead to the emergence or selection of strains with reduced susceptibility. These observations are consistent with the suggestion that it is more difficult to select for resistance against compounds that exert their antibacterial action by acting on multiple cellular targets and the cell membrane. This present study provides some indication of the difficulty in rapidly selecting for resistance against EPA, but future investigations will use further bacterial species and strains, undertake more passages, and employ more incremental sub-inhibitory PUFA concentrations.

To conclude, the data in this present study provide support for the development of EPA as a possible new antibacterial agent due to its favorable potency against Gram-positive pathogens, lack of rapid selection of bacterial strains with reduced susceptibility or resistance, and bactericidal mechanism of action.

4. Materials and Methods

4.1. Reagents and Bacteria

EPA (>99% purity) and culture media were purchased from Sigma-Aldrich Ltd. (Poole, Dorset, UK). An EPA stock was made in ethanol (≥99.5%) to 20 mg/mL and stored at −20 °C. All other solutions and media were made with ultrapure deionized water (Option 3; Elga, High Wycombe, Bucks, UK) and were sterilized by autoclaving at 121 °C for 15 min or by filtration (polyethersulphone, 0.22 μm; Millipore, Watford, UK). *S. aureus* Newman (gifted by Dr. Angelika Gründling, Imperial College London, UK) and *B. cereus* NCIMB 9373 were resuscitated on MH agar at 37 °C from 15% glycerol (*v/v*) stocks kept at −70 °C, and maintained thereafter at 4 °C.

4.2. Preparation of Bacterial Suspensions

Typically, bacterial suspensions were prepared from cultures that were inoculated with 3–5 colonies into 5 mL MH broth in universal bottles and incubated (37 °C, 150 rpm) until late exponential phase (determined by measuring the absorbance at 600 nm of the culture and comparing to growth curves constructed for each species; approximately 12 h). Next, bacterial cells were harvested by centrifugation ($2000\times g$, 10 min, 4 °C), washed twice with PBS (for 1 L: 8 g of NaCl, 0.2 g of KCl, 1.78 g of $Na_2HPO_4 \cdot 2H_2O$, 0.24 g of KH_2PO_4; pH 7.4), and re-suspended to the desired CFU/mL in MH broth or PBS. The CFU/mL of the suspensions were checked by serially diluting 10 μL in PBS (in duplicate), plating on MH agar, incubating overnight (37 °C, 24 h), and performing CFU counts.

4.3. Assessing Antibacterial Potency

MIC values were determined by broth micro-dilution according to Clinical and Laboratory Standards Institute protocol [15]. Briefly, EPA in MH broth was serially diluted in flat-bottomed 96-well microtitre plates to final well concentrations of 4, 8, 16, 32, 64, 128, 256, and 512 mg/L (100 μL per well). Five microliters of bacterial suspension at 1×10^7 CFU/mL (prepared in PBS as described in Section 4.2) was used to inoculate each well. Plates were sealed with Parafilm, incubated (37 °C, 24 h), and the lowest concentration that prevented bacterial growth visible to the naked eye was determined to be the MIC. Control wells containing either MH broth only or the volume of ethanol equal to the greatest volume in a test well were also inoculated; a non-inoculated MH broth well was also included. MBC was determined according to Clinical and Laboratory Standards Institute protocol [16] by plating 20 μL from each well showing no visible growth at 24 h on to MH agar and incubating these plates for colonies to form (37 °C, 24 h). Contents from individual wells were spread across a quarter of an agar plate and, as no attempt was made to wash away residual EPA, the effect of carryover preventing the formation of colonies cannot be dismissed. The lowest concentration of EPA that killed ≥99.9% of the initial inoculum was determined to be the MBC. MIC and MBC values were determined from duplicated series of wells.

4.4. Killing Kinetics

The times required for EPA at 128 mg/L to kill 5×10^5 CFU/mL from exponential phase cultures of *B. cereus* and *S. aureus* were determined. Briefly, 200 μL of EPA at 128 mg/L in MH was prepared and dispensed into an Eppendorf tube, while control wells received an equal volume of carrier solvent (2.56 μL ethanol). Then, 10 μL of a cell suspension (at 1×10^7 CFU/mL and prepared as described in

Section 4.2) was added to each tube, before the contents were mixed by inversion and then incubated statically at 37 °C. At 15 min, 30 min, 1 h, 2 h, and 4 h, the CFU/mL in each tube was determined by serial dilution and plating of 10 μL of suspension as described in Section 4.2. Geometric means and standard errors of these values were calculated from quadruplicate trials. The experiment was repeated in PBS to determine whether active bacterial growth was necessary for the killing activity of EPA.

4.5. Selection of Bacterial Strains with Reduced Susceptibility to EPA

To investigate the possibility to select experimentally for strains with reduced susceptibility to EPA quickly, bacteria were serially passaged in the presence of sub-inhibitory concentrations of this fatty acid. Briefly, an MIC plate for each bacterium was prepared as described in Section 4.3, except that the final well concentrations of EPA were 16, 32, 64, and 128 mg/L. After 24–48 h incubation, 5 μL from the well showing growth at the greatest concentration of EPA was used to inoculate the wells of a new MIC plate set up such that the greatest EPA concentration was double that of the concentration in the well from which the inoculum was taken. Meanwhile, the remainder of the well contents were stored at −70 °C in a cryogenic tube with 15% glycerol (*v/v*). This process was continued for 13 passages for each species of bacterium. Finally, the MICs against EPA of each passage isolate and the original parent strain were determined as described in Section 4.3, once the cultures had been recovered from cryogenic stocks by culturing on MH agar (37 °C, 24 h) as described in Section 4.1.

4.6. Leakage of A260-Absorbing Material from Bacterial Cells

To assess bacterial membrane perturbation and increasing membrane permeability, leakage of 260-nm absorbing material was quantified from *B. cereus* and *S. aureus* cells in suspension after exposure to increasing concentrations of EPA for 30 min, according to a protocol modified from Carson et al. [17]. These measurements were performed concomitant with bacterial viability assessments by plating cell suspensions on agar. For this, EPA (solubilized in ethanol) was added to PBS and made up to 90 μL to give 16, 32, 64, 128, 256, and 512 mg/L (final volume concentrations after addition of inoculum below), while control tubes contained 90 μL PBS with ethanol concentrations corresponding to each of the respective EPA-containing tubes. Two negative control tubes contained PBS only (no EPA or ethanol). For 10 μL from each tube, A260 was determined on a NanoDrop 1000 spectrophotometer (ThermoScientific, Wilmington, DE, USA) and each of these values for the tube solutions served as the 'blank' when readings were to be taken again from each tube at 30 min. After this, to each tube was added 10 μL of bacterial suspension at 1.5×10^{10} CFU/mL, which had been prepared in PBS as described in Section 4.2, except that the cells were derived from a 500-mL shake flask culture (37 °C, 150 rpm, 24 h). Immediately, one of the negative control tubes was sampled to determine CFU/mL and the A260 of sterile-filtered supernatant. CFU/mL was determined by serial dilution and plating of 10 μL of suspension as described in Section 4.2, while the remaining 80 μL of the suspension was centrifuged (2000× *g*, 10 min). Then the supernatant was collected and passed through a 0.22-μm syringe filter (4 mm; Sterlitech, Washington, DC, USA), before the A260 of this filtrate was measured on the NanoDrop against its respective 'blank'. Meanwhile, all other tubes were incubated (37 °C, 30 min). After incubation, the CFU/mL of unfiltered suspension and A260 of filtered suspension were determined for each tube as described above, however CFU/mL counts were performed only for the remaining negative control, each EPA-containing tube, and the tube containing the greatest volume of carrier solvent. Note that as the control incubations in the presence of carrier solvent (ethanol) showed the presence of only negligible quantities of A260-absorbing material in cell-free filtrates (data not shown), these A260 values were not subtracted from the A260 value of each respective EPA treatment reading. This experiment was repeated twice more, though CFU/mL was determined for only one of these further trials.

Acknowledgments: No external funding was received for this study.

Author Contributions: Andrew P. Desbois and Phuc Nguyen Thien Le conceived and designed the experiments; Phuc Nguyen Thien Le performed the experiments; Andrew P. Desbois and Phuc Nguyen Thien Le analyzed the data; Andrew P. Desbois contributed reagents/materials/analysis tools; Andrew P. Desbois and Phuc Nguyen Thien Le wrote the paper.

Conflicts of Interest: The authors declare no conflict of interest.

References

1.   Desbois, A.P. Potential applications of antimicrobial fatty acids in medicine, agriculture and other industries. *Recent Pat. Antiinfect. Drug Discov.* **2012**, *7*, 111–122. [CrossRef] [PubMed]

2.   Desbois, A.P.; Lawlor, K.C. Antibacterial activity of long-chain polyunsaturated fatty acids against *Propionibacterium acnes* and *Staphylococcus aureus*. *Mar. Drugs* **2013**, *11*, 4544–4557. [CrossRef] [PubMed]

3.   Desbois, A.P. Antimicrobial properties of eicosapentaenoic acid (C20:5n-3). In *Marine Microbiology: Bioactive Compounds and Biotechnological Applications*; Kim, S.K., Ed.; Wiley-VCH Verlag GmbH & Co. KGaA: Hoboken, NJ, USA, 2013; pp. 351–367, ISBN 978-3-527-33327-1.

4.   Sun, M.; Zhou, Z.; Dong, J.; Zhang, J.; Xia, Y.; Shu, R. Antibacterial and antibiofilm activities of docosahexaenoic acid (DHA) and eicosapentaenoic acid (EPA) against periodontopathic bacteria. *Microb. Pathog.* **2016**, *99*, 196–203. [CrossRef] [PubMed]

5.   Crossley, K.B.; Jefferson, K.K.; Archer, G.L.; Fowler, V.G. (Eds.) *Staphylococci in Human Disease*, 2nd ed.; John Wiley & Sons Ltd.: Chichester, UK, 2009; ISBN 978-1-4443-0847-1.

6.   Bottone, E.J. *Bacillus cereus*, a volatile human pathogen. *Clin. Microbiol. Rev.* **2010**, *23*, 382–398. [CrossRef] [PubMed]

7.   Brogden, N.K.; Mehalick, L.; Fischer, C.L.; Wertz, P.W.; Brogden, K.A. The emerging role of peptides and lipids as antimicrobial epidermal barriers and modulators of local inflammation. *Skin Pharmacol. Physiol.* **2012**, *25*, 167–181. [CrossRef] [PubMed]

8.   Parsons, J.B.; Yao, J.; Frank, M.W.; Jackson, P.; Rock, C.O. Membrane disruption by antimicrobial fatty acids releases low-molecular-weight proteins from *Staphylococcus aureus*. *J. Bacteriol.* **2012**, *194*, 5294–5304. [CrossRef] [PubMed]

9.   Wang, Y.D.; Peng, K.C.; Wu, J.L.; Chen, J.Y. Transgenic expression of salmon delta-5 and delta-6 desaturase in zebrafish muscle inhibits growth of *Vibrio alginolyticus* and affects fish immunomodulatory activity. *Fish Shellfish Immunol.* **2014**, *39*, 223–230. [CrossRef] [PubMed]

10.   Porter, E.; Ma, D.C.; Alvarez, S.; Faull, K.F. Antimicrobial lipids: Emerging effector molecules of innate host defense. *World J. Immunol.* **2015**, *5*, 51–61. [CrossRef]

11.   Mullen, A.; Loscher, C.E.; Roche, H.M. Anti-inflammatory effects of EPA and DHA are dependent upon time and dose-response elements associated with LPS stimulation in THP-1-derived macrophages. *J. Nutr. Biochem.* **2010**, *21*, 444–450. [CrossRef] [PubMed]

12.   Shingel, K.I.; Faure, M.P.; Azoulay, L.; Roberge, C.; Deckelbaum, R.J. Solid emulsion gel as a vehicle for delivery of polyunsaturated fatty acids: Implications for tissue repair, dermal angiogenesis and wound healing. *J. Tissue Eng. Regen. Med.* **2008**, *2*, 383–393. [CrossRef] [PubMed]

13.   Nakatsuji, T.; Kao, M.C.; Fang, J.Y.; Zouboulis, C.C.; Zhang, L.; Gallo, R.L.; Huang, C.M. Antimicrobial property of lauric acid against *Propionibacterium acnes*: Its therapeutic potential for inflammatory acne vulgaris. *J. Investig. Dermatol.* **2009**, *129*, 2480–2488. [CrossRef] [PubMed]

14.   Desbois, A.P.; Smith, V.J. Antibacterial free fatty acids: Activities, mechanisms of action and biotechnological potential. *Appl. Microbiol. Biotechnol.* **2010**, *85*, 1629–1642. [CrossRef] [PubMed]

15.   Clinical and Laboratory Standards Institute. *Methods for Dilution Antimicrobial Susceptibility Tests for Bacteria That Grow Aerobically*; Approved Standard M07-A8; Clinical and Laboratory Standards Institute: Wayne, PA, USA, 2009.

16.   Clinical and Laboratory Standards Institute. *Methods for Determining Bactericidal Activity of Antimicrobial Agents*; Approved Guideline M26-A Vol. 19 No. 18; Clinical and Laboratory Standards Institute: Wayne, PA, USA, 1999.

17. Carson, C.F.; Mee, B.J.; Riley, T.V. Mechanism of action of *Melaleuca alternifolia* (tea tree) oil on *Staphylococcus aureus* determined by time-kill, lysis, leakage, and salt tolerance assays and electron microscopy. *Antimicrob. Agents Chemother.* **2002**, *46*, 1914–1920. [CrossRef] [PubMed]

18. Shin, S.Y.; Bajpai, V.K.; Kim, H.R.; Kang, S.C. Antibacterial activity of eicosapentaenoic acid (EPA) against foodborne and food spoilage microorganisms. *LWT Food Sci. Technol.* **2007**, *40*, 1515–1519. [CrossRef]

19. Desbois, A.P.; Mearns-Spragg, A.; Smith, V.J. A fatty acid from the diatom *Phaeodactylum tricornutum* is antibacterial against diverse bacteria including multi-resistant *Staphylococcus aureus* (MRSA). *Mar. Biotechnol.* **2009**, *11*, 45–52. [CrossRef] [PubMed]

20. Borst, P.; Loos, J.A.; Christ, E.J.; Slater, E.C. Uncoupling activity of long chain fatty acids. *Biochim. Biophys. Acta* **1962**, *62*, 509–518. [CrossRef]

21. Boyaval, P.; Corre, C.; Dupuis, C.; Roussel, E. Effects of free fatty acids on propionic acid bacteria. *Lait* **1995**, *75*, 17–29. [CrossRef]

22. Wojtczak, L.; Więckowski, M.R. The mechanisms of fatty acid induced proton permeability of the inner mitochondrial membrane. *J. Bioenerg. Biomembr.* **1999**, *31*, 447–455. [CrossRef] [PubMed]

23. Greenway, D.L.A.; Dyke, K.G.H. Mechanism of the inhibitory action of linoleic acid on the growth of *Staphylococcus aureus*. *J. Gen. Microbiol.* **1979**, *115*, 233–245. [CrossRef] [PubMed]

24. Sheu, C.W.; Freese, E. Effects of fatty acids on growth and envelope proteins of *Bacillus subtilis*. *J. Bacteriol.* **1972**, *111*, 516–524. [PubMed]

25. Galbraith, H.; Miller, T.B. Physiological effects of long chain fatty acids on bacterial cells and their protoplasts. *J. Appl. Bacteriol.* **1973**, *36*, 647–658. [CrossRef] [PubMed]

26. Thormar, H.; Isaacs, C.E.; Brown, H.R.; Barshatzky, M.R.; Pessolano, T. Inactivation of enveloped viruses and killing of cells by fatty acids and monoglycerides. *Antimicrob. Agents Chemother.* **1987**, *31*, 27–31. [CrossRef] [PubMed]

27. Quinn, P.J. Membranes as targets of antimicrobial lipids. In *Lipids and Essential Oils as Antimicrobial Agents*; Halldor, T., Ed.; John Wiley and Sons, Ltd.: Hoboken, NJ, USA, 2011; pp. 1–24, ISBN 978-0-470-74178-8.

28. Arouri, A.; Mouritsen, O.G. Membrane-perturbing effect of fatty acids and lysolipids. *Prog. Lipids Res.* **2013**, *52*, 130–140. [CrossRef] [PubMed]

29. Clarke, S.R.; Mohamed, R.; Bian, L.; Routh, A.F.; Kokai-Kun, J.F.; Mond, J.J.; Tarkowski, A.; Foster, S.J. The *Staphylococcus aureus* surface protein IsdA mediates resistance to innate defenses of human skin. *Cell Host Microbe* **2007**, *1*, 199–212. [CrossRef] [PubMed]

30. Kenny, J.G.; Ward, D.; Josefsson, E.; Jonsson, I.M.; Hinds, J.; Rees, H.H.; Lindsay, J.A.; Tarkowski, A.; Horsburgh, M.J. The *Staphylococcus aureus* response to unsaturated long chain free fatty acids: Survival mechanisms and virulence implications. *PLoS ONE* **2009**, *4*, e4344. [CrossRef] [PubMed]

31. Kohler, T.; Weidenmaier, C.; Peschel, A. Wall teichoic acid protects *Staphylococcus aureus* against antimicrobial fatty acids from human skin. *J. Bacteriol.* **2009**, *191*, 4482–4484. [CrossRef] [PubMed]

32. Chamberlain, N.R.; Mehrtens, B.G.; Xiong, Z.; Kapral, F.A.; Boardman, J.L.; Rearick, J.I. Correlation of carotenoid production, decreased membrane fluidity, and resistance to oleic acid killing in *Staphylococcus aureus* 18Z. *Infect. Immun.* **1991**, *59*, 4332–4337. [PubMed]

33. Xiong, Z.; Kapral, F.A. Carotenoid pigment levels in *Staphylococcus aureus* and sensitivity to oleic acid. *J. Med. Microbiol.* **1992**, *37*, 192–194. [CrossRef] [PubMed]

34. Marounek, M.; Skivanová, E.; Rada, V. Susceptibility of *Escherichia coli* to C2-C18 fatty acids. *Folia Microbiol.* **2003**, *48*, 731–735. [CrossRef]

35. Petschow, B.W.; Batema, R.P.; Ford, L.L. Susceptibility of *Helicobacter pylori* to bactericidal properties of medium-chain monoglycerides and free fatty acids. *Antimicrob. Agents Chemother.* **1996**, *40*, 302–306. [PubMed]

36. Obonyo, M.; Zhang, L.; Thamphiwatana, S.; Pornpattananangkul, D.; Fu, V.; Zhang, L. Antibacterial activities of liposomal linolenic acids against antibiotic-resistant *Helicobacter pylori*. *Mol. Pharm.* **2012**, *9*, 2677–2685. [CrossRef] [PubMed]

37. Sun, C.Q.; O'Connor, C.J.; Roberton, A.M. Antibacterial actions of fatty acids and monoglycerides against *Helicobacter pylori*. *FEMS Immunol. Med. Microbiol.* **2003**, *36*, 9–17. [CrossRef]

38. Lacey, R.W.; Lord, V.L. Sensitivity of staphylococci to fatty acids: Novel inactivation of linolenic acid by serum. *J. Med. Microbiol.* **1981**, *14*, 41–49. [CrossRef] [PubMed]

marine drugs

MDPI

Article

Lipophilic Fraction of Cultivated *Bifurcaria bifurcata* R. Ross: Detailed Composition and In Vitro Prospection of Current Challenging Bioactive Properties

Sónia A. O. Santos [1,*], Stephanie S. Trindade [1], Catia S. D. Oliveira [1], Paula Parreira [2], Daniela Rosa [2], Maria F. Duarte [2,3], Isabel Ferreira [4,5], Maria T. Cruz [4,5], Andreia M. Rego [6], Maria H. Abreu [6], Silvia M. Rocha [7] and Armando J. D. Silvestre [1]

[1] CICECO—Aveiro Institute of Materials and Department of Chemistry, University of Aveiro, 3810-193 Aveiro, Portugal; sst@ua.pt (S.S.T.); cs.oliveira@ua.pt (C.S.D.O.); armsil@ua.pt (A.J.D.S.)
[2] Centro de Biotecnologia Agrícola e Agro-Alimentar do Alentejo (CEBAL)/Instituto Politécnico de Beja (IPBeja), 7801-908 Beja, Portugal; parreira@i3s.up.pt (P.P.); daniela.rosa@cebal.pt (D.R.); fatima.duarte@cebal.pt (M.F.D.)
[3] ICAAM—Instituto de Ciências Agrárias e Ambientais Mediterrânicas, Universidade de Évora, Pólo da Mitra, 7006-554 Évora, Portugal
[4] CNC—Center for Neuroscience and Cell Biology, University of Coimbra, 3004-504 Coimbra, Portugal; isabelcvf@gmail.com (I.F.); trosete@ff.uc.pt (M.T.C.)
[5] FFUC—Faculty of Pharmacy, University of Coimbra, 3000-548 Coimbra, Portugal
[6] ALGAplus—Prod. e Comerc. De Algas e Seus Derivados, Lda., 3830-196 Ílhavo, Portugal; amrego@algaplus.pt (A.M.R.); htabreu@algaplus.pt (M.H.A.)
[7] QOPNA and Department of Chemistry, University of Aveiro, 3810-193 Aveiro, Portugal; smrocha@ua.pt
* Correspondence: santos.sonia@ua.pt; Tel.: +351-234-370-711; Fax: +351-234-370-084

Received: 20 September 2017; Accepted: 23 October 2017; Published: 1 November 2017

Abstract: Macroalgae have been seen as an alternative source of molecules with promising bioactivities to use in the prevention and treatment of current lifestyle diseases. In this vein, the lipophilic fraction of short-term (three weeks) cultivated *Bifurcaria bifurcata* was characterized in detail by gas chromatography–mass spectrometry (GC-MS). *B. bifurcata* dichloromethane extract was composed mainly by diterpenes (1892.78 ± 133.97 mg kg^{-1} dry weight (DW)), followed by fatty acids, both saturated (550.35 ± 15.67 mg kg^{-1} DW) and unsaturated (397.06 ± 18.44 mg kg^{-1} DW). Considerable amounts of sterols, namely fucosterol (317.68 ± 26.11 mg kg^{-1} DW) were also found. In vitro tests demonstrated that the *B. bifurcata* lipophilic extract show antioxidant, anti-inflammatory and antibacterial activities (against both Gram-positive and Gram-negative bacteria), using low extract concentrations (in the order of µg mL^{-1}). Enhancement of antibiotic activity of drug families of major clinical importance was observed by the use of *B. bifurcata* extract. This enhancement of antibiotic activity depends on the microbial strain and on the antibiotic. This work represents the first detailed phytochemical study of the lipophilic extract of *B. bifurcata* and is, therefore, an important contribution for the valorization of *B. bifurcata* macroalgae, with promising applications in functional foods, nutraceutical, cosmetic and biomedical fields.

Keywords: *Bifurcaria bifurcata*; macroalgae; lipids; lipophilic compounds; diterpenes; antioxidant activity; anti-inflammatory activity; antibacterial activity

1. Introduction

Currently there is a remarkable demand for new bioactive molecules to treat diseases, caused by the modern lifestyle and the growing exposure to industrial pollutants and environmental toxins.

Antioxidant compounds have been widely researched, due to their protective action against damage induced by oxidative stress, which is the major cause of diseases like cancer, diabetes, asthma, cardiovascular, or neurodegenerative problems [1], and of skin damage. Current lifestyle diseases have been also associated with inflammatory processes [2], which have also encouraged the scientific community to search for new molecules with anti-inflammatory capacity. At the same time, the emergence and spread of antibiotic-resistant bacteria strains is a growing public health problem [3], which has led the scientific community to search for new antibacterial substitutes or combination of drugs that might overcome this resistance [4].

Nature has always been an important source of molecules with biological properties as such, or after chemical modification [5]. In recent years, marine resources have become alternative sources of several value-added compounds [6,7]. Macroalgae, due to their high biological and chemical diversity and fast-growing properties, are one of the most explored marine resources [8,9]. In fact, approximately 28.5 million tons of macroalgae (brown, red or green) were produced worldwide in 2014, from both capture and aquaculture [10], with an estimated increase of 15%/year. This resource has been mainly explored for food consumption or for the production of specialty products, such as alginic acid, carrageenan, agar, or colorants [7]. In this context, the exploitation of macroalgae has been gaining increasing attention from several research groups around the world [11–13] and notably in Portugal [14–18]. In fact, the exploitation of marine resources is also one of the critical sectors for Portuguese development, given the extent and richness of its Exclusive Economic Zone (EEZ), being included in the Portuguese Research and Innovation Strategy for Smart Specialization for 2014–2020 [19].

Amongst the macroalgae species (close to 10,000), only a limited number have been the object of extensive studies due to their unique composition and consequent diversity of biological activities or health benefits [20]. Species belonging to the brown Dictyotaceae and Sargassaceae families have been studied in detail due to the presence of diterpenes, which have a wide range of biological properties, such as antimicrobial or antitumoral [21]. Among those families, *Bifurcaria bifurcata*, which can only be found in the northeastern Atlantic coasts, from Morocco to northwestern Ireland, is known to biosynthesize components rarely found in other macroalgae species, namely linear diterpenes [22]. In fact, the presence of these components in *B. bifurcata* has been widely reported [23–27]. In addition, several in vitro effects of *B. bifurcata* extracts, such as antimicrobial [28], antimitotic [29], or antiproliferative activity [30], have been attributed to the presence of diterpenes, although this assumption has never been adequately complemented by a detailed study of the composition of the tested extracts. In fact, the relationship between the chemical composition (limited to the fatty acids profile) and the biological properties (namely, antimicrobial and antioxidant activity) of *B. bifurcata* extracts has been addressed in only one study [31].

It has been suggested that the diterpenic composition of *B. bifurcata* is largely dependent on abiotic factors and especially on its geographic origin [32]. Furthermore, there are no studies regarding the lipophilic composition (and the diterpenic fraction in particular) of *B. bifurcata* cultivated in nutrient-rich waters (Integrated Multi-Trophic Aquaculture systems (IMTA)) (or even from the Portuguese coast), where, due to the totally different biotic and abiotic growth conditions, substantial differences in composition might be anticipated.

In addition, most of the published studies using wild *B. bifurcata* have neglected the abundance of diterpenic components, and the few studies that reported their quantification have focused on specific fractions, not considering the lipophilic fraction as a whole [29,33,34]. The abundance of other important classes of lipophilic compounds are also unknown, namely sterols, long-chain aliphatic alcohols and even fatty acids, for which only a single study can be found [31]. The limited number of compounds identified in these studies can be overcome using gas chromatography coupled with mass spectrometry (GC-MS), which allows the detailed identification and quantification of complex mixtures of compounds found in the lipophilic extracts [35,36].

In this study the lipophilic fraction of *B. bifurcata* cultivated for a short period (three weeks) in a land-based IMTA system was detailed characterized by GC-MS analysis. In addition, in vitro assays were performed to evaluate the antioxidant, antibacterial and anti-inflammatory activities of this fraction. Experiments were also conducted aiming to access the synergistic antibiotic–*B. bifurcata* lipophilic extract bacterial growth inhibition by determining the minimum inhibitory concentration (MIC—the lowest concentration where no bacterial growth was detected).

2. Results and Discussion

2.1. Lipophilic Composition

The dichloromethane extraction yield of *B. bifurcata* accounted for 3.92 ± 0.09% (*w/w*), which is quite similar to the *n*-hexane extraction yield obtained for *B. bifurcata* from both aquaculture and wild collected in Portuguese coast [37]. Additionally, the dichloromethane extraction yield was considerably higher than those observed previously for other Phaeophyta species [15] or for Chlorophyta or Rhodophyta macroalgae [14].

The chemical composition of the *B. bifurcata* dichloromethane extract was studied in detail by GC-MS analysis. With the exception of the diterpenes family, all the lipophilic components were identified as their trimethylsilyl derivatives, after derivatization of the dichloromethane extract. The identification of the main lipophilic extractives and the corresponding quantification is summarized in Table 1. In general, the extract was mainly composed of diterpenes, free fatty acids (C14–C22), long-chain aliphatic alcohols (C14–C28), and sterols, among other components. To the best of our knowledge this is the first study reporting the analysis of the lipophilic fraction of *B. bifurcata* by GC-MS. Recently, Alves et al. [31] analyzed the dichloromethane extract of *B. bifurcata* by GC with flame ionization detector; however, only the presence of fatty acids was reported.

Table 1. Compounds identified in *B. bifurcata* dichloromethane extract expressed in mg g^{-1} of extract and in mg kg^{-1} of dry macroalgae [1].

Rt (min)	Compound	mg g^{-1} of Extract	mg kg^{-1} of Dry Macroalgae
	Fatty acids [2]	**24.18 ± 0.56**	**947.88 ± 21.94**
	Saturated	14.04 ± 0.40	550.35 ± 15.67
31.0	Tetradecanoic acid	1.94 ± 0.04	76.23 ± 1.69
33.5	Pentadecanoic acid	0.24 ± 0.02	9.45 ± 0.60
36.0	Hexadecanoic acid	10.11 ± 0.43	396.43 ± 17.01
38.2	Heptadecanoic acid	0.30 ± 0.02	11.81 ± 0.78
40.5	Octadecanoic acid	1.05 ± 0.09	41.01 ± 3.49
44.7	Docosanoic acid	0.24 ± 0.01	9.28 ± 0.48
48.5	Tetracosanoic acid	0.16 ± 0.01	6.15 ± 0.20
	Unsaturated	10.13 ± 0.47	397.06 ± 18.44
35.3	Hexadec-9-enoic acid	1.08 ± 0.07	42.24 ± 2.79
39.5	Octadeca-9,12-dienoic acid	1.04 ± 0.05	40.74 ± 2.08
39.6	Octadeca-9,12,15-trienoic acid	1.47 ± 0.07	57.81 ± 2.70
39.9	Octadec-9-enoic acid	5.13 ± 0.26	200.95 ± 10.32
42.9	Eicosa-5,8,11,14,17-pentaenoic acid	0.91 ± 0.07	35.86 ± 2.57
43.3	Eicosa-5,8,11-trienoic acid	0.02 ± 0.00	0.75 ± 0.03
43.4	Eicosa-11,14-dienoic acid	0.07 ± 0.01	2.86 ± 0.05
	ω-hydroxyacids	0.01 ± 0.00	0.47 ± 0.02
55.3	22-Hydroxydocosanoic acid	0.01 ± 0.00	0.47 ± 0.02
	Long-chain aliphatic alcohols [2]	**0.43 ± 0.03**	**17.03 ± 1.29**
29.0	Tetradecan-1-ol	0.10 ± 0.00	3.78 ± 0.02
34.1	Hexadecan-1-ol	0.15 ± 0.01	5.89 ± 0.38
38.7	Octadecan-1-ol	0.09 ± 0.00	3.41 ± 0.19
58.9	Octacosan-1-ol	0.12 ± 0.00	4.78 ± 0.17
	Sterols [2]	**10.37 ± 0.67**	**406.45 ± 26.19**
58.6	Cholesterol	0.19 ± 0.01	7.29 ± 0.24
62.0	Desmosterol	1.11 ± 0.03	43.56 ± 1.10
63.1	Fucosterol	8.10 ± 0.67	317.68 ± 26.11
63.4	Campesterol	0.97 ± 0.04	37.92 ± 1.61

Table 1. *Cont.*

Rt (min)	Compound	mg g^{-1} of Extract	mg kg^{-1} of Dry Macroalgae
	Monoglycerides [2]	**0.89 ± 0.03**	**34.99 ± 1.10**
47.8	1-Monohexadecanoin	0.66 ± 0.03	25.82 ± 1.01
50.9	1-Monooctadecenoin	0.13 ± 0.00	5.17 ± 0.17
51.5	1-Monoeicosa-tetraenoin	0.10 ± 0.01	4.00 ± 0.31
	Diterpenes [3]	**48.29 ± 3.42**	**1892.78 ± 133.97**
26.8	Neophytadiene	1.53 ± 0.08	59.86 ± 3.29
32.4	Phytol	0.87 ± 0.07	34.02 ± 3.25
34.0	*Trans*-geranylgeraniol	1.70 ± 0.16	66.53 ± 2.93
34.5	6,7,9,10,11,12,14,15-Tetradehydrophytol	2.41 ± 0.88	94.35 ± 6.29
36.5	6-Hydroxy-13-oxo-7,7',10,11-didehydrophytol	16.27 ± 2.41	637.84 ± 34.41
40.7	1-Acetyl-10,13-dioxo-6,7,11,11',14,15-tridehydrophytol	25.51 ± 4.04	1000.17 ± 94.47
	Other terpenic compounds [3]	**0.28 ± 0.00**	**10.79 ± 0.18**
26.6	6,10,14-Trimethyl-2-pentadecanone	0.28 ± 0.00	10.79 ± 0.18
	Total	**84.44 ± 4.14**	**3309.93 ± 162.25**

[1] Results are the average of the concordant values obtained from the triplicated extracts each injected twice (less than 5% variation between injections of the same aliquot and between triplicated dichloromethane extracts); [2] Compounds identified as trimethylsilyl derivatives; [3] Compounds identified in the method without derivatization.

2.1.1. Diterpenes and Other Terpenoids

Several diterpenes (Figure 1) and other terpenoids were identified in *B. bifurcata* dichloromethane extract, namely neophytadiene, phytol, *trans*-geranylgeraniol, 6,7,9,10,11,12,14,15-tetradehydrophytol, 6-hydroxy-13-oxo-7,7',10,11-didehydrophytol, 1-acetyl-10,13-dioxo-6,7,11,11',14,15-tridehydrophytol and 6,10,14-trimethyl-2-pentadecanone.

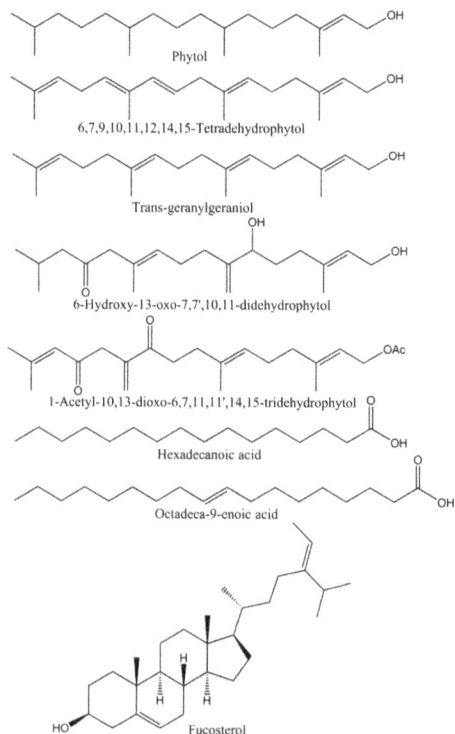

Figure 1. Diterpenes and other major lipophilic compounds detected in *B. bifurcata* dichloromethane extract.

The fragmentation patterns of these components depend on the number and position of the double bonds, and on the nature of the substituent groups [23,25,26]. As an example, the mass spectra of neophytadiene and *trans*-geranylgeraniol are presented in Figure 2.

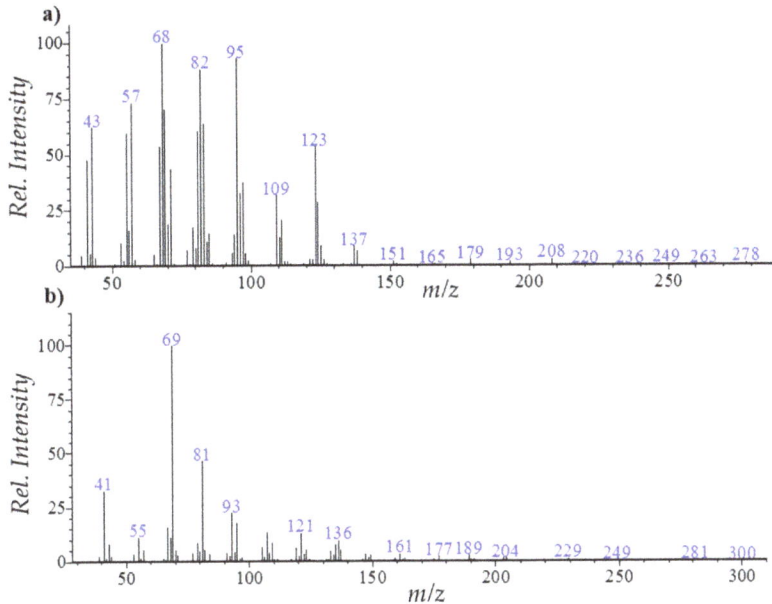

Figure 2. Mass spectra of (**a**) neophytadiene and (**b**) *trans*-geranylgeraniol.

The mass spectrum of neophytadiene presents a molecular ion $[M]^+$ at m/z 278 and major product ions at m/z 43 ($[C_3H_7]^+$), 57 ($[C_4H_9]^+$), 68, 82, 95 ($[C_7H_{11}]^+$), 109 and 123 ($[C_9H_{15}]^+$). Similarly, *trans*-geranylgeraniol presents a molecular ion $[M]^+$ at m/z 290, and characteristic product ions at m/z 69 ($[C_5H_9]^+$), 81 ($[C_6H_9]^+$), 121 ($[C_{11}H_{19}]^+$) and 272 ($M-H_2O]^+$).

Diterpenes, accounting for 1892.78 $\pm$ 133.97 mg kg^{-1} DW, correspond to about 57% of the total amount of lipophilic compounds detected in *B. bifurcata* extract, with 1-acetyl-10,13-dioxo-6,7,11,11',14,15-tridehydrophytol and 6-hydroxy-13-oxo-7,7',10,11-didehydrophytol as the major components of this family, accounting for 1000.17 $\pm$ 94.47 and 637.84 $\pm$ 34.41 mg kg^{-1} DW, respectively. 1-acetyl-10,13-dioxo-6,7,11,11',14,15-tridehydrophytol and 6-hydroxy-13-oxo-7,7',10,11-didehydrophytol were already described as constituent of *B. bifurcata* collected in Brittany (France) [25,26], but no information was reported regarding its quantification. Considerable amounts of 6,7,9,10,11,12,14,15-tetradehydrophytol, *trans*-geranylgeraniol and neophytadiene were also found, and minor amounts of phytol were also detected. 6,7,9,10,11,12,14,15-Tetradehydrophytol was previously identified in *B. bifurcata* collected in Morocco [23], though no information has been provided concerning its abundance, while, to the best of our knowledge, phytol and neophytadiene are reported, for the first time, as constituents of *B. bifurcata*. In opposition, geranylgeraniol has been one of the most reported constituents of this macroalga species [26,29,38]. Additionally, the abundance of this component in the *B. bifurcata* dichloromethane extract is in the range of those (0.0–1.5 mg g^{-1} DW) reported in a seasonal variation study of the same species collected in Brittany (France) [39]. Notwithstanding, only traces of this diterpene were found in a geographical variation study of the diterpene composition of *B. bifurcata* collected in Spain [33].

Despite the high abundance verified for all the detected diterpenes, these were not reported in some of the studies regarding the seasonal and geographical variation of diterpenes in *B. bifurcata* [29,33,34].

Diterpenes have been associated with a wide number of health benefits [40]. In particular, those from macroalgae have been reported to have antioxidant, antimicrobial [41], antifungal [42], anti-viral [43], or antitumoral activities [44].

Minor amounts (10.79 ± 0.18 mg kg^{-1} DW) of 6,10,14-trimethyl-2-pentadecanone were found in *B. bifurcata* dichloromethane extract. This terpenoid and its derivatives have been described to be produced by macroalgae from Cystoseiraceae family, namely *Cystophora moniliformis* [45,46].

2.1.2. Fatty Acids

Fatty acids were the second most abundant family of compounds detected in the lipophilic fraction of *B. bifurcata*, corresponding to 947.88 ± 21.94 mg kg^{-1} DW. Hexadecanoic acid (Figure 1) was the most abundant fatty acid (396.43 ± 17.01 mg kg^{-1} DW), followed by tetradecanoic (76.23 ± 1.69 mg kg^{-1} DW) and octadecanoic (41.01 ± 3.49 mg kg^{-1} DW) acids. The high abundance of hexadecanoic and tetradecanoic acid in *B. bifurcata* was already reported [31], however, in considerably higher amounts than those found in this study.

Considerable amounts of unsaturated fatty acids were found in *B. bifurcata* (397.06 ± 18.44 mg kg^{-1} DW), representing about 42% of the total fatty acids. Octadec-9-enoic acid (Figure 1) was the most abundant unsaturated fatty acid, accounting for 200.95 ± 10.32 mg kg^{-1} DW, followed by the ω-3 octadeca-9,12,15-trienoic acid (57.81 ± 2.70 mg kg^{-1} DW) and the hexadec-9-enoic acid (42.24 ± 2.79 mg kg^{-1} DW). Significant amounts of the ω-3 fatty acid eicosa-5,8,11,14,17-pentaenoic acid were also found (35.86 ± 2.57 mg kg^{-1} DW), despite being 10 times lower than those reported previously by Alves et al. [31] for *B. bifurcata* collected in Peniche coast, Portugal. Notwithstanding, the high abundance of this component contributes to the low ω-6/ω-3 ratio verified for this macroalga (~0.46), which is considerably lower than the maximum (4:1) recommended for a diet in order to prevent the development of inflammatory processes [47]. Additionally, this ratio is significantly lower than those verified for other macroalgae species [15,48], which were considered promising to be incorporated in healthy diets. The high abundance of such components, widely associated with the prevention or delay of chronic diseases, such as cancer, cardiovascular, or coronary problems [49], highlights the potential of valorization of *B. bifurcata* also in food and/or nutraceutical industry. However, such exploitation requires that more detailed studies should be performed, namely evaluating their side effects and toxicity.

2.1.3. Long-Chain Aliphatic Alcohols

Minor amounts of long-chain aliphatic alcohols (LCAA) were found in *B. bifurcata* dichloromethane extract, accounting for 17.03 ± 1.29 mg kg^{-1} DW. To the best of our knowledge this is the first study reporting the presence of these components in *B. bifurcata*. Hexadecan-1-ol was the most abundant LCAA detected, followed by octadecan-1-ol and tetradecan-1-ol.

2.1.4. Sterols

Significant amounts of sterols were found in the dichloromethane extract of *B. bifurcata* (406.45 ± 26.19 mg kg^{-1} DW). Other studies reported the total sterols content in this macroalga collected in different countries [33,39]; however, only fucosterol has been described, with no further detailed characterization. In fact, fucosterol (Figure 1) was the major sterol detected in *B. bifurcata* dichloromethane extract, accounting for 317.68 ± 26.11 mg kg^{-1} DW. Notwithstanding, this value was lower than the previous reported contents (1400–5900 mg kg^{-1} DW) [33,39]. The bioactivities assigned to this phytosterol have been widely described in literature, such as its capacity to modulate cholesterol levels [50], its anti-inflammatory potential [51], or its dermo-protective effect [52]. Considerable amounts of desmosterol (43.56 ± 1.10 mg kg^{-1} DW) and campesterol (37.92 ± 1.61 mg kg^{-1} DW) were also found, which is in accordance with results observed for other Phaeophyta macroalgae [15]. Minor amounts of cholesterol were also detected.

2.1.5. Monoglycerides

Monoglycerides were also detected in *B. bifurcata* dichloromethane extract, representing only a small fraction of the total lipophilic compounds identified (34.99 $\pm$ 1.10 mg kg^{-1} DW). To our knowledge, this is also the first study reporting the monoglycerides profile of *B. bifurcata*. 1-Monohexadecanoin was the most abundant compound detected from this family, representing almost 74% of the total monoglycerides. Minor amounts of 1-monooctadecenoin (5.17 $\pm$ 0.17 mg kg^{-1} DW) and 1-monoeicosa-tetraenoin (4.00 $\pm$ 0.31 mg kg^{-1} DW) were also found.

2.2. Antioxidant Activity

The antioxidant activity of *B. bifurcata* dichloromethane extract was evaluated by both DPPH and ABTS in vitro assays. Table 2 presents the obtained results, expressed as the amount of extract needed to decrease the DPPH$^{\bullet}$ and ABTS$^{+\bullet}$ concentrations by 50% (IC$_{50}$), as well as in mg of ascorbic acid (AAE) and trolox (TE) equivalents per g of dry weight. The IC$_{50}$ values for ascorbic acid and BHT (for DPPH assay) and for trolox (for ABTS assay) were also estimated and are reported in Table 2 for comparative purposes. *B. bifurcata* lipophilic extract showed antioxidant activity against both radicals, although a higher activity was observed against ABTS$^{+\bullet}$, with the respective IC$_{50}$ accounting for 116.25 $\pm$ 2.54 μg mL^{-1}, representing 23.10 $\pm$ 0.51 mg of trolox equivalents g^{-1} DW.

Table 2. Antioxidant activity of *B. bifurcata* dichloromethane extract expressed as IC$_{50}$ values (μg mL^{-1}) and in mg of ascorbic acid (AAE)/trolox equivalents (TE) g^{-1} of dry weight.

Heading	DPPH Assay		ABTS Assay	
	IC$_{50}$	mg AAE g^{-1} DW	IC$_{50}$	mg TE g^{-1} DW
B. bifurcata	365.57 $\pm$ 10.04	11.18 $\pm$ 0.30	116.25 $\pm$ 2.54	23.10 $\pm$ 0.51
Ascorbic acid	4.08 $\pm$ 0.05			
BHT	14.32 $\pm$ 0.69			
Trolox			2.68 $\pm$ 0.07	

To our knowledge, this is the first study reporting the antioxidant activity of a *B. bifurcata* extract against ABTS$^{+\bullet}$. In addition, the IC$_{50}$ value determined for DPPH assay (365.57 $\pm$ 10.04 μg mL^{-1}) was quite similar to that described before for a dichloromethane extract of *B. bifurcata* collected from rock pools in Portuguese coast [31]. Notwithstanding, *B. bifurcata* lipophilic extracts showed IC$_{50}$ values against both radicals significantly lower than those determined for ascorbic acid or trolox (Table 2).

2.3. Anti-Inflammatory Activity

In order to study the capacity of *B. bifurcata* dichloromethane extract to modulate nitric oxide production, an in vitro model of inflammation consisting of macrophages stimulated with LPS was performed. Concomitantly, cell viability was evaluated by the resazurin-based assay (Figure 3) in order to select concentrations with bioactivity and without cytotoxicity. *B. bifurcata* lipophilic extract showed a slightly but statistically significant cytotoxic effect at 100 μg mL^{-1}, when compared to LPS-treated cells.

The effect of *B. bifurcata* extract on NO production was analyzed by measuring the accumulation of nitrites in the culture medium, using the Griess assay. RAW 264.7 cells were stimulated with LPS in the presence or absence of *B. bifurcata* extract for 24 h. *B. bifurcata* extract inhibited LPS-induced NO production in a concentration-dependent manner (Figure 4). This extract markedly inhibited LPS-induced NO production to 6% and 40% at 50 and 25 μg mL^{-1}, respectively.

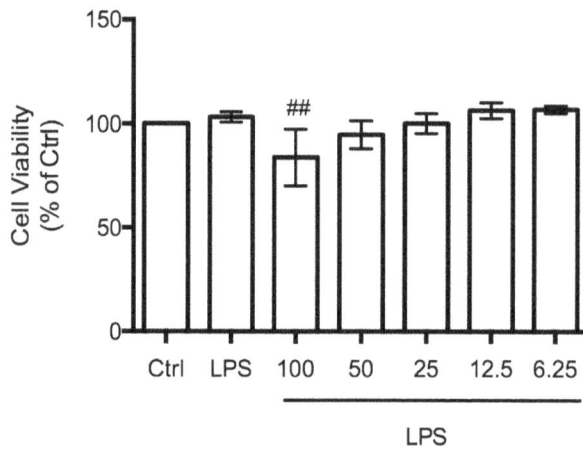

Figure 3. Effect of *B. bifurcata* dichloromethane extract at 6.25, 12.5, 25, 50, and 100 µg mL^{-1} on RAW 264.7 cell viability. Evaluation of statistical significance was performed using one-way ANOVA with *Dunnett's* multiple comparison test; ## $p < 0.01$ compared to control (Ctrl).

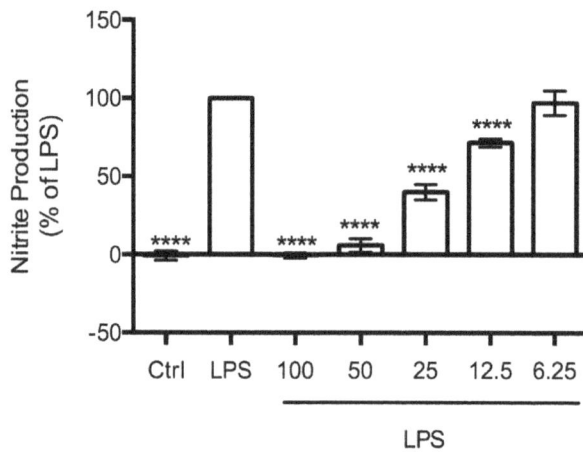

Figure 4. Inhibitory effect of *B. bifurcata* dichloromethane extract on nitric oxide production evoked by LPS in RAW 264.7 cells. *B. bifurcata* dichloromethane extract was used in the concentration range of 6.25–100 µg mL^{-1}, **** $p < 0.0001$ compared to LPS.

The decrease in LPS-induced NO production at 50 µg mL^{-1} to levels similar to those observed in untreated cells reveals the remarkable anti-inflammatory potential of this extract. In fact, the anti-inflammatory activity of the extract is considerable higher than reported for other macroalgae extracts, namely from *Saccharina japonica* [53] or *Undaria pinnatifida* [54]. To our knowledge, this is the first study reporting the anti-inflammatory activity of *B. bifurcata*.

2.4. Antibacterial Activity

The antibacterial activity of *B. bifurcata* lipophilic extract against *Staphylococcus aureus, Escherichiacoli, Pseudomonas aeruginosa*, and *Staphylococcus epidermidis* were evaluated. Therapeutic strategies to counteract infections caused by these bacterial strains have been one of the main concerns, particularly in health care settings. In fact, *S. aureus, E. coli*, and *P. aeruginosa* present the higher rates of antibiotic resistance, while *S. epidermidis* has been considered an opportunistic pathogen, being the most common source of infections on indwelling medical devices [55]. Thus, in this study, the antimicrobial efficacy of four antibiotics representing drug families of major clinical importance, such as aminoglycosides (Gent: Gentamicin), tetracyclines (Tetra: Tetracycline), macrocyclics (Rif: Rifampicin), and β-lactams antibiotics, such as aminopenicillins, (Amp: Ampicillin) were evaluated in combination with the *B. bifurcata* lipophilic extract.

Results regarding the antibacterial activity of the *B. bifurcata* dichloromethane extract are presented in Table 3. Activity against *S. aureus* ATCC®6538 (MIC = 1024 μg mL^{-1}), *S. aureus* ATCC®43300 (MIC = 2048 μg mL^{-1}) and *E. coli* (MIC = 2048 μg mL^{-1}) was observed. No growth inhibition of *S. epidermidis* and *P. aeruginosa* PAO1 was verified in the range of concentrations tested (MIC > 2048 μg mL^{-1}). *B. bifurcata* extract showed inhibition against both Gram-negative and Gram-positive bacteria, in opposition to that observed by Alves et al. [31], which only verified activity of a *B. bifurcata* dichloromethane extract against Gram-negative bacteria. Additionally, these authors have not reported growth inhibition against the same *E. coli* strain. Notwithstanding the fact that the extraction conditions in both studies were not the same, the macroalgae origins are distinct, which may alter the metabolite composition and therefore their bioactivities, highlighting the importance of controlling the growth conditions of macroalgae in order to maximize their valorization.

Table 3. Antibacterial activity of *B. bifurcata* dichloromethane extract and synergistic effects with different antibiotics expressed in MIC (μg mL^{-1}).

Bacteria	Ext	Rif	Rif+ Ext	Gent	Gent+ Ext	Amp	Amp+ Ext	Tetra	Tetra + Ext
					MIC (μg mL^{-1})				
E. coli ATCC®25922	2048	32	16	>256	<2	32	128	18	<2
Staphylococcus aureus ATCC®43300	2048	16	<2	>256	16	128	256	>256	<2
Staphylococcus aureus ATCC®6538	1024	64	64	32	16	256	512	16	8
Pseudomonas aeruginosa PAO1	>2048	32	-	2	-	16	-	16	-
Staphylococcus epidermidis	>2048	32	-	512	-	>2048	-	32	-

Key: Ext: Extract; Rif: Rifampicin; Gent: Gentamicin; Amp: Ampicillin; Tetra: Tetracycline.

The synergism of *B. bifurcata* dichloromethane extract with antibiotics was also evaluated against the same bacterial panel. This approach has been suggested as a promising tool to overcome the bacterial resistance that has been developed to most of the antibiotic classes recommended for their treatment [3,4,56]; to our knowledge, no similar study has previously been reported for this macroalga. The synergetic activity (antibiotic + extract) screening was performed against the bacterial strains for which a new antibiotic MIC was possible to determine, namely for *E. coli* ATCC®25922, *S. aureus* ATCC®43300 and *S. aureus* ATCC®6538 (Table 3). Interestingly, the combination of the extract with gentamicin or tetracycline drastically decreased the antibiotic MIC against the three strains under study, enabling it to be effective at considerably lower concentrations. Similar behavior was observed for both *E. coli* ATCC®25922 and *S. aureus* ATCC®43300 when the extract was combined with rifampicin. These observations suggest that the use of *B. bifurcata* lipophilic fraction may be considered within a coadjutant/synergistic approach to conventional antibiotherapy, enabling a superior therapeutic potential against the above-mentioned bacteria.

Concerning the interaction between *B. bifurcata* dichloromethane extract and ampicillin, an antagonism effect was observed, with MIC values increasing considerably. This may be linked to alterations in the antibiotic structure induced by the presence of the extract in solution, leading to its partial inactivation.

3. Material and Methods

3.1. Sample

Bifurcaria bifurcata R. Ross was cultivated in the land-based aquaculture system of ALGAplus, Lda, at Ria de Aveiro (Portugal, 40°36'43" N, 8°40'43" W). The company operates under the IMTA concept, using solely filtered nutrient-rich water from a fishpond to produce seaweed biomass. The culture starting material was obtained at Aguda beach (Portugal, 41°2′38″ N, 8°39′10″ W) in May 2014 and grown by vegetative propagation during three weeks at constant conditions. The biomass was washed with seawater to remove salts, epiphytes, and/or microorganisms and dried at 25 °C until it reached a total moisture content of 12%. The samples were transformed into flakes (1–2 mm), packed, and stored in hermetic bags in the company's storage room.

3.2. Lipophilic Compounds Extraction

Three aliquots (20 g) of lyophilized macroalgae samples were Soxhlet extracted with dichloromethane (Sigma Chemical Co., Madrid, Spain) for 9 h. Solvent was evaporated to dryness, lipophilic extracts weighted and the results were expressed in percent of dry weight material (% DW). Dichloromethane was selected as a fairly specific solvent for lipophilic extractives isolation for analytical purposes only.

3.3. GC-MS Analysis

The analysis of the lipophilic extracts followed two distinct methodologies: the first suitable for trimethylsilyl (TMS) derivatizable compounds and the second for diterpenic compounds, as described below.

3.3.1. Analysis of Trimethylsilyl Derivatizable Compounds

Before GC-MS analysis, two aliquots of each dried extract (20 mg each) and an accurate amount of internal standard (tetracosane, 0.25–0.50 mg, Sigma Chemical Co., Madrid, Spain) were dissolved in 250 µL of pyridine (Sigma Chemical Co.).

The compounds containing hydroxyl and carboxyl groups were converted into TMS ethers and esters, respectively, by adding 250 µL of *N,O*-bis(trimethylsilyl)trifluoroacetamide (Sigma Chemical Co.) and 50 µL of trimethylchlorosilane (Sigma Chemical Co.), standing the mixture at 70 °C for 30 min [57]. The derivatized extracts were analyzed by GC-MS following previously described methodologies [15,35] on a GCMS-QP2010 Ultra (Shimadzu, Kyoto, Japan), equipped with a DB–1 J&W capillary column (30 m × 0.32 mm inner diameter, 0.25 µm film thickness). The chromatographic conditions were as follows: initial temperature, 80 °C for 5 min; temperature gradient, 4 °C min^{-1}; final temperature, 260 °C; temperature gradient, 2 °C min^{-1}; final temperature, 285 °C for 8 min; injector temperature, 250 °C; transfer-line temperature, 290 °C; split ratio, 1:33.

Compounds were identified as TMS derivatives by comparing their mass spectra with two commercial GC-MS spectral libraries (Wiley 275 and U.S. National Institute of Science and Technology (NIST14)), their characteristic retention times obtained under the described experimental conditions [15,35], and by comparing their mass spectra fragmentation profiles with published data or by injection of standards.

For semi-quantitative analysis, GC-MS was calibrated with pure reference compounds, representative of the major lipophilic extractive families (cholesterol, hexadecanoic acid and nonadecan-1-ol (Sigma Chemical Co.)) relative to tetracosane. The respective response factors were calculated as an average of six GC-MS runs. Each one of the three extracts prepared from macroalgae was injected in duplicate (*n* = 6). The presented results are the average of the concordant values obtained (less than 5% variation between injections of the same aliquot and between triplicated extracts of the same macroalgae).

3.3.2. Analysis of diterpenes

Diterpenes were identified and quantified by the analysis of the *B. bifurcata* extract without derivatization. About 10 mg of extract were dissolved in 1100 µL of dichloromethane with 0.8 mg of internal standard (*n*-hexadecane (Supelco, Bellefonte, PA, USA)).

The extracts were injected in the same GC-MS equipment as described above and the chromatographic conditions were as follows: initial temperature, 80 °C for 5 min; temperature gradient, 5 °C min^{-1}, final temperature 200 °C, temperature gradient, 2 °C min^{-1}, final temperature 240 °C; temperature gradient, 5 °C min^{-1}, final temperature 285 °C for 8 min; injector temperature, 250 °C; transfer-line temperature, 290 °C; split ratio, 1:40. All other conditions were the same as described above. Diterpenes were identified by comparing their mass spectra fragmentation profile with library (Wiley 275 and U.S. National Institute of Science and Technology (NIST14)) and with the characteristic fragmentation pathway described in literature for these components [23,25,26].

Semi-quantitative analysis was carried out determining the response factor (an average of six GC-MS runs) of a representative standard, namely phytol (Sigma Chemical Co.), relative to *n*-hexadecane. Each one of the three aliquots were injected in duplicate ($n = 6$).

3.4. Antioxidant Activity

3.4.1. DPPH Assay

The antioxidant activity of the *B. bifurcata* lipophilic extract was measured by their hydrogen-donating or radical scavenging ability using the stable free radical 2,2-diphenyl-1-picrylhydrazyl (DPPH$^{\bullet}$) (Sigma Chemical Co.), following a previously described procedure [35]: 0.25 mL of DPPH$^{\bullet}$ (0.8 mM in methanol (Fluka Chemie, Madrid, Spain)) were added to 1.00 mL of the aqueous solution of the *B. bifurcata* dichloromethane extract and 2.75 mL of methanol. The extract concentration ranged between 61.9 and 556.9 µg mL^{-1}. After 30 min of incubation in the dark, at room temperature, the absorbance was determined at 517 nm, using a Shimadzu UV-1800 spectrophotometer (Kyoto, Japan). Standards of ascorbic acid (Fluka Chemie) and 3,5-di-*tert*-4-butylhydroxytoluene (BHT) (Sigma Chemical Co. were used with concentrations ranging from 0.7 to 5.6 µg mL^{-1} and 5.0 to 75.8 µg mL^{-1}. Duplicate measurements of three extracts were carried out ($n = 6$). The antioxidant activity was expressed in IC$_{50}$ values, defined as the inhibitory concentration of the extract required to decrease the initial DPPH radical concentration by 50%, as well as in g of ascorbic acid equivalents per kg of dry weight (g AAE kg^{-1} DW).

3.4.2. ABTS Assay

The ABTS assay is based on the scavenging of the 2,2′-azino-bis(3-ethylbenzothiazoline-6-sulphonic acid) radical cation, ABTS$^{+\bullet}$ (Sigma Chemical Co.) converting it into a colorless product. In this test, ABTS$^{+\bullet}$ cation was generated by reacting 50 mL of ABTS 7 mM solution with 25 mL of potassium persulfate (Fluka Chemie) 2.45 mM, following a methodology described before with minor modifications [35]. This mixture was then incubated in the dark, at room temperature, for 16 h. Before usage, the ABTS$^{+\bullet}$ solution was diluted with methanol to obtain an absorbance of 0.700 ± 0.02 at 734 nm. 30 µL of *B. bifurcata* dichloromethane extract or trolox (Aldrich Chemical Co., Madrid, Spain), used as reference, were mixed with 3 mL of ABTS$^{+\bullet}$ solution, obtaining final concentrations of 47.9–239.5 µg mL^{-1} and 1.0–5.0 µg mL^{-1}, respectively. The absorbance was measured at 734 nm using a Shimadzu UV-1800 spectrophotometer (Kyoto, Japan). Duplicate measurements of three extracts were performed.

The antioxidant activity was expressed as IC$_{50}$ values (extract or trolox concentration providing 50% of ABTS$^{+\bullet}$ inhibition) as well as in mg of trolox equivalents per g of dry weight (mg TE g^{-1} DW).

3.5. Anti-Inflammatory Activity

Test solutions of *B. bifurcata* dichloromethane extract (100 mg mL^{-1}) were prepared in ethanol and diluted in a culture medium (prepared as described below). Ethanol concentrations ranged from 0.007 to 0.1% (*v/v*), corresponding to extract concentrations between 6.25 and 100 µg mL^{-1}.

3.5.1. Cell Culture

Raw 264.7, a mouse leukaemic monocyte macrophage cell line from American Type Culture Collection (ATCC TIB-71), kindly supplied by Dr. Otília Vieira (Centro de Neurociências e Biologia Celular, Universidade de Coimbra, Coimbra, Portugal), was cultured in Dulbecco's Modified Eagle Medium supplemented with 10% non-inactivated fetal bovine serum, 100 U mL^{-1} penicillin, and 100 µg mL^{-1} streptomycin at 37 °C in a humidified atmosphere of 95% air and 5% CO_2. During the experiments, cells were monitored by microscope observation in order to detect any morphological changes.

3.5.2. Determination of Cell Viability

Assessment of metabolically active cells was performed using a resazurin (a nonfluorescent blue dye) based assay [58]. Briefly, cell duplicates were plated at a density of 0.1×10^6/well, in a 96 well plate and allowed to stabilize overnight. Following this period, cells were either maintained in culture medium (control) or pre-incubated with *B. bifurcata* extract diluted in culture medium for 1 h, and later activated with 50 ng mL^{-1} of the *Toll-like* receptor 4 agonist lipopolysaccharide (LPS) for 24 h. After the treatments, resazurin solution (50 µM in culture medium) was added to each well and incubated at 37 °C for 1 h, in a humidified atmosphere of 95% air and 5% CO_2. Viable cells are able to reduce resazurin into resorufin (fluorescent pink) and, hence, their number correlates with the magnitude of dye reduction. Quantification of resorufin was performed on a Biotek Synergy HT plate reader (Biotek, Winooski, VT, USA) at 570 nm, with a reference wavelength of 620 nm.

3.5.3. Measurement of Nitrite Production by Griess Reagent

The production of nitric oxide (NO) was measured by the accumulation of nitrite in the culture supernatants of cells treated with or without *B. bifurcata* dichloromethane extract (6.25–100 µg mL^{-1}) in the presence or absence of LPS, using a colorimetric reaction with the Griess reagent [59]. Briefly, 170 µL of culture supernatants were diluted with equal volumes of the Griess reagent [0.1% (*w/v*) N-(1-naphthyl)-ethylenediamine dihydrochloride and 1% (*w/v*) sulphanilamide containing 5% (*w/v*) H3PO4] and maintained during 30 min, in the dark. The absorbance at 550 nm was measured in a Biotek Synergy HT plate reader. Culture medium was used as blank and nitrite concentration was determined from a regression analysis using serial dilutions of sodium nitrite as standard (0.5–50 µM).

- Statistical Analysis

Statistical analysis was performed using GraphPad Prism 6.0 for Mac OS X (GraphPad Software, San Diego, CA, USA). For each experiment, the results are expressed as mean ± SD of, at least, three independent experiments. Evaluation of statistical significance was performed using one-way ANOVA with *Dunnett's* multiple comparison test. Values of $p < 0.05$ were considered statistically significant.

3.6. Antibacterial Activity

3.6.1. Bacterial Strains

Bacterial stocks of Gram-positive, *Staphylococcus aureus* ATCC®6538, *S. aureus* ATCC®43300 and *S. epidermidis* and Gram-negative *Escherichia coli* ATCC®25922 and *Pseudomonas aeruginosa* PAOI were kept in Brucella Broth (Fluka Chemie, Madrid, Spain) with 20% (*v/v*) glycerol (Sigma, Madrid, Spain) at −80 °C.

Pre-inoculum was prepared by stocks defrosting at room temperature, suspension in Mueller Hinton Broth (MHB; Liofilchem, Roseto degli Abruzzi, Italy) followed by a 6 h incubation at 37 °C, 220 rpm. Afterwards, bacteria were streaked onto Mueller Hinton Agar (MHA; Liofilchem, Roseto degli Abruzzi, Italy) and incubated overnight at 37 °C. Then, bacteria were harvested with sterile peptone water (Liofilchem), washed twice by centrifugation at 2700 rpm and bacterial pellet suspended in MHB.

3.6.2. Growth Kinetics

For each strain, growth curves were performed in three independent experiments in order to determine the respective exponential phase. Colonies from MHA were harvested with sterile peptone water and bacterial pre-inoculum was prepared as mentioned above. The initial optical density of each bacterial strain was adjusted to 0.04 in the referred media (λ = 600 nm). T-flasks (Starsted, Numbrecht, Germany) containing media and bacterial inoculums were incubated with agitation (220 rpm) at 37 °C. At different time points, samples were taken and the optical density was measured at λ = 600 nm.

The correlation between the obtained optical density values with the number of colony-forming units per mL (CFU mL^{-1}) was accessed. For CFU mL^{-1} determination, bacteria were harvested from liquid media at different time points; serial dilutions were done in peptone water (10^{-2} until 10^{-7}) and 10 µL of each dilution plated in MHA. Incubation was done as previously described and the number of CFUs was determined after 24 h. Experiments were performed twice and in triplicate (n = 6 for each experiment).

3.6.3. Minimal Inhibitory Concentration (MIC) Determination

MIC assays were performed in accordance to the Clinical and Laboratory Standards Institute (CLSL) guidelines [60]. Briefly, bacterial cells in exponential growth phase (about 2 h of incubation, previously defined in Section 3.6.2) were suspended in MHB (Fluka Chemie) and the *B. bifurcata* dichloromethane extract was dissolved in dimethyl sulfoxide (DMSO; AppliChem, Darmstadt, Germany) to a final stock concentration of 50 mg mL^{-1}. The antibacterial performance of the extract was screened using the microbroth dilution method in a range of concentrations from 8 to 2048 µg mL^{-1} in 96 well-plates (Starsted, Numbrecht, Germany) [61]. The MIC was qualitatively determined after 24 h of incubation, at 37 °C, by addition of 3-(4,5-dimethylthiazol-2-yl)-2,5-diphenyltetrazolium bromide (MTT, Calbiochem® supplied by Millipore, Madrid, Spain) with slight adaptations to the protocol described by Ellof et al. [62]. The use of MTT, metabolized to a formazan chromophore by the viable cells, was required for MIC determination due to the coloration of the extract that turned the naked eye MIC visualization impracticable. The following controls were performed: (i) solvent control: bacterial cultures with 4% (v/v) of DMSO, which represents the maximum amount of solvent added to the cultures, to infer its effect on bacterial growth; (ii) pure cultures (only bacterial inoculum); and (iii) culture media. Experiments were performed twice and in triplicate (n = 6 for each experiment).

3.6.4. Synergistic Assays

The synergistic potential of the extract was accessed by conjugation of the extract with antibiotics, namely: rifampicin. gentamicin, ampicillin, and tetracycline (all supplied by Sigma, Madrid, Spain) in a concentration range between 2 and 512 µg mL^{-1}; against the same bacterial panel. *B. bifurcata* extract was used at the respective MIC concentration, calculated as described above. Antibiotics MIC, when used alone or in extract + antibiotic combination were determined as mentioned above. Experiments were performed twice and in triplicate (n = 6 for each experiment).

4. Conclusions

The complete study of the lipophilic fraction of *B. bifurcata* from a Portuguese aquaculture system was characterized in detail by GC-MS. *B. bifurcata* dichloromethane extract was composed mainly of diterpenes (1892.78 ± 133.97 mg kg^{-1} DW), followed by fatty acids (947.88 ± 21.94 mg kg^{-1} DW). Considerable amounts of sterols (406.45 ± 26.19 mg kg^{-1} DW) were also found, with fucosterol accounting for 317.68 ± 26.11 mg kg^{-1} DW. The sterols, long-chain aliphatic alcohols, and monoglycerides profile of *B. bifurcata* is reported for the first time. Additionally, two diterpenes, phytol and neophytadiene, are also reported for the first time as constituents of

B. bifurcata. This extract showed relevant antioxidant activities against both DPPH and ABTS radicals. Additionally, antibacterial activity was shown to be strain-dependent, with activity being verified against both Gram-positive (*S. aureus*) and Gram-negative (*E. coli*) strains. *B. bifurcata* extract was shown to increase antibiotic antimicrobial activity, with this effect being microbial strain- and antibiotic-dependent. This study provides the basis for further exploitation of this alga's biomolecules for current antimicrobial chemotherapeutics, showing the practical utility of marine natural products to sensitize pathogenic bacteria, increasing their susceptibility. In addition, the studied extract showed promising anti-inflammatory activity, which is reported for the first time. It is important to point out that all biological activities were expressed using low extract concentrations, on the order of μg mL^{-1}. This study is, therefore, an important contribution for the valorization of *B. bifurcata* macroalga from aquaculture systems, with promising applications in the functional food, nutraceutical, cosmetic, and biomedical fields, obviously considering the development of eco-friendly methodologies to extract this lipophilic fraction.

Acknowledgments: The authors wish to thank to FCT-Portugal (Fundação para a Ciência e Tecnologia) and POPH/FSE for the postdoctoral grant to S.A.O.S. (SFRH/BPD/84226/2012), PhD grant to I.F. (SFRH/BD/110717/2015), for CICECO-Aveiro Institute of Materials, POCI-01-0145-FEDER-007679 (FCT Ref. UID/CTM/50011/2013), QOPNA research unit (FCT UID/QUI/00062/2013) financed by national funds through the FCT/MEC and when applicable co-financed by FEDER under the PT2020 Partnership Agreement and CNC financed by the European Regional Development Fund (ERDF), through the Centro 2020 Regional Operational Program: project CENTRO-01-0145-FEDER-000012-HealthyAging2020, the COMPETE 2020—Operational Program for Competitiveness and Internationalization, and the Portuguese national funds via FCT—Fundação para a Ciência e a Tecnologia, I.P.: project POCI-01-0145-FEDER-007440. The work was also supported by FEDER funds through the Operational Program Competitiveness Factors—COMPETE and national funds by FCT under the strategic project UID/NEU/04539/2013. M.F.D. also acknowledges FCT for the financial support under the Project UID/AGR/00115/2013. Part of the work of ALGAplus team was funded by the project SEACOLORS, LIFE13 ENV/ES/000445; the team also wishes to thank intern Raquel Silva.

Author Contributions: S.A.O.S. contributed to the GC-MS analysis, collected and interpreted all the data, and wrote the manuscript. S.S.T. and C.S.D.O. performed the extractions, extracts derivatization, GC-MS analysis, and antioxidant activity tests. P.P., D.R. and M.F.D. performed the antibacterial activity analysis and helped in the data interpretation. I.F. and M.T.C. performed the anti-inflammatory activity analysis and helped in the data interpretation. M.H.A. and A.M.R. contributed the cultivation of the algae and information on the processes for obtaining biomass. Also, they provided information on previous results from this species at ALGAplus and reviewed the manuscript; S.M.R. and A.J.D.S. (together with S.A.O.S.) were responsible for the conception and design of the work, also performing a critical revision of the manuscript. All authors revised the article and approved the version to be submitted.

Conflicts of Interest: The authors declare no conflict of interest. The funding sponsors had no role in the design of the study; in the collection, analyses, or interpretation of data; in the writing of the manuscript, and in the decision to publish the results.

References

1. Reuter, S.; Gupta, S.C.; Chaturvedi, M.M.; Aggarwal, B.B. Oxidative stress, inflammation, and cancer: How are they linked? *Free Radic. Biol. Med.* **2010**, *49*, 1603–1616. [CrossRef] [PubMed]

2. Coussens, L.M.; Werb, Z. Inflammation and cancer. *Nature* **2002**, *420*, 860–867. [CrossRef] [PubMed]

3. Davies, J.; Davies, D. Origins and evolution of antibiotic resistance. *Microbiol. Mol. Biol. Rev.* **2010**, *74*, 417–433. [CrossRef] [PubMed]

4. Hemaiswarya, S.; Kruthiventi, A.K.; Doble, M. Synergism between natural products and antibiotics against infectious diseases. *Phytomedicine* **2008**, *15*, 639–652. [CrossRef] [PubMed]

5. Mishra, B.B.; Tiwari, V.K. Natural products: An evolving role in future drug discovery. *Eur. J. Med. Chem.* **2011**, *46*, 4769–4807. [CrossRef] [PubMed]

6. Blunt, J.W.; Copp, B.R.; Keyzers, R.A.; Munro, M.H.G.; Prinsep, M.R. Marine natural products. *Nat. Prod. Rep.* **2014**, *31*, 160–258. [CrossRef] [PubMed]

7. Carvalho, L.G.; Pereira, L. Review of marine algae as source of bioactive metabolites. In *Marine Algae Bodiversity, Taxonomy, Environmental Assessment, and Biotechnology*; Pereira, L., Neto, J.M., Eds.; CRC Press: Boca Raton, FL, USA, 2014; pp. 195–227.

8. Plaza, M.; Cifuentes, A.; Ibanez, E. In the search of new functional food ingredients from algae. *Trends Food Sci. Technol.* **2008**, *19*, 31–39. [CrossRef]
9. Van Hal, J.W.; Huijgen, W.J.J.; López-Contreras, A.M. Opportunities and challenges for seaweed in the biobased economy. *Trends Biotechnol.* **2014**, *32*, 231–233. [CrossRef] [PubMed]
10. FAO Yearbook of Fishery Statistics. Available online: http://www.fao.org/3/a-i5716t.pdf (accessed on 4 April 2017).
11. Martins, C.D.L.; Ramlov, F.; Nocchi Carneiro, N.P.; Gestinari, L.M.; dos Santos, B.F.; Bento, L.M.; Lhullier, C.; Gouvea, L.; Bastos, E.; Horta, P.A.; et al. Antioxidant properties and total phenolic contents of some tropical seaweeds of the Brazilian coast. *J. Appl. Phycol.* **2012**, *25*, 1179–1187. [CrossRef]
12. Fleurence, J.; Gutbier, G.; Mabeau, S.; Leray, C. Fatty acids from 11 marine macroalgae of the French Brittany coast. *J. Appl. Phycol.* **1994**, *6*, 527–532. [CrossRef]
13. Tabarsa, M.; Rezaei, M.; Ramezanpour, Z.; Robert Waaland, J.; Rabiei, R. Fatty acids, amino acids, mineral contents, and proximate composition of some brown seaweeds. *J. Phycol.* **2012**, *48*, 285–292. [CrossRef] [PubMed]
14. Santos, S.A.O.; Vilela, C.; Freire, C.S.R.; Abreu, M.H.; Rocha, S.M.; Silvestre, A.J.D. Chlorophyta and Rhodophyta macroalgae: A source of health promoting phytochemicals. *Food Chem.* **2015**, *183*, 122–128. [CrossRef] [PubMed]
15. Santos, S.A.O.; Oliveira, C.S.D.; Trindade, S.S.; Abreu, M.H.; Rocha, S.S.M.; Silvestre, A.J.D. Bioprospecting for lipophilic-like components of five Phaeophyta macroalgae from the Portuguese coast. *J. Appl. Phycol.* **2016**, 1–8. [CrossRef]
16. Lopes, G.; Sousa, C.; Bernardo, J.; Andrade, P.B.; Valentão, P.; Ferreres, F.; Mouga, T. Sterol profiles in 18 macroalgae of the Portuguese coast. *J. Phycol.* **2011**, *47*, 1210–1218. [CrossRef] [PubMed]
17. Andrade, P.B.; Barbosa, M.; Matos, R.P.; Lopes, G.; Vinholes, J.; Mouga, T.; Valentão, P. Valuable compounds in macroalgae extracts. *Food Chem.* **2013**, *138*, 1819–1828. [CrossRef] [PubMed]
18. Melo, T.; Alves, E.; Azevedo, V.; Martins, A.S.; Neves, B.; Domingues, P.; Calado, R.; Abreu, M.H.; Domingues, M.R. Lipidomics as a new approach for the bioprospecting of marine macroalgae—Unraveling the polar lipid and fatty acid composition of *Chondrus crispus*. *Algal Res.* **2015**, *8*, 181–191. [CrossRef]
19. Fundação para a Ciência e a Tecnologia Estratégia de Investigação e Inovação para uma Especialização Inteligente 2014–2020. Available online: https://www.fct.pt/gabestudosestrategia/ENEI/docs/ENEI_Julho2014_aposconsulta_VF_completa.pdf (accessed on 27 February 2017).
20. Gupta, S.; Abu-Ghannam, N. Bioactive potential and possible health effects of edible brown seaweeds. *Trends Food Sci. Technol.* **2011**, *22*, 315–326. [CrossRef]
21. Bourgougnon, N.; Stiger-Pouvreau, V. Chemodiversity and Bioactivity within Red and Brown Macroalgae Along the French coasts, Metropole and Overseas Departements and Territories. In *Handbook of Marine Macroalgae*; John Wiley & Sons, Ltd.: Chichester, UK, 2011; pp. 58–105. ISBN 9781119977087.
22. Muñoz, J.; Culioli, G.; Köck, M. Linear diterpenes from the marine brown alga *Bifurcaria bifurcata*: A chemical perspective. *Phytochem. Rev.* **2013**, *12*, 407–424. [CrossRef]
23. Culioli, G.; Daoudi, M.; Ortalo-magne, A.; Valls, R.; Piovetti, L. (*S*)-12-Hydroxygeranylgeraniol-derived diterpenes from the brown alga *Bifurcaria bifurcata*. *Phytochemistry* **2001**, *57*, 529–535. [CrossRef]
24. Valls, R.; Piovetti, L.; Banaigs, B.; Archavlis, A.; Pellegrini, M. (*S*)-13-hydroxygeranylgeraniol-derived furanoditerpenes from *Bifurcaria bifurcata*. *Phytochemistry* **1995**, *39*, 145–149. [CrossRef]
25. Ortalo-Magné, A.; Culioli, G.; Valls, R.; Pucci, B.; Piovetti, L. Polar acyclic diterpenoids from *Bifurcaria bifurcata* (Fucales, Phaeophyta). *Phytochemistry* **2005**, *66*, 2316–2323. [CrossRef] [PubMed]
26. Culioli, G.; Daoudi, M.; Mesguiche, V.; Valls, R.; Piovetti, L. Geranylgeraniol-derived diterpenoids from the brown alga *Bifurcaria bifurcata*. *Phytochemistry* **1999**, *52*, 1447–1454. [CrossRef]
27. Göthel, Q.; Muñoz, J.; Köck, M. Formyleleganolone and bibifuran, two metabolites from the brown alga *Bifurcaria bifurcata*. *Phytochem. Lett.* **2012**, *5*, 693–695. [CrossRef]
28. Biard, J.; Verbist, J.; Letourneux, Y.; Floch, R. Cétols Diterpeniques à Activité Antimicrobienne de *Bifurcaria bifurcata*. *Planta Med.* **1980**, *40*, 288–294. [CrossRef] [PubMed]
29. Valls, R.; Banaigs, B.; Piovetti, L.; Archavlis, A.; Artaud, J. Linear diterpene with antimitotic activity from the brown alga *Bifurcaria bifurcata*. *Phytochemistry* **1993**, *34*, 1585–1588. [CrossRef]

30. Moreau, D.; Thomas-Guyon, H.; Jacquot, C.; Jugé, M.; Culioli, G.; Ortalo-Magné, A.; Piovetti, L.; Roussakis, C. An extract from the brown alga *Bifurcaria bifurcata* induces irreversible arrest of cell proliferation in a non-small-cell bronchopulmonary carcinoma line. *J. Appl. Phycol.* **2006**, *18*, 87–93. [CrossRef]

31. Alves, C.; Pinteus, S.; Simões, T.; Horta, A.; Silva, J.; Tecelão, C.; Pedrosa, R. *Bifurcaria bifurcata*: A key macro-alga as a source of bioactive compounds and functional ingredients. *Int. J. Food Sci. Technol.* **2016**, *51*, 1638–1646. [CrossRef]

32. Le Lann, K.; Rumin, J.; Cérantola, S.; Culioli, G.; Stiger-Pouvreau, V. Spatiotemporal variations of diterpene production in the brown macroalga *Bifurcaria bifurcata* from the western coasts of Brittany (France). *J. Appl. Phycol.* **2014**, *26*, 1207–1214. [CrossRef]

33. Daoudi, M.; Bakkas, S.; Culioli, G.; Ortalo-Magné, A.; Piovetti, L.; Guiry, M.D. Acyclic diterpenes and sterols from the genera Bifurcaria and Bifurcariopsis (Cystoseiraceae, Phaeophyceae). *Biochem. Syst. Ecol.* **2001**, *29*, 973–978. [CrossRef]

34. Valls, R.; Mesguiehe, V.; Pellegrini, M.; Pellegrini, L.; Banaigs, B. Variation de la composition en diterpènes de *Bifurcaria bifurcata* sur les côtes atlantiques françaises. *Acta Bot. Gall.* **1995**, *142*, 119–124. [CrossRef]

35. Touati, R.; Santos, S.A.O.; Rocha, S.M.; Belhamel, K.; Silvestre, A.J.D. *Retama sphaerocarpa*: An unexploited and rich source of alkaloids, unsaturated fatty acids and other valuable phytochemicals. *Ind. Crops Prod.* **2015**, *69*, 238–243. [CrossRef]

36. Vilela, C.; Santos, S.A.O.; Oliveira, L.; Camacho, J.F.; Cordeiro, N.; Freire, C.S.R.; Silvestre, A.J.D. The ripe pulp of *Mangifera indica* L.: A rich source of phytosterols and other lipophilic phytochemicals. *Food Res. Int.* **2013**, *54*, 1535–1540. [CrossRef]

37. Silva, R.G. *Estudos em Bifurcaria bifurcata: Aquacultura, Atividade Antioxidante e Antifungica*; Universidade de Coimbra: Coimbra, Portugal, 2014.

38. Culioli, G.; Mesguiche, V.; Piovetti, L.; Valls, R. Geranylgeraniol and geranylgeraniol-derived diterpenes from the brown alga *Bifurcaria bifurcata* (Cystoseiraceae). *Biochem. Syst. Ecol.* **1999**, *27*, 665–668. [CrossRef]

39. Culioli, G.; Ortalo-Magné, A.; Richou, M.; Valls, R.; Piovetti, L. Seasonal variations in the chemical composition of *Bifurcaria bifurcata* (Cystoseiraceae). *Biochem. Syst. Ecol.* **2002**, *30*, 61–64. [CrossRef]

40. González-Burgos, E.; Gómez-Serranillos, M.P. Terpene compounds in nature: A review of their potential antioxidant activity. *Curr. Med. Chem.* **2012**, *19*, 5319–5341. [CrossRef] [PubMed]

41. Gouveia, V.; Seca, A.M.L.; Barreto, M.C.; Pinto, D.C.G.A. Di- and sesquiterpenoids from Cystoseira genus: Structure, intra-molecular transformations and biological activity. *Mini Rev. Med. Chem.* **2013**, *13*, 1150–1159. [CrossRef] [PubMed]

42. Manzo, E.; Ciavatta, M.L.; Bakkas, S.; Villani, G.; Varcamonti, M.; Zanfardino, A.; Gavagnin, M. Diterpene content of the alga Dictyota ciliolata from a Moroccan lagoon. *Phytochem. Lett.* **2009**, *2*, 211–215. [CrossRef]

43. Abrantes, J.L.; Barbosa, J.; Cavalcanti, D.; Pereira, R.C.; Frederico Fontes, C.L.; Teixeira, V.L.; Moreno Souza, T.L.; Paixão, I.C.P. The effects of the diterpenes isolated from the Brazilian brown algae Dictyota pfaffii and Dictyota menstrualis against the herpes simplex type-1 replicative cycle. *Planta Med.* **2010**, *76*, 339–344. [CrossRef] [PubMed]

44. Awad, N.E. Bioactive Brominated Diterpenes from the Marine Red Alga *Jania rubens* (L.) Lamx. *Phyther. Res.* **2004**, *279*, 275–279. [CrossRef] [PubMed]

45. Ravi, B.; Murphy, P.; Lidgard, R.; Warren, R.; Wells, R. C18 terpenoid metabolites of the brown alga *Cystophora moniliformis*. *Aust. J. Chem.* **1982**, *35*, 171–182. [CrossRef]

46. Reddy, P.; Urban, S. Linear and cyclic C18 terpenoids from the southern Australian marine brown alga *Cystophora moniliformis*. *J. Nat. Prod.* **2008**, *71*, 1441–1446. [CrossRef] [PubMed]

47. Wall, R.; Ross, R.P.; Fitzgerald, G.F.; Stanton, C. Fatty acids from fish: The anti-inflammatory potential of long-chain omega-3 fatty acids. *Nutr. Rev.* **2010**, *68*, 280–289. [CrossRef] [PubMed]

48. Kumari, P.; Kumar, M.; Gupta, V.; Reddy, C.R.K.; Jha, B. Tropical marine macroalgae as potential sources of nutritionally important PUFAs. *Food Chem.* **2010**, *120*, 749–757. [CrossRef]

49. Chen, B.; McClements, D.J.; Decker, E.A. Design of foods with bioactive lipids for improved health. *Annu. Rev. Food Sci. Technol.* **2013**, *4*, 35–56. [CrossRef] [PubMed]

50. Hoang, M.-H.; Jia, Y.; Jun, H.; Lee, J.H.; Lee, B.Y.; Lee, S.-J. Fucosterol is a selective liver X receptor modulator that regulates the expression of key genes in cholesterol homeostasis in macrophages, hepatocytes, and intestinal cells. *J. Agric. Food Chem.* **2012**, *60*, 11567–11575. [CrossRef] [PubMed]

51. Jung, H.A.; Jin, S.E.; Ahn, B.R.; Lee, C.M.; Choi, J.S. Anti-inflammatory activity of edible brown alga *Eisenia bicyclis* and its constituents fucosterol and phlorotannins in LPS-stimulated RAW264.7 macrophages. *Food Chem. Toxicol.* **2013**, *59*, 199–206. [CrossRef] [PubMed]

52. Hwang, E.; Park, S.-Y.; Sun, Z.; Shin, H.-S.; Lee, D.-G.; Yi, T.H. The protective effects of fucosterol against skin damage in UVB-irradiated human dermal fibroblasts. *Mar. Biotechnol.* **2014**, *16*, 361–370. [CrossRef] [PubMed]

53. Islam, M.N.; Ishita, I.J.; Jin, S.E.; Choi, R.J.; Lee, C.M.; Kim, Y.S.; Jung, H.A.; Choi, J.S. Anti-inflammatory activity of edible brown alga *Saccharina japonica* and its constituents pheophorbide a and pheophytin a in LPS-stimulated RAW 264.7 macrophage cells. *Food Chem. Toxicol.* **2013**, *55*, 541–548. [CrossRef] [PubMed]

54. Hwang, J.-H.; Oh, Y.-S.; Lim, S.-B. Anti-inflammatory activities of some brown marine algae in LPS-stimulated RAW 264.7 cells. *Food Sci. Biotechnol.* **2014**, *23*, 865–871. [CrossRef]

55. Otto, M. *Staphylococcus epidermidis*—The "accidental" pathogen. *Nat. Rev. Microbiol.* **2009**, *7*, 555–567. [CrossRef] [PubMed]

56. Lewis, K. Platforms for antibiotic discovery. *Nat. Rev. Drug Discov.* **2013**, *12*, 371–387. [CrossRef] [PubMed]

57. Freire, C.S.R.; Silvestre, A.J.D.; Neto, C.P.; Cavaleiro, J.A.S. Lipophilic extractives of the inner and outer barks of *Eucalyptus globulus*. *Holzforschung* **2002**, *56*, 372–379. [CrossRef]

58. O'Brien, J.; Wilson, I.; Orton, T.; Pognan, F. Investigation of the Alamar Blue (resazurin) fluorescent dye for the assessment of mammalian cell cytotoxicity. *Eur. J. Biochem.* **2000**, *267*, 5421–5426. [CrossRef] [PubMed]

59. Green, L.C.; Wagner, D.A.; Glogowski, J.; Skipper, P.L.; Wishnok, J.S.; Tannenbaum, S.R. Analysis of nitrate, nitrite, and [15N]nitrate in biological fluids. *Anal. Biochem.* **1982**, *126*, 131–138. [CrossRef]

60. CLSI. *Performance Standards for Antimicrobial Susceptibility Testing; Twenty-Fifth Informational Supplement*; Clinical and Laboratory Standards Institute: Wayne, PA, USA, 2015; ISBN 1562387855.

61. Wiegand, I.; Hilpert, K.; Hancock, R.E.W. Agar and broth dilution methods to determine the minimal inhibitory concentration (MIC) of antimicrobial substances. *Nat. Protoc.* **2008**, *3*, 163–175. [CrossRef] [PubMed]

62. Ellof, J. A sensitive and quick microplate method to determine the minimal inhibitory concentration of plant extracts for bacteria. *Planta Med.* **1998**, *64*, 711–713. [CrossRef] [PubMed]

marine drugs

MDPI

Article

Chemical Profiling and Bioactivity of Body Wall Lipids from *Strongylocentrotus droebachiensis*

Alexander N. Shikov [1,*], Into Laakso [2], Olga N. Pozharitskaya [1], Tuulikki Seppänen-Laakso [3], Anna S. Krishtopina [1], Marina N. Makarova [1], Heikki Vuorela [2] and Valery Makarov [1]

[1]   Saint-Petersburg Institute of Pharmacy, Leningrad Region, Vsevolozhsky District, Kuzmolovo P 245, 188663 Saint-Petersburg, Russia; olgapozhar@mail.ru (O.N.P.); annakrishtopina@list.ru (A.S.K.); mmn2410@yandex.ru (M.N.M.); makarov.vg@doclinika.ru (V.M.)

[2]   Division of Pharmaceutical Biosciences, Faculty of Pharmacy, University of Helsinki, P.O. Box 56 (Viikinkaari 5E), FI-00014 Helsinki, Finland; into.laakso@helsinki.fi (I.L.); heikki.vuorela@helsinki.fi (H.V.)

[3]   VTT Technical Research Centre of Finland Ltd., P.O. Box 1000 (Tietotie 2), FI-02044 VTT Espoo, Finland; tuulikki.seppanen-laakso@vtt.fi

*   Correspondence: spbpharm@mail.ru; Tel.: +7-812-603-2432

Received: 31 August 2017; Accepted: 15 November 2017; Published: 24 November 2017

Abstract: The lipids from gonads and polyhydroxynaphthoquinone pigments from body walls of sea urchins are intensively studied. However, little is known about the body wall (BW) lipids. Ethanol extract (55 °C) contained about equal amounts of saturated (SaFA) and monounsaturated fatty acids (MUFA) representing 60% of total fatty acids, with myristic, palmitic and eicosenoic acids as major SaFAs and MUFAs, respectively. Non-methylene-interrupted dienes (13%) were composed of eicosadienoic and docosadienoic acids. Long-chain polyunsaturated fatty acids (LC-PUFA) included two main components, n6 arachidonic and n3 eicosapentaenoic acids, even with equal concentrations (15 μg/mg) and a balanced n6/n3 PUFA ratio (0.86). The UPLC-ELSD analysis showed that a great majority of the lipids (80%) in the ethanolic extract were phosphatidylcholine (60 μg/mg) and phosphatidylethanolamine (40 μg/mg), while the proportion of neutral lipids remained lower than 20%. In addition, alkoxyglycerol derivatives—chimyl, selachyl, and batyl alcohols—were quantified. We have assumed that the mechanism of action of body wall lipids in the present study is via the inhibition of MAPK p38, COX-1, and COX-2. Our findings open the prospective to utilize this lipid fraction as a source for the development of drugs with anti-inflammatory activity.

Keywords: sea urchin; body wall lipids; inhibition of p38 MAPK; COX; GC-MS; UPLC-ELSD

1. Introduction

Sea urchins have wide distribution and they play an important role in ecosystem of both shallow and deeper waters of the ocean. A number of species of sea urchin are intensively utilized in food, pharmaceutical, and cosmetic industries. In 2015, the total world production of edible sea urchin was 71,229 tons [1]. Green sea urchin, *Strongylocentrotus droebachiensis*, is an edible species of the phylum *Echinodermata*, which is a typical inhabitant of the polar region of Russia, including the Barents Sea. The gonads are delicacies in many parts of the world and considered as highly valued seafood. Several studies have indicated that the gonads of *S. droebachiensis* are rich in important bioactive compounds like polyunsaturated fatty acids (PUFA), phospholipids, tocopherols, sterols [2–4], carotenoids [2,5], and amino acids [6,7]. Extract of gonad tissue has also revealed effective anti-inflammatory and antidiabetic properties [4].

After removal of gonads, the residual shells and spines (body wall, BW) of sea urchins which account for more than 40% of total body weight are discarded as waste. Previous studies have also shown that sea urchin BW contain polyhydroxynaphthoquinones. These pigments have evoked renewed interest as a promising source for the development of drugs. A number of bioactivities have been found, for example antiallergic [8], antidiabetic [9], antihypertensive [10], anti-inflammatory [11], antioxidant [12–14], cardioprotective [15], and hypocholesterolemic [16] effects.

Total concentration of pigments in sea urchin body wall is quite low (1.2–1.6 mg/g) [17], however, a purification method has been recently reported in order to improve the yield of shell pigments [18]. The inner layer of the body wall, on the contrary, is covered with a biomembrane, consisting of lipids. However, their profile has not been described yet. Marine lipids are unique sources of essential fatty acids, phospholipids, sterols, and alkoxyglycerols, and they have a broad pharmacological activity.

The aim of this study was to analyze the ethanolic extract of lipids underlying the body wall (BW) of green sea urchin by using gas chromatography-mass spectrometry (GC-MS) and ultra-performance liquid chromatography (UPLC-ELSD). In addition, the anti-inflammatory potential of BW lipids was investigated in vitro.

2. Results

2.1. Fatty Acid Composition of Body Wall (BW) Lipids

The ethanolic (95%) extract, which was prepared at 55 °C for 3 h, contained about equal amounts of saturated (SaFA) and monounsaturated fatty acids (MUFA) representing 60% of total fatty acids. Myristic, palmitic, and eicosenoic acids were the main SaFAs and MUFAs, respectively (Table 1 and Figure 1). The principal MUFA is unusual and this isomer is suggested to be 20:1n15. Among non-methylene-interrupted dienes (NMID, 13%), eicosadienoic (20:2) and docosadienoic (22:2) acids were characteristic. They are typical constituents of sea urchin among which 20:2Δ5,11 often appears as the most abundant isomer. In addition, n12 and n5 isomers of 18:1, 20:3n9 and cyclopropaneoctanoic and -decanoic acid 2-octyl methyl esters accounted 10 µg/mg.

The composition of long-chain polyunsaturated fatty acids (LC-PUFA) was characterized by two major components, n6 arachidonic (20:4n6; AA) and n3 eicosapentaenoic acids (20:5n3; EPA), with equal concentrations (15 µg/mg) and a balanced n6/n3 PUFA ratio (0.86). These LC-PUFAs possess important properties, since they act as the precursors of eicosanoids.

A very low amount of docosahexaenoic acid (22:6n3, DHA) was typical for BW lipid extract (Table 1), as well as the high proportion of free (22%) vs. esterified fatty acids. Quantitatively, free AA covered 1/3 of total AA and free EPA 1/4 of total EPA, respectively, reflecting decomposition of bound fatty acids during extraction procedure. The relatively high proportion of LC-PUFAs, n6 AA, and n3 EPA (20%), would suggest that the extract is rich in phospholipids.

Table 1. The concentrations (µg/mg; mean ± SD, *n* = 3) and relative amounts (%) of bound and free fatty acids in ethanolic extract of BW lipids of sea urchin determined by GC-MS.

Fatty Acids	Bound Fatty Acids as FAME	%	FFA	%
C10:0–C13:0	0.4 ± 0.1	0.3	-	-
C14:0	15.1 ± 0.7	9.6	3.7 ± 0.1	8.4
C15:0	1.5 ± 0.1	0.9	0.5 ± 0.1	1.1
C16:0	21.9 ± 1.2	13.9	5.0 ± 0.2	11.4
C18:0	4.0 ± 0.2	2.5	2.1 ± 0.1	4.8
C19:0	1.0 ± 0.1	0.6	-	-
C20:0	0.6 ± 0.1	0.4	-	-
Σ SaFAΣ	44.5 ± 2.1	28.2	11.3 ± 0.3	25.7
C14:1n5	1.1 ± 0.1	0.7	-	-
C16:1n9	0.9 ± 0.1	0.6	-	-
C16:1n7	7.8 ± 0.6	4.9	1.9 ± 0.1	4.3
C16:1n5	4.5 ± 0.2	2.8	-	-
C18:1n9	3.0 ± 0.2	1.9	1.4 ± 0.1	3.2

Table 1. *Cont.*

Fatty Acids	Bound Fatty Acids as FAME	%	FFA	%
C18:1n7	4.2 ± 0.2	2.7	2.9 ± 0.1	6.6
C20:1n15	16.7 ± 0.8	10.6	-	-
C20:1n9	5.2 ± 0.3	3.3	8.2 ± 0.4	18.6
C20:1n7	1.3 ± 0.1	0.8	-	-
C22:1n9	4.3 ± 0.2	2.7	-	-
Σ MUFA	49.0 ± 1.3	31.0	14.4 ± 0.4	32.7
20:2Δ5,11	12.2 ±1.2	7.7	-	-
20:2Δ5,13	3.0 ± 0.5	1.9	-	-
22:2Δ7,13	0.9 ± 0.2	0.6	-	-
22:2Δ7,15	4.7 ± 0.4	3.0	-	-
Σ NMID	20.9 ± 1.9	13.2	-	-
C18:2n6	1.5 ± 0.1	0.9	5.2 ± 0.2	11.8
C20:2n6	1.9 ± 0.1	1.2	-	-
C20:3n6	0.9 ± 0.1	0.6	-	-
C20:4n6	15.8 ± 0.6	10.0	7.3 ± 0.1	16.6
Σ n6 PUFA	20.1 ± 1.0	12.7	12.5 ± 0.8	28.4
C18:3n3	1.2 ± 0.1	0.8	-	-
C18:4n3	2.3 ± 0.1	1.5	-	-
C20:3n3	2.3 ± 0.2	1.5	-	-
C20:4n3	0.6 ± 0.1	0.4	-	-
C20:5n3	15.2 ± 0.7	9.6	5.8 ± 1.0	13.2
C22:5n3	0.2 ± 0.1	0.1	-	-
C22:6n3	1.7 ± 0.1	1.1	-	-
Σ n3 PUFA	23.5 ± 1.1	14.9	5.8 ± 1.0	13.2
Σ Fatty acids	158.0 ± 5.3	100.0	44.0 ± 1.8	100.0
n6/n3 PUFA	0.86		2.16	

FAME, fatty acid methyl ester; FFA, free fatty acid; SaFA, saturated fatty acid; MUFA, mono-unsaturated fatty acid; NMID, non-methylene-interrupted diene; PUFA, polyunsaturated fatty acid.

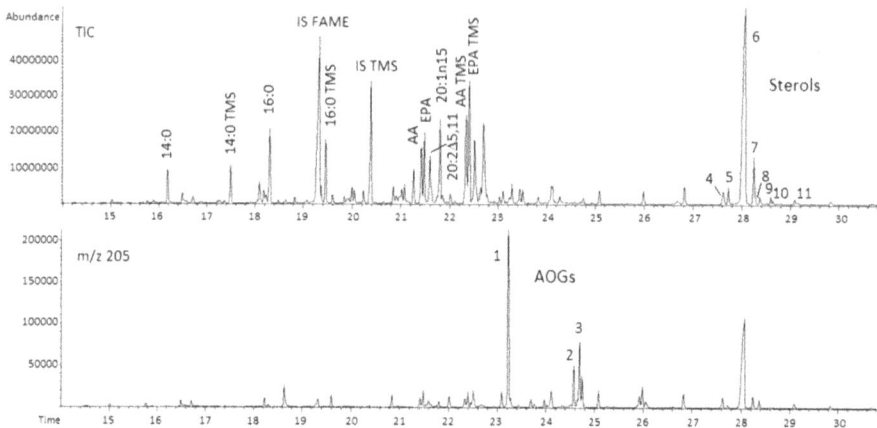

Figure 1. GC-MS analysis of transesterified and trimethylsilylated (TMS) fatty acids from ethanolic extract of BW lipids of sea urchin. Total ion (TIC) and extracted ion chromatogram (*m/z* 205) shows TMS derivatives of alkylglycerols (AOG; (1) 16:0-AOG, (2) 18:1-AOG, and (3) 18:0-AOG) and sterols (peaks 4–11, TIC): (4–5) unidentified sterols, (6) cholesterol, (7) desmosterol, (8) cholecalciferol as shoulder, (9) campesterol, (10) stigmasterol, and (11) clionasterol. Heptadecanoic acid (as FAME and TMS derivative) was used as internal standard (IS). Other peaks represent FAMEs and TMS ethers.

2.2. Sterols and Alkoxyglycerols (AOG)

Sterol and AOG samples were analyzed by GC-MS as TMS-derivatives. BW lipid extract contained mainly cholesterol (50 µg/mg) (Figure 1 and Table 2). Minor non-cholesterol sterols included desmosterol, campesterol, stigmasterol, and clionasterol (gamma-sitosterol) which is a common constituent in oyster, for example [19]. For GC-MS analyses from abundant AOG sources, purified samples from unsaponifiable fraction have been used. Because of low concentration, extracted ion chromatogram (m/z 205) of silylated AOG was taken to confirm the peak location. This fragment is formed after cleavage between carbons 1 and 2 of the glycerol moiety [20]. Quantified AOG derivatives were chimyl (C16:0), selachyl (C18:1) and batyl alcohols (C18:0).

Table 2. Sterol and alkoxyglycerol (AOG) content (µg/mg; mean ± SD, n = 3) of ethanolic extract of BW lipids of sea urchin. Sterols and AOGs were determined as TMS ethers by GC-MS.

Sterols and AOGs	µg/mg
Cholesterol	50.3 ± 2.8
Non-cholesterol sterols *	10.6 ± 0.5
C16:0-AOG	1.0 ± 0.1
C18:1-AOG	0.4 ± 0.1
C18:0-AOG	0.3 ± 0.1
Σ Alkoxyglycerols	1.7 ± 0.2

* Non-cholesterol sterols include desmosterol, campesterol, stigmasterol, clionasterol, and two unidentified sterols (Figure 1).

2.3. Lipid Classes

The UPLC-ELSD analyses from the ethanol extract of BW lipids, shown in Figure 2 and Table 3, demonstrate high abundance of phospholipids (PL, 80%), especially those of phosphatidylcholine (PC, 60 µg/mg) and phosphatidylethanolamine (PE, 40 µg/mg). The proportion of neutral lipids remained less than 20%. It is clear that ethanol extracts polar PLs better than neutral lipids (NL) like triacylglycerols and cholesteryl esters. The fatty acid profile with relatively high content of PUFAs indicates that the fatty acids have mostly originated from PLs. Lysophosphatidylcholine (LPC) eluted late as a broad peak and covered about 8% of total lipids.

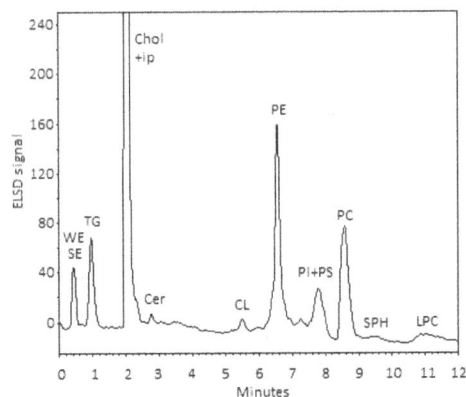

Figure 2. UPLC-ELSD chromatogram of BW lipids of ethanol (95%) extract of sea urchin. UPLC, ultra-performance liquid chromatography; ELSD, evaporative light scattering detector; WE, wax ester; SE, steryl ester; TG, triacylglycerol; Chol, cholesterol; ip, impurity; Cer, ceramide; CL, cardiolipin; PE, phosphatidylethanolamine; PI, phosphatidylinositol; PS, phosphatidylserine; PC, phosphatidylcholine; SPH, sphingomyelin; LPC, lysophosphatidylcholine.

Table 3. Concentration of major lipid classes (µg/mg, mean ± SD; *n* = 3) and their relative amounts (%) in ethanolic extract of BW lipids of sea urchin. Analyses were carried out by UPLC-ELSD.

Lipid Classes	µg/mg	%
WE + SE	14.6 ± 0.7	8.4
TG	15.9 ± 1.0	9.1
Σ Neutral lipids	30.5 ± 1.6	17.5
PE	38.6 ± 1.1	22.2
PI + PS	27.8 ± 0.1	16.0
PC	63.3 ± 1.7	36.4
LPC	13.5 ± 0.5	7.8
Σ Phospholipids	143.2 ± 0.7	82.5
Σ Total lipids	173.7 ± 2.3	100.0

WE, wax ester; SE, steryl ester; TG, triacylglycerol; PE, phosphatidylethanolamine; PS, phosphatidylserine; PI, phosphatidylinositol; PC, phosphatidylcholine; LPC, lysophosphatidylcholine.

2.4. Bioactivity of Body Wall (BW) Lipids

Inflammation is a part of the body's normal response to infection and injury, extreme or inappropriate inflammation, on the contrary, is linked to the pathobiology of several diseases [21]. The anti-inflammatory effect of *S. droebachiensis* lipids of the extract was assessed in the human mononuclear U937 cells stimulated with lipopolysaccharide (LPS). The stimulation of U937 resulted in direct activation of MAPK p38. The results (Table 4) revealed that BW lipids dose-dependently inhibited MAPK p38. The most effective dose of the lipid extract was 0.033 µg/mL. In addition, BW lipids were clearly more potent than a specific MAPK p38 inhibitor SB203580 (1.88 µg/mL) at the doses of 0.0037–0.1 µg/mL. The COX-1 and COX-2 isoenzymes were inhibited by BW lipids dose-dependently with IC_{50} = 15.7 µg/mL and 21 µg/mL, respectively.

Table 4. Effect of body wall (BW) lipids on the phosphorylation of MAPK p38 in the human mononuclear U937 cells Mean ± SEM, (*n* = 6).

Sample, Concentration	Percentage of MAPK p38 (%)
Intact cells (no stimulation with LPS)	23.0 ± 1.2
Control cells stimulated with LPS (1 µg/mL)	100
SB203580 (1.88 µg/mL) + LPS	30.0 ± 1.7
BWL (10 µg/mL) + LPS	59.0 ± 1.3
BWL (5 µg/mL) + LPS	53.0 ± 1.9
BWL (1 µg/mL) + LPS	49.0 ± 0.9
BWL (0.5 µg/mL) + LPS	38.0 ± 1.6
BWL (0.1 µg/mL) + LPS	17.0 ± 1.5
BWL (0.033 µg/mL) + LPS	12.0 ± 0.5
BWL (0.011 µg/mL) + LPS	21.0 ± 1.7
BWL (0.0037 µg/mL) + LPS	27.0 ± 0.7
BWL (0.0012 µg/mL) + LPS	38.0 ± 1.6
BWL (0.0004 µg/mL) + LPS	52.0 ± 1.2

3. Discussion

The fatty acid composition of the ethanol extract of body wall (BW) lipids of sea urchin (Table 1 and Figure 1) was consistent with literature including non-methylene interrupted dienes (NMID) and unusual cyclopropane derivatives [2,22,23]. The principal PUFAs were n6 arachidonic (AA) and n3 eicosapentaenoic acids (EPA). The proportion docosahexanoic acid (22:6n3, DHA) in BW lipids, on the contrary, was very low as has been reported also by others [2,22]. In our previous study, gonads of sea urchin contained about 50 µg/mg of EPA and DHA [4]. Analysis of lipid classes showed two major phospholipids—i.e., phosphatidylcholine and phosphatidylethanolamine. This data is in general agreement with the phospholipids profile of gonads of *S. droebachiensis* [7].

Fats in human diet are responsible for severe health problems because of long-term intake of unbalanced proportions of SaFA, MUFA, n6, and n3 PUFA. The n6 and n3 PUFA intake favors too much n6 PUFA. This presupposes an adequate intake of PUFA precursors (n6 linoleic and n3 alpha-linolenic acids) to form long-chain LC-PUFAs (n6-AA and n3-EPA), which, in turn, act as eicosanoid precursors which determine the balance and effects of eicosanoids in the body [24]. Arachidonic acid is converted to thromboxane-type eicosanoids via cyclo-oxygenase enzyme, while EPA is converted to prostacycline, antagonizing the conversion of AA to eicosanoids. This would enhance anti-aggregatory and anti-inflammatory conditions [25].

From a biological point of view, the biomembrane covering the inner layer of the body wall plays an important protective role for survival of the sea urchin. Damage of the body wall will follow with inflammation. Mitogen-activated protein kinases (MAPKs) are among the most important molecules in the signaling pathways among which MAPK p38 signaling plays an essential role in regulating cellular processes, especially inflammation [26]. In our study, we observed 88% of MAPK p38 inhibition by body wall (BW) lipids at a very low dose of 0.033 µg/mL (Table 4). The inhibition of MAPK p38 might be attributed to the different active compounds of BW lipids. Ait-Said et al. [27] established that EPA, unlike DHA, failed to inhibit nuclear factor-κB (NF-κB) activation, and suppressed MAPK p38 phosphorylation in IL-1β stimulated human pulmonary microvascular endothelial cells. The anti-inflammatory properties of EPA in LPS-stimulated BV2 microglia cells were mediated by downregulation of NF-κB and MAPKs such as ERK, p38, JNK, and Akt activation [28].

Cyclooxygenase-2 (COX-2) isoenzyme could be induced by a wide range of proinflammatory agents. Prostaglandin-dependent amplification of COX-2 is hypothesized to be an important part of sustained proliferative and chronic inflammatory conditions [29]. EPA as well as DHA effectively inhibited COX-2 expression in LPS-stimulated HUVEC endothelial cells [30]. Recently, we have reported that lipid rich fraction from gonads of *S. droebachiensis* inhibited COX-2 with IC_{50} = 49 µg/mL, but was not effective against COX-1 isoform [4]. It is important to note that, in our current study, body wall lipids were more effective and inhibited COX-2 in lower dose and inhibited COX-1 with IC_{50} = 15.7 µg/mL.

Low amounts of ether-bonded alkoxyglycerols were also found in ethanolic extract of BW lipids (Figure 2 and Table 3). The total amount of AOG was less than 0.2% which would correspond about the level in human milk and plasma lipids, for example [31]. By using the present UPLC-ELSD method, however, it was not possible to detect glycerophospholipid-based alkyl- and alkenylacyl lipids, since their analysis first requires the separation of phospholipid subclasses. These lipids, such as glycerophosphatidylethanolamine and -choline, are known to have important activities.

Biological activity of alkoxyglycerols has been known already more than half a century. Some of the activities, such as anti-inflammatory effects and protection against radiation damage, are still under study together with the more recent interest in the possible cell-signaling properties of phospholipid-based ether-bonded compounds. These lipids have shown multiple pharmacological activities such as anti-cancer [32], wound healing [33], and immunostimulatory effects [34]. It has been demonstrated that AOG differentially modulate LPS-mediated MAPK and NF-κB signaling in adipocytes and that they do not activate signaling in the absence of LPS. Saturated alkyl chain increased LPS-mediated activation of the MAPK signaling, which could cause the expression of inflammatory genes. Conversely, unsaturated alkyl chain decreased LPS-mediated activation of the MAPK and NF-κB signaling [35].

Taking into account all these aspects, we can hypothesize that the lipids containing n3 PUFA, especially EPA, could contribute anti-inflammatory activity. Besides the content of neutral lipid alkoxyglycerols, it is necessary to confirm the occurrence of alkyl- and alkenylacyl lipids and fatty acid compositions of individual phospholipid classes in BW lipid extracts.

4. Material and Methods

4.1. Sample Preparation and Extraction of Lipids

Green sea urchins, *Strongylocentrotus droebachiensis*, were harvested in the Barents Sea in 2016 by divers. After removal of gonads, coelomic fluid, and internal organs, the shells with spines were washed in cold tap water, dried at 4 °C for two days and stored in a dark place. The shells with spines (20 g) were ground and macerated with 160 mL of 95% ethanol for 3 h with constant stirring at 55 °C. The extract was filtered and evaporated into dryness by rotary evaporator (IKA RV 10; IKA®-Werke GmbH & Co. KG, Staufen, Germany). The yield of the body wall lipids was 1.2%.

4.2. Chemical Analyses

4.2.1. Analysis of Fatty Acids, Sterols and Alkoxyglycerols by GC-MS

Analyses of bound and free (FFA) fatty acids were carried out by using the method described previously [4]. Shortly, the lipid extract, spiked with internal standards (IS) TG(17:0/17:0/17:0) and FFA 17:0, were transesterified with 0.5 N sodium methoxide at 45 °C for 5 min. After acidification, fatty acid methyl esters (FAMEs) as well as FFAs were extracted with petroleum ether. The method enabled the esterification of bound fatty acids from neutral lipids and also from phospholipids [36]. The mixtures of FAMEs and FFAs were used as reference substances.

The analyses were performed on an Agilent 7890A GC mounted with Gerstel MPS injection system and an Agilent 5975C mass selective detector. The column was an Agilent FFAP silica capillary column (25 m × 0.2 mm × 0.3 μm) and helium was used as the carrier gas. The oven temperature raised from 70 °C to 235 °C, with a total run time of 30 min. The temperatures of the injector and MS source were 220 and 230 °C, respectively, and the data were collected in EI mode (70 eV) at a mass range of m/z 40–600.

After analyzing the composition of FAMEs by GC-MS, the same samples were derivatized to determine the contents of FFAs, cholesterol, and minor sterols. Analyses of alkoxyglycerols (AOG) were done according to a previous method [31]. Samples were evaporated, re-dissolved in dichloromethane, and silylated with MSTFA [*N*-Methyl-*N*-(trimethylsilyl)-trifluoroacetamide] (Pierce, Rockford, IL, USA) at 80 °C for 20 min. TMS-derivatives were analyzed on an Rtx-5-ms column (15 m × 0.25 mm × 0.25 μm) (Restek, Bellefonte, PA, USA). The split ratio was 20:1 and the oven temperature was programmed to go from 70 (1 min) to 270 °C at a rate of 10 °C/min, the total run time was 30 min. The data was collected by MSD ChemStation software (Agilent Technologies, Inc., Santa Clara, CA, USA). Identification of the compounds was based on retention times of reference substances, library comparisons (The Wiley® Registry of Mass Spectral Data, John Wiley and Sons, Inc., New York, NY, USA; NIST 08 spectral library, National Institute of Standards and Technology, Gaithersburg, MD, USA) and on literature data.

4.2.2. Analysis of Lipid Classes by UPLC-ELSD

The same lipid extract as above (without derivatization) was analyzed on a Waters Acquity™ H-class UPLC (ultra-performance liquid chromatograph) equipped with an evaporative light scattering detector (ELSD) by modifying previous conditions [37]. Separation of the lipid classes was carried out on a Waters Spherisorb silica column (3 μm, 100 × 2.1 mm I.D.). The gradient solvent system consisted of (A) iso-octane-tetrahydrofurane (99:1), (B) 2-propanol-dichloromethane (3:2) and (C) 2-propanol-water (1:1) with an analysis time of 20 min. The temperature of the drift tube was 40 °C and air flow 50 psi. The multigradient system started from 100% A, the proportion of A decreased to 32%, that of B increased to 52% and simultaneously that of C (containing water) increased to 16%. Stable retention times were obtained by keeping continuous cycle running. The flow rate was 0.800 mL/min and the injection volume 2 μL. The temperature of the sample manager was 10 °C.

4.3. Biological Assays

4.3.1. Cell Lines and Cell Culture

The human mononuclear U937 cells were purchased from the Russian Collection of Cell Culture (Institute of Cytology of Russian Academy of Science, Saint-Petersburg, Russia), and maintained at 37 °C in a humidified 95% air and 5% CO_2 in RPMI1640 supplemented with 2 mM glutamine, 10% heat-inactivated FBS, 100 U/mL penicillin, and 100 µg/mL streptomycin. BW lipids were dissolved in dimethyl sulfoxide (DMSO) as a stock solution at a 10 mg/mL concentration, and the stock solution was then diluted with the medium to the desired concentration prior to use. Cells derived from the freeze-down batch were thawed, grown, and seeded (106 cells per well) onto 12-well tissue culture plates and cultured in medium for 24 h. The cells were then stimulated with 1 µg/mL *Escherichia coli* LPS (Sigma-Aldrich, St. Louis, MO, USA) at 37 °C for 1 h. After that the cells were treated with SB203580 (Sigma-Aldrich, St. Louis, MO, USA) and various concentrations of BWL at 37 °C for 1 h.

4.3.2. Western Blotting

Cells were washed in cold (4 °C) phosphate-buffered saline (PBS; 0.5 mol/L sodium phosphate, pH 7.5) and separated by centrifugation (Hermle Labortechnik, Germany) at 1500 rpm^{-1} for 5 min at 4 °C, harvested by gentle scraping, and used to prepare total protein or nuclear extracts. Cells were treated with lysis buffer—1 mol/L Tris-HCl pH 7.5, 1.5 mol/L NaCl, 10% Triton X-100, 0.2 mol/L Na_3VO_4, 1 mol/L NaF, 0.2 mol/L EDTA, phenylmethylsulphonyl fluoride (PMSF), Abcam's protease inhibitor cocktail, and Abcam's phosphatase inhibitor cocktail—for 20 min at 4 °C. The lysates were then clarified by centrifugation at 15,000 rpm^{-1} for 15 min at 4 °C and the supernatant was collected.

The protein concentrations of the extracts were determined using the [38] with an XMark spectrophotometer (Bio-Rad, Hercules, CA, USA). For Western blot analysis, 40 µg of protein were desaturated by boiling with Laemmli buffer (5 min at 100 °C) and subjected to 4–14% SDS-polyacrylamide gels, and transferred to nitrocellulose membrane membranes (Bio-Rad) by electroblotting. The membranes were blocked with 5% non-fat dry milk in PBS with Tween 20 buffer (PBS-T) (Tris-HCl (pH 7.5), 1.5 mol NaCl, and 0.1% Tween 20) for 1 h at room temperature. Membranes were then incubated overnight at 4 °C with the primary antibodies, probed with enzyme-linked secondary antibodies, and visualized using a chemiluminescent detection with LumiGLO® reagent (Cell Signaling Technology, Danvers, MA, USA) according to the manufacturer's instructions. After detection, the membranes were scanned (Epson Perfection V330 Photo, Seiko Epson Corporation, Nagano, Japan) and processed with Scion Image software (Alpha 4.0.3.2, Scion, Fredrick, MD, USA). The band intensities were used for calculations. Phospho-p38 MAPK antibody, rabbit, p38 MAPK XP rabbit mAb, β-actin rabbit mAb, and anti-rabbit IgG, HRP-linked antibody were from Cell Signaling Technology (Danvers, MA, USA).

4.3.3. Assessment of Cyclooxygenase Activity

Inhibition of human recombinant cyclooxygenase COX-1 and COX-2 (Cayman Chemical, Ann Arbor, MI, USA) was assessed according to the manufacturer's instructions. Indomethacin (1 µg/mL) from Sigma (St. Louis, MO, USA) was used as reference. The BWL was dissolved in DMSO prior to analysis.

4.4. Statistical Analysis

Data were analyzed using Statistica version 10.0. All biological assay data are presented as mean ± SEM or mean ± SD. Differences among groups were evaluated by one way analysis of variance (ANOVA) and post-hoc Tukey's test. In all comparisons, $p < 0.05$ was accepted as statistically significant.

5. Conclusions

To the best of our knowledge, this is the first time when the profile of body wall lipids is reported. It can be assumed that the mechanism of action of body wall lipids in the present study is via the inhibition of MAPK p38, COX-1, and COX-2. Our findings open the potential to utilize this lipid fraction as a source for the development of drugs with anti-inflammatory activity. Further- more, the anti-inflammatory properties of the lipid extract may be useful for ameliorating neuro- degenerative diseases, as well as suppressing LPS-induced shock.

Acknowledgments: This study was done without financial support. The cost to publish in open access will be pay from the budget of Division of Pharmaceutical Biosciences, Faculty of Pharmacy, University of Helsinki.

Author Contributions: A.N.S., O.N.P., V.M. and H.V. conceived and designed the experiments; A.N.S., O.N.P., A.S.K., I.L. and T.S.-L. performed the experiments; A.N.S., I.L. and T.S.-L. analyzed the data; M.M., V.M. and H.V. contributed to data interpretation and discussion; A.N.S., O.N.P., I.L. and T.S.-L. wrote the paper with all others contributing in editing and revision.

Conflicts of Interest: The authors declare no conflict of interest.

References

1. FAO. Global Production Statistics (1950–2015). Available online: http://www.fao.org/fishery/statistics/global-production/query/en (accessed on 21 June 2017).
2. Liyana-Pathirana, C.; Shahidi, F.; Whittick, A.; Hooper, R. Lipid and lipid soluble components of gonads of green sea urchin (*Strongylocentrotus droebachiensis*). *J. Food Lipids* **2002**, *9*, 105–126. [CrossRef]
3. Kalogeropoulos, K.; Mikellidi, A.; Nomikos, T.; Chiou, A. Screening of macro- and bioactive micro-constituents of commercial finfish and sea urchin eggs. *LWT-Food Sci. Technol.* **2012**, *46*, 525–531. [CrossRef]
4. Pozharitskaya, O.N.; Shikov, A.N.; Laakso, I.; Seppänen-Laakso, T.; Makarenko, I.E.; Faustova, N.M.; Makarova, M.N.; Makarov, V.G. Bioactivity and chemical characterization of gonads of green sea urchin *Strongylocentrotus droebachiensis* from Barents Sea. *J. Funct. Foods* **2015**, *17*, 227–234. [CrossRef]
5. Matsuno, T.; Tsushima, M. Carotenoids in sea urchins. *Dev. Aquac. Fish. Sci.* **2001**, *32*, 115–138.
6. Dincer, T.; Cakli, S. Chemical composition and biometrical measurements of the Turkish sea urchin (*Paracentrotus lividus*, Lamarck, 1816). *Crit. Rev. Food Sci. Nutr.* **2007**, *27*, 21–26. [CrossRef]
7. Shikov, A.N.; Laakso, I.; Pozharitskaya, O.N.; Makarov, V.G.; Hiltunen, R. Phospholipids and amino-acid composition of eggs of sea urchin from Barents Sea. *Planta Med.* **2012**, *78*, 1146. [CrossRef]
8. Pozharitskaya, O.N.; Shikov, A.N.; Makarova, M.N.; Ivanova, S.A.; Kosman, V.M.; Makarov, V.G.; Bazgier, V.; Berka, K.; Otyepka, M.; Ulrichová, J. Antiallergic effects of pigments isolated from green sea urchin (*Strongylocentrotus droebachiensis*) shells. *Planta Med.* **2013**, *79*, 1698–1704. [CrossRef] [PubMed]
9. Kovaleva, M.A.; Ivanova, S.A.; Makarova, M.N.; Pozharitskaya, O.N.; Shikov, A.N.; Makarov, V.G. Effect of a complex preparation of sea urchin shells on blood glucose level and oxidative stress parameters in type II diabetes model. *Eksp. Klin. Farmakol.* **2013**, *76*, 27–30. [PubMed]
10. Agafonova, I.G.; Bogdanovich, R.N.; Kolosova, N.G. Assessment of nephroprotective potential of histochrome during induced arterial hypertension. *Bull. Exp. Biol. Med.* **2015**, *160*, 223–227. [CrossRef] [PubMed]
11. Talalaeva, O.S.; Mishchenko, N.P.; Bryukhanov, V.M.; Zverev, Y.F.; Fedoreyev, S.A.; Lampatov, V.V.; Zharikov, A.Y. The influence of histochrome on exudative and proliferative phases of the experimental inflammation. *Bull. Sib. Branch RAMS* **2012**, *32*, 28–31.
12. Kuwahara, R.; Hatate, H.; Yuki, T.; Murata, H.; Tanaka, R.; Hama, Y. Antioxidant property of polyhydroxylated naphthoquinone pigments from shells of purple sea urchin *Anthocidaris crassispina*. *LWT-Food Sci. Technol.* **2009**, *42*, 1296–1300. [CrossRef]
13. Pozharitskaya, O.N.; Ivanova, S.A.; Shikov, A.N.; Makarov, V.G. Evaluation of free radical-scavenging activity of sea urchin pigments using HPTLC with post-chromatographic derivatization. *Chromatographia* **2013**, *76*, 1353–1358. [CrossRef]
14. Powell, C.; Hughes, A.D.; Kelly, M.S.; Conner, S.; McDougall, G.J. Extraction and identification of antioxidant polyhydroxynaphthoquinone pigments from the sea urchin, *Psammechinus miliaris*. *LWT-Food Sci. Technol.* **2014**, *59*, 455–460. [CrossRef]

15. Anufriev, V.P.; Novikov, V.L.; Maximov, O.B.; Elyakov, G.B.; Levitsky, D.O.; Lebedev, A.V.; Sadretdinov, S.M.; Shvilkin, A.V.; Afonskaya, N.I.; Ruda, M.Y.; et al. Synthesis of some hydroxynaphthazarins and their cardioprotective effects under ischemia-reperfusion in vivo. *Bioorg. Med. Chem. Lett.* **1998**, *8*, 587–592. [CrossRef]

16. Lakeev, Y.V.; Kosykh, V.A.; Kosenkov, E.I.; Novikov, V.L.; Lebedev, A.V.; Repin, V.S. Effect of natural and synthetic antioxidants (polyhydroxynaphthaquinones) on cholesterol metabolism in cultured rabbit hepatocytes. *Bull. Exp. Biol. Med.* **1992**, *114*, 1611–1614. [CrossRef]

17. Amarowicz, R.; Synowiecki, J.; Shahidi, F. Chemical composition of shells from red (*Strongylocentrotus franciscanus*) and green (*Strongylocentrotus droebachiensis*) sea urchin. *Food Chem.* **2012**, *133*, 822–826. [CrossRef]

18. Krishtopina, A.S.; Urakova, I.N.; Pozharitskaya, O.N.; Razboeva, E.V.; Kosman, V.M.; Makarov, V.G.; Shikov, A.N. Optimization of polyhydroxynaphtoquinone extraction from shells of *Strongylocentrotus droebachiensis* sea urchins. *Pharm. Chem. J.* **2017**, *51*, 407–410. [CrossRef]

19. Phillips, K.M.; Ruggio, D.M.; Exler, J.; Patterson, K.Y. Sterol composition of shellfish species commonly consumed in the United States. *Food Nutr. Res.* **2012**, *56*, 19831. [CrossRef] [PubMed]

20. Bordier, C.G.; Sellier, N.; Foucault, A.P.; Le Goffic, F. Purification and characterization of deep sea shark *Centrophorus squamosus* liver oil 1-O-alkylglycerol ether lipids. *Lipids* **1996**, *31*, 521–528. [CrossRef] [PubMed]

21. Calder, P.C. n-3 polyunsaturated fatty acids, inflammation, and inflammatory diseases. *Am. J. Clin. Nutr.* **2006**, *83*, S1505–S1519.

22. Takagi, T.; Eaton, C.A.; Ackman, R.G. Distribution of fatty acids in lipids of the common Atlantic sea urchin *Strongylocentrotus droebachiensis*. *Can. J. Fish. Aquat. Sci.* **1980**, *37*, 195–202. [CrossRef]

23. Castell, J.D.; Kennedy, E.J.; Robinson, S.M.C.; Parsons, G.J.; Blair, T.J.; Gonzales-Duran, E. Effect of dietary lipids on fatty acid composition and metabolism in juvenile green sea urchin (*Strongylocentrotus droebachiensis*). *Aquaculture* **2004**, *242*, 417–436. [CrossRef]

24. Lands, W.E.M. Diets could prevent many diseases. *Lipids* **2003**, *38*, 317–321. [CrossRef] [PubMed]

25. Lands, B. A critique of paradoxes in current advice on dietary lipids. *Prog. Lipid Res.* **2008**, *47*, 77–106. [CrossRef] [PubMed]

26. Yang, Y.; Kim, S.C.; Yu, T.; Yi, Y.S.; Rhee, M.H.; Sung, G.H.; Yoo, B.C.; Cho, J.Y. Functional roles of p38 mitogen-activated protein kinase in macrophage-mediated inflammatory responses. *Mediat. Inflamm.* **2014**. [CrossRef] [PubMed]

27. Ait-Said, F.; Elalamy, I.; Werts, C.; Gomard, M.T.; Jacquemin, C.; Couetil, J.P.; Hatmi, M. Inhibition by eicosapentaenoic acid of IL-1β-induced PGHS-2 expression in human microvascular endothelial cells: Involvement of lipoxygenase-derived metabolites and p38 MAPK pathway. *Biochim. Biophys. Acta* **2003**, *1631*, 77–84. [CrossRef]

28. Moon, D.O.; Kim, K.C.; Jin, C.Y.; Han, M.H.; Park, C.; Lee, K.J.; Park, Y.M.; Choi, Y.H.; Kim, G.Y. Inhibitory effects of eicosapentaenoic acid on lipopolysaccharide-induced activation in BV2 microglia. *Int. Immunopharmacol.* **2007**, *7*, 222–229. [CrossRef] [PubMed]

29. Bagga, D.; Wang, L.; Farias-Eisner, R.; Glaspy, J.A.; Reddy, S.T. Differential effects of prostaglandin derived from ω-6 and ω-3 polyunsaturated fatty acids on COX-2 expression and IL-6 secretion. *Proc. Natl. Acad. Sci. USA* **2003**, *100*, 1751–1756. [CrossRef] [PubMed]

30. Lee, S.; Kim, H.J.; Chang, K.C.; Baek, J.C.; Park, J.K.; Shin, J.K.; Choi, W.J.; Lee, J.H.; Paik, W.Y. DHA and EPA down-regulate COX-2 expression through suppression of NF-κB activity in LPS-treated human umbilical vein endothelial cells. *Korean J. Physiol. Pharmacol.* **2009**, *13*, 301–307. [CrossRef] [PubMed]

31. Seppänen-Laakso, T.; Laakso, I.; Hiltunen, R. Rapid determination of unsubstituted alkylglyceryl ethers by gas chromatography-mass spectrometry. *J. Chromatogr.* **1990**, *530*, 94–101. [CrossRef]

32. Pugliese, P.T.; Jordan, K.; Cederberg, H.; Brohult, J. Some biological actions of alkylglycerols from shark liver oil. *J. Altern. Complement. Med.* **1998**, *4*, 87–99. [CrossRef] [PubMed]

33. Pedrono, F.; Martin, B.; Leduc, C.; Le Lan, J.; Saïag, B.; Legrand, P.; Moulinoux, J.P.; Legrand, A.B. Natural alkylglycerols restrain growth and metastasis of grafted tumors in mice. *Nutr. Cancer* **2004**, *48*, 64–69. [CrossRef] [PubMed]

34. Iannitti, T.; Palmieri, B. An update on the therapeutic role of alkylglycerols. *Mar. Drugs* **2010**, *8*, 2267–2300. [CrossRef] [PubMed]

35. Zhang, M.; Sun, S.; Tang, N.; Cai, W.; Qian, L. Oral administration of alkylglycerols differentially modulates high-fat diet-induced obesity and insulin resistance in mice. *Evid.-Based Complement. Altern.* **2013**, *2013*, 834027. [CrossRef] [PubMed]

36. Seppänen-Laakso, T.; Laakso, I.; Hiltunen, R. Analysis of fatty acids by gas chromatography, and its relevance to research on health and nutrition. *Anal. Chim. Acta* **2002**, *465*, 39–62. [CrossRef]

37. Seppänen-Laakso, T.; Laakso, I.; Vanhanen, H.; Kiviranta, K.; Lehtimäki, T.; Hiltunen, R. Major human plasma lipid classes determined by quantitative high-performance liquid chromatography, their variation and associations with phospholipid fatty acids. *J. Chromatogr. B* **2001**, *754*, 437–445. [CrossRef]

38. Bradford, M.M. A rapid and sensitive method for the quantification of microgram quantities of protein utilizing the principle of protein-dye binding. *Anal. Biochem.* **1976**, *72*, 248–254. [CrossRef]

marine drugs

MDPI

Article

The Marine Fungi *Rhodotorula* sp. (Strain CNYC4007) as a Potential Feed Source for Fish Larvae Nutrition

M. Barra [1,2], A. Llanos-Rivera [1], F. Cruzat [1,2,3], N. Pino-Maureira [1,2,3] and
R. R. González-Saldía [1,2,3,*]

[1] Marine Biotechnology Unit, Department of Oceanography, Faculty of Natural and Oceanographic Sciences, Universidad de Concepción, Casilla 160-C, 4030000 Concepción, Chile; mariajosbarra@udec.cl (M.B.); alllanos@udec.cl (A.L.-R.); fecruzat@udec.cl (F.C.); napino@udec.cl (N.P.-M.)
[2] Center for Oceanographic Research COPAS Sur-Austral, Universidad de Concepción, Casilla 160-C, 4030000 Concepción, Chile
[3] Doctoral Program in Aquatic Living Resources (MaReA), Faculty of Natural and Oceanographic Sciences, University of Concepción, Casilla 160-C, 4030000 Concepción, Chile
*   Correspondence: rogonzal@udec.cl; Tel.: +56-41-2204520

Received: 31 August 2017; Accepted: 16 November 2017; Published: 1 December 2017

Abstract: Fish oil is used in the production of feed for cultured fish owing to its high polyunsaturated fatty acid content (PUFA). The over-exploitation of fisheries and events like "El Niño" are reducing the fish oil supply. Some marine microorganisms are considered potentially as alternative fatty acid sources. This study assesses a strain of *Rhodotorula* sp. (strain CNYC4007; 27% docosahexaenoic acid (DHA) of total fatty acids), as feed for fish larvae. The total length and ribonucleic acid (RNA)/deoxyribonucleic acid (DNA) ratio of *Danio rerio* larvae was determined at first feeding at six and 12 days old (post-yolk absorption larvae). Larvae fed with microencapsulated *Rhodotorula* sp. CNYC4007 had a significantly higher RNA/DNA ratio than control group (C1). At six days post-yolk absorption group, the RNA/DNA ratio of larvae fed with *Rhodotorula* sp. bioencapsulated in *Brachionus* sp. was significantly higher than control group fed with a commercial diet high in DHA (C2-DHA). Finally, at 12 days post-yolk absorption, the RNA/DNA ratio was significantly higher in larvae fed with *Rhodotorula* sp. CNYC4007 and C2-DHA (both bioencapsulated in *Artemia* sp. nauplii) than in control group (C1). These results suggest that *Rhodotorula* sp. CNYC4007 can be an alternative source of DHA for feeding fish at larval stage, providing a sustainable source of fatty acids.

Keywords: *Rhodotorula*; DHA; EPA; *Danio rerio*; larvae nutrition; RNA/DNA ratio

1. Introduction

Marine resources for fishmeal and fish oil have been exploited beyond maximum sustainable yields [1], and the rate of exploitation is increasing by 8.8% per year [2]. One of the fish oil uses is the production of feed for aquaculture [3], an area that has expanded significantly, resulting in the rising demand for fish oil, which has consequently increased the price of this product [4]. Fish oil is used because of its high nutritional value and essential polyunsaturated fatty acid content (PUFA), among which are docosahexaenoic acid (C22:6, DHA), eicosapentaenoic acid (C20:5, EPA) and docosapentaenoic acid (C22:5, DPA), [5]. Freshwater fish generally have sufficient elongase and desaturase activities to produce these fatty acids from the 18C precursor. On the contrary, marine fish have a very limited capacity to synthesize these fatty acids [6], hence a strict requirement for long-chain PUFA, eicosapentaenoic, docosahexaenoic and arachidonic acids (essential fatty acid; EFA). Therefore, when these fatty acids are not synthesized in fish they should be incorporated into the diets of the larval-to-adult stages of the species that are of commercial interest [7,8]. Their presence is important for normal growth, reproduction, metabolic functions, feeding efficiency and immunocompetence [9].

Specifically, DHA has a high biological value during larval development and is selectively incorporated in neural tissue, contributing to pigmentation and visual acuity [10]. Lipids are maternally supplied and sustain both embryonic and yolk sac larvae development, but when the yolk absorption is completed and first feeding is started an exogenous source of PUFA is required [3]. Then, at the time of first feeding the larvae should receive a diet that covers all the nutritional requirements, so for aquaculture species the research focus has been to improve the nutritional content of artificial diets for larvae [7,11], as diets lacking in nutritional content result in low survival rates [12].

Fatty acids have also been encapsulated in rotifers (*Brachionus* sp.) and *Artemia* sp. nauplii [13,14], these organisms allow for the passing of essential nutrients to the larvae [15,16]. Morphological and histological parameters, as well as molecular ratios like the ribonucleic acid (RNA)/deoxyribonucleic acid (DNA) ratio, have been used to determine if effectively delivered nutrients improve the nutritional condition of the target species [17]. Theoretically, RNA levels in cells vary in relation to protein synthesis [18], while DNA concentrations remain constant even during starvation [19]. Therefore, the RNA/DNA ratio is a potential indicator of cellular protein synthesis and growth [17].

The declining supply of traditional sources of essential fatty acids and their high cost has led to the search for alternative sources [20], among which are vegetable seed oils [21] that contain alpha-linolenic acid (ALA, 18:3 ω-3), a precursor of DHA and EPA that is assimilated, but not necessarily transformed, into fatty acids [22]. However, it has not been fully possible to enhance these oils for their use in aquaculture as anti-nutritional components have been found in them that affect lipid homeostasis and energy metabolism [23], along with weakening immune response [24]. Another alternative to fish oil is single-cell oils (SCO) from one-cell organisms like bacteria, algae, fungoid protists and marine fungi, which have great potential because of their high essential long-chain unsaturated fatty acid content [25,26]. Marine fungi has also been described as producers of carotenoids, omega-3 fatty acids, including DHA and EPA [27], and omega-6 acids like arachidonic acid (20:04 ω6, AA) [28].

Given their characteristics, strains of marine microorganisms constitute as alternative sources that can partially or totally replace fish oil for the nutrition of fish [25,29,30]. For example, *Schizochytrium* sp., a strain that produces a high level of DHA (28% of its dry weight), has been highly studied because it has been proven with juvenile *Salmo salar* that the complete replacement of fish oil with this strain does not negatively affect growth, immune response, and digestibility of nutrients [25]. Its use with other marine species under intensive cultivation like *Sparus aurata* [31] has also been tested and with positive results. *Schizochytrium* sp. has therefore been proposed as a sustainable source of fatty acids [25].

A strain of *Rhodotorula* sp. (CNYC4007) from the Humboldt Current System off Central Chile was isolated and grown in a bioreactor. The strain, which has high levels of DHA, EPA and carotenoids [27], could be considered as a suitable source of omega-3 for the aquaculture demand. In reference to the above the aim of this study was to evaluate the use of strain CNYC4007 for feeding *Danio rerio* larvae. The zebrafish (*D. rerio*) has been widely studied in various fields of biology [32], and in recent years has been proposed as a possible model in studies of nutrition and growth in fish [33], especially as a model in aquaculture research [34]. However, despite potential limitations due to species-specific differences in fatty acid metabolism, zebrafish is a useful tool in the initial screening of new supplies for larval nutrition with lower cost and rearing times than the use of aquaculture species [35]. To date, research has been provided on the anti-nutritional aspects of some alternative ingredients [36] and individualized the requirements for mineral and trace element requirements that have been established for adequate larval development and growth in zebrafish that could benefit other fish species [16]. Therefore, we used zebrafish larvae as a model in the first step to determine the potential use of *Rhodotorula* sp. (CNYC4007) in larval nutrition according to the usual feeding protocols implemented in larval rearing.

2. Results

2.1. RNA/DNA Ratio as a Proxy for the Nutritional Condition of Danio rerio

The total DNA content per individual did not present a significant difference among live larvae up to the 5th day five of the experiment, when the larvae were 10 days old (Kruskal-Wallis test, $p = 0.0565$). As well, no significant differences were observed for the same parameter among the treatments during the days of the assay (Figure 1). However, there was 100% mortality rate by 6th day of the larvae in starvation (11 days old), while there was 100% survival rate among the larvae that were fed. The quantity of RNA in larval tissue (Figure 1) progressively increased over the course of the treatment with the standard feed, presenting significant differences throughout the assay (one-way analysis of variance (ANOVA), F-statistic = 31.09, p-value = 4.18×10^{-5}). In contrast, RNA content in the tissue of the larvae under the starvation treatment decreased over the course of the assay. Comparing RNA content when the larvae were 10 days old (Figure 1), a significant difference between larvae fed with the standard feed (C1) and those in starvation (one-way analysis of variance (ANOVA), F-statistic = 11.6, p-value = 0.00362) was found. The average of the RNA/DNA ratio of the treatment with the standard feed (Figure 2) when larvae were 10 days old (5th day of the treatment) was 0.81 ± 0.12, and the starvation treatment had a lower average (0.71 ± 0.06). This ratio presents significant differences between the treatments (two-way ANOVA, F-statistic = 5.873, p-value = 0.0210).

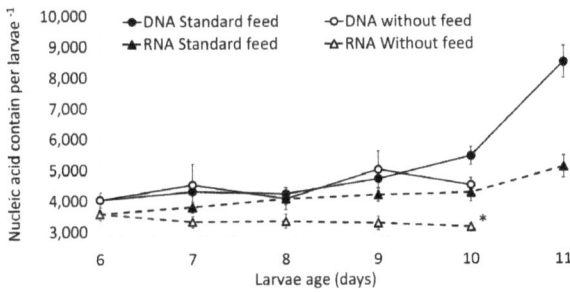

Figure 1. Average nucleic acid content (ng) versus larvae age of *Danio rerio* fed with standard feed compared to treatment with larvae in starvation. Each point represents the average of four larvae per treatment. * $p < 0.05$.

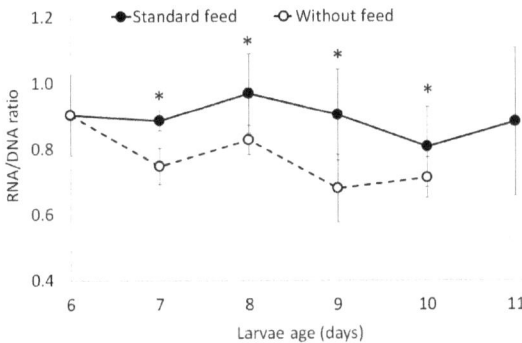

Figure 2. Ribonucleic acid (RNA)/deoxyribonucleic acid (DNA) ratio versus test of larvae of *Danio rerio* fed with standard feed compared to that of larvae in starvation. The daily average is based on four larvae per treatment. * $p < 0.05$.

2.2. Assay with Six-Day-Old Larvae (First-Feeding Larvae, One Day Post-Yolk Absorption)

There were significant differences among the three treatments (one-way ANOVA, F-statistic = 1809, p-value = 2×10^{-16}) by the final day of the assay (Figure 3). A significantly greater average larval length (7.7 ± 0.3 mm) was obtained with the control feed (control group) than with the microencapsulated *Rhodotorula* sp. CNYC4007 and with the flour lyophilized of *Rhodotorula* sp. CNYC4007 (7.0 ± 0.2 mm and 7.1 ± 0.2 mm, respectively; Tukey test; p-value = 0.000125 for both treatments). There were no significant differences in the total length between larvae fed on flour and those fed on microencapsulated *Rhodotorula* sp. CNYC4007 (Tukey test, p = 0.3641). With respect to the molecular analysis, the three treatments did not present any significant differences over the course of the assay (Kruskal-Wallis p = 0.148), except for the final day (Figure 4) when the RNA/DNA ratios in the larvae in both the treatments with microencapsulated *Rhodotorula* sp. CNYC4007 and with lyophilized flour (which had the same ratios; 0.8 ± 0.14) were higher than that of the larvae in the control group (0.53 ± 0.09).

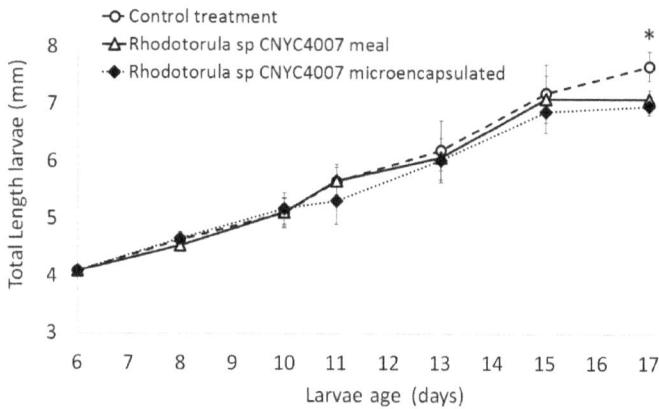

Figure 3. Total length versus larvae age of *Danio rerio* at first feeding larvae. The daily average is based on six larvae per treatment. * p < 0.05.

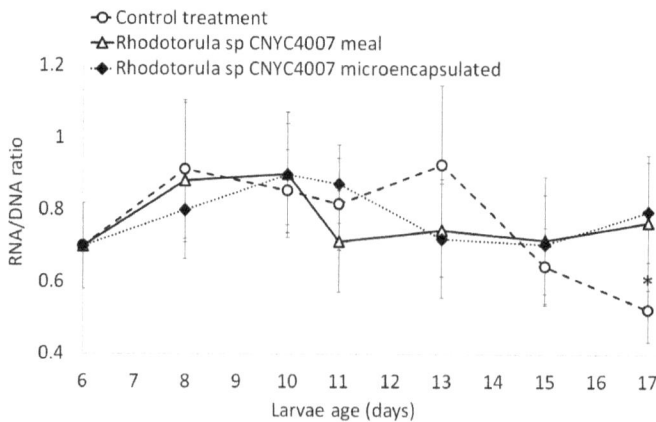

Figure 4. RNA/DNA ratio versus larvae age of *Danio rerio* at first feeding larvae. The daily average daily is based on six larvae per treatment. * p < 0.05.

2.3. Feeding Six-Day-Old Larvae (Post-Yolk Absorption) with Encapsulation in Brachionus sp.

There were no significant differences in the total length between the larvae (Table 1) fed with rotifers without enrichment (control treatment), rotifers enriched with C2-DHA (positive control) and rotifers enriched with *Rhodotorula* sp. CNYC4007 (one-way ANOVA, F-statistic = 0.89, p-value = 0.41), nor were significant differences observed in the lengths of larvae from the different treatments on the final day of the assay (one-way ANOVA, F-statistic = 0.34, p-value = 0.72). Similarly, there were no significant differences in the RNA/DNA ratios of the larvae from the three treatments (Kruskal-Wallis p = 0.148), except for 7th day of the assay when larvae are 12 days old (Table 2), the RNA/DNA ratios of larvae fed with *Brachionus* sp. (control treatments) and *Brachionus* sp. enriched with C2-DHA (positive control) were significantly lower than those of larvae fed with *Rhodotorula* sp. CNYC4007 (Tukey test, p = 0.0017 and p = 0.0087, respectively). There were no significant differences (Tukey test, p = 0.6803) between the treatments with *Brachionus* sp. (control treatment) and *Brachionus* sp. enriched with C2-DHA (Table 2).

Table 1. Total larval length (average ± SD) for assay with six-day-old larvae of *Danio rerio* after post-yolk absorption. Larvae were fed with bioencapsulated *Brachionus* sp. rotifers and *Artemia* sp. nauplii. The daily average is based on six larvae per treatment. C1: without enrichment; C2-DHA: feed rich in docosahexaenoic acid (DHA) treatment; CNYC4007: *Rhodotorula* sp. CNYC4007 treatment.

Larvae Age (Days)	Total Length Larvae (mm)					
	Brachionus sp.			*Artemia* sp. nauplii		
	C1	C2-DHA	CNYC4007	C1	C2-DHA	CNYC4007
6	5.0 ± 0.2	5.0 ± 0.2	5.0 ± 0.2	n/a	n/a	n/a
8	5.2 ± 0.3	5.5 ± 0.4	5.1 ± 0.2	n/a	n/a	n/a
10	5.6 ± 0.1	5.6 ± 0.3	5.6 ± 0.3	n/a	n/a	n/a
12	5.6 ± 0.1	5.7 ± 0.2	5.8 ± 0.2	5.6 ± 0.3	5.6 ± 0.3	5.6 ± 0.3
14	n/a	n/a	n/a	5.6 ± 0.3	5.7 ± 0.3	5.8 ± 0.4
16	n/a	n/a	n/a	5.6 ± 0.2	6.0 ± 0.3	6.0 ± 0.4
18	n/a	n/a	n/a	6.6 ± 0.1	6.9 ± 0.3	7.0 ± 0.3
20	n/a	n/a	n/a	9.5 ± 0.4	10.2 ± 0.3	10.1 ± 0.7
22	n/a	n/a	n/a	9.9 ± 0.8	10.6 ± 0.3	10.6 ± 0.5

Table 2. RNA/DNA ratio for assay with six-day-old larvae of *Danio rerio* after post-yolk absorption. Larvae were fed with bioencapsulated *Brachionus* sp. rotifers and *Artemia* sp. nauplii. The daily average is based on six larvae per treatment. C1: without enrichment; C2-DHA: feed rich in DHA treatment; CNYC4007: *Rhodotorula* sp. CNYC4007 treatment.

Larvae Age (Days)	RNA/DNA Ratio					
	Brachionus sp.			*Artemia* sp. nauplii		
	C1	C2-DHA	CNYC4007	C1	C2-DHA	CNYC4007
6	1.1 ± 0.2	1.1 ± 0.2	1.1 ± 0.2	n/a	n/a	n/a
8	1.3 ± 0.2	1.1 ± 0.3	1.4 ± 0.3	n/a	n/a	n/a
10	1.2 ± 0.2	1.5 ± 0.1	1.5 ± 0.3	n/a	n/a	n/a
12	1.0 ± 0.2	1.1 ± 0.1	1.4 ± 0.2 *	0.8 ± 0.2	0.8 ± 0.2	0.8 ± 0.2
14	n/a	n/a	n/a	1.1 ± 0.1	0.9 ± 0.2	1.1 ± 0.3
16	n/a	n/a	n/a	1.0 ± 0.1	1.1 ± 0.2	1.1 ± 0.3
18	n/a	n/a	n/a	1.0 ± 0.2	1.1 ± 0.1	1.2 ± 0.2
20	n/a	n/a	n/a	0.8 ± 0.1	1.1 ± 0.1	1.1 ± 0.1
22	n/a	n/a	n/a	0.8 ± 0.1 *	1.1 ± 0.1	1.0 ± 0.1

* = p-value < 0.05; n/a = not available.

2.4. Experiments with 12-Day-Old Larvae

There were no significant differences in the total length of these larvae (Table 1) used in the treatment with the enriched *Artemia* sp. nauplii (one-way ANOVA, F-statistic = 0.713, p-value = 0.493), either throughout or on the final day of the assay (one-way ANOVA, F-statistic = 0.869, p-value = 0.43).

However, by the final day there were significant differences in the RNA/DNA ratio (Table 2) between the control group and the group treatment with nauplii enriched with C2-DHA and with the strain CNYC4007 (Tukey test, *p*-value = 0.0003 and *p*-value = 0.005, respectively). Finally, there were no significant differences on the final day of assay between the treatment with *Artemia* sp. nauplii enriched with C2-DHA and that with nauplii enriched with CNYC4007 (Tukey test, *p*-value = 0.39).

2.5. DHA and EPA Enrichment of Artemia sp. nauplii

There was a significant high DHA concentration (Figure 5) in nauplii enriched with the strain CNYC4007 (*Rhodotorula* sp.) compared to those without enrichment or enriched with C2-DHA (one-way ANOVA, *p*-value = 0.0036). There were no significant differences in DHA concentrations (*p*-value > 0.05) between nauplii without enrichment (control treatment) and those fed with C2-DHA. The same was observed with EPA concentrations, which were significantly high for nauplii fed with the strain CNYC4007 (*Rhodotorula* sp.) compared to those of nauplii without enrichment or enriched with C2-DHA (one-way ANOVA, *p*-value = 0.00017). There were no significant differences in EPA concentrations (*p*-value > 0.05) between *Artemia* sp. nauplii without enrichment (control treatment) and those fed with C2-DHA.

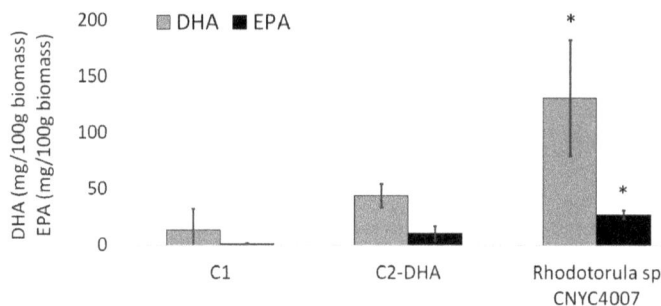

Figure 5. Average DHA and eicosapentaenoic acid (EPA) concentrations (± S.D.) in *Artemia* sp. nauplii without enrichment (C1), fed feed rich in DHA (C2-DHA) and fed with the strain *Rhodotorula* sp. CNYC4007. * *p*-value < 0.05.

3. Discussion

The use of strains of marine microorganisms in commercial fish feed and other applications, such as a source of food for human consumption, requires in vivo validation. The first step is to rule out toxic effects and then verify the nutritional contribution to the species under study. The condition index of Fulton [37] is generally used for organisms that are sufficiently large to be easily weighed and measured. However, with fish larvae the determination of difference in a very small body mass tends to lack precision. The present work studied the effect of the strain CNYC4007 of the marine basidiomycete *Rhodotorula* sp. [27] in the RNA/DNA ratio, which was used as a proxy for the nutritional state of *Danio rerio* larvae. The larvae were fed in their different stages of development and the strain was administered in three forms (flour, microencapsulated, and bioencapsulated).

The results show the efficacy of the RNA/DNA method to evaluate the nutritional condition of larvae in the experiments carried out, given that it effectively distinguished between the conditions of the larvae that were fed and those in starvation (Figures 1 and 2). This method is relevant and has been used above all with animals at the size scale of zooplankton [22,38] and fish larvae [17,39,40]. This ratio is considered a metabolic index and has been used as a measurement of individual growth rate and nutritional state [18,41]. The increase in RNA concentrations in larvae fed with standard feed (Figure 1) suggests there was protein synthesis and with this a progressive increase in larval size throughout the assay, as has been observed with other species like *Sardina pilchardus*, *Engraulis encrasicolus*,

Atherina presbyter and *Paralichthys orbignyanus* in different development stages [42]. This pattern concurs with that described by Chung and Segnini [43], who found that RNA levels were higher in rainbow trout larvae that were fed, as opposed to those in starvation. In the present work, average DNA levels per individual were observed to be constant during technical standardization assays (Figure 1). It has been described that DNA levels do not vary under stress, whether caused by starvation or environmental conditions [44]. Coincident with our results, Ben Khemis et al. [17] proposed that RNA/DNA is a useful molecular tool to evaluate larval condition in aquaculture.

In all of the assays the RNA/DNA ratio in the larval stages of *Danio rerio* ranged between 0.7 and 2. These values were similar to those obtained in *Oncorhynchus mykiss* (1.5–2.5) [45]. It was observed that the ratio was sensitive to changes in the diet composition; this concurs with what has been described in rainbow trout, where the RNA/DNA ratio of larvae varied in response to distinct protein compositions included in their diet (40–50%) [45]. There were no differences among the feeding treatments when only considering larval length, although there were differences in nutritional state.

The first feeding assay with *Danio rerio* larvae indicates that the strain *Rhodotorula* sp. CNYC4007 can be used effectively for this purpose. Larvae fed with flour and microcapsules had similar lengths (Figure 3) and were in a similar nutritional condition (Figure 4), indicating that the two forms of administrating feed can be considered as appropriate. The microencapsulated feed contained stabilizers (maltodextrine and capsule) that protected the fatty acids against moisture, light and other environmental factors [46], which because of storage or contact with water can deteriorate the feed [47]. Furthermore, the preparation of microcapsules only required 4% of the microorganisms biomass required for the flour lyophilized with *Rhodotorula* sp. CNYC4007 and its physical properties allowed the feed to remain longer on the surface of the water, which facilitates feeding the larvae, in contrast to the flour, which decants rapidly. Thus, the microencapsulation method to administer *Rhodotorula* sp. (CNYC4007) is recommended for first-feeding the larvae because it is more stable, offers better protection of fatty acids and is more bioavailable than the lyophilized flour method (100% of flour), the latter thus requiring a large quantity of microorganisms to achieve the same effect.

As the larvae grow their mouth size and nutritional requirements increase, because of which the quantity and nutritional composition of the feed must be modified [48]. Because of this, in aquaculture, especially with marine species, live prey such as rotifers and *Artemia* sp. nauplii are used. However, these have low levels of polyunsaturated fatty acids. Rotifers have 4.19% arachidonic acid and 2.29% docosapentaenoic acid of total fatty acids, with the absence of DHA [16], and *Artemia* sp. nauplii have low levels of EPA with the absence of DHA [47]. Consequently, it is necessary to enrich these organisms for use as larval feed [13]. With larvae at six days post-yolk absorption (six days old), *Rhodotorula* sp. CNYC4007 was bioencapsulated in rotifers. The results indicate that the strain is not toxic for *Brachionus* sp., *Artemia* sp. nauplii or zebrafish larvae (0% mortality in the assay). In the case of larvae, its presence in the digestive tract was confirmed and larvae presented a better molecular condition than larvae fed on rotifers without enrichment. This concurs with the results obtained with bioencapsulation in rotifers of *Schizochytrium mangrovei*, a marine fungoid used to enrich live prey [14]. By comparing the fatty acid content of rotifers with and without enrichment, the authors observed that the rotifers absorbed significant levels of the DHA present in the lyophilized of *S. mangrovei* [14]. Barclay and Zeller [49] also found the DHA content in rotifers fed with *Schizochytrium* sp. was significantly higher than that of control rotifers fed with beer yeast. *Schizochytrium* sp. is one of the ingredients included in the formulation of feed, termed in this study as C2-DHA, which was considered as a positive control in experiments in the present work. Both *Schizochytrium mangrovei* [14] and the strain *Rhodotorula* sp. CNYC4007, used in this experiment, have high percentages of DHA (31.53% and 27%, respectively, of total fatty acids).

The incorporation of the bioencapsulated strain CNYC4007 in feed resulted in significantly higher concentrations of polyunsaturated fatty acids in *Artemia* sp. nauplii than in nauplii fed with bioencapsulated C2-DHA (Figure 5). Consequently, *Rhodotorula* sp. CNYC407, as well as *Schizochytrium* sp. [50], represent alternative species for enriching nauplii, even though the percentages of DHA and EPA as part of the total of fatty acids differ between the two species (27% and 43% of DHA; 7.2% and 2.8% of EPA for *Rhodotorula* sp. and *Schizochytrium* sp., respectively). With larvae at 12 days post-yolk absorption, the RNA/DNA ratio of larvae in the control treatment was lower than that of larvae in the treatment with *Artemia* sp. nauplii enriched with feed rich in DHA (C2-DHA) and with the strain *Rhodotorula* sp. CNYC4007. The bioencapsulation of highly unsaturated fatty acids of the omega-3 series in *Artemia* sp. nauplii has been shown to improve growth and survival of marine fish larvae [51], which concurs with what was observed with the enrichment of *Artemia* sp. with the strain studied in this work.

Essential fatty acid requirements vary qualitatively and quantitatively with environmental origin and during the ontogeny of fish, with early developmental stages and broodstock being critical periods [3]. Specifically, as there is evidence that n-3 HUFA (highly unsaturated fatty acids) and DHA may be more important and, possibly, essential in the larvae of some species of freshwater fish compared with adults. In marine species, larvae are characterized by having a greater requirement for n-3 HUFA than juvenile and pre-adult fish, although there are relatively few species where the requirements at larval and juvenile stages can be directly compared. Notwithstanding the foregoing, scarce data exists in relation to EFA requirements (expressed as % dry diet), in larvae and early juvenile fish, with values that ranged between 1 to 5% [3], described the fatty acid requirements (n-3/n-6) in zebrafish larvae using larval growth as a proxy, and concluded that a diet with a low n-3/n-6 ratio maximizes growth. This finding is coincident with the requirement of typical warm water species, namely, a higher demand for n-6 PUFA, for example, tilapia.

While this study is based mainly on the use of the RNA/DNA ratio as a proxy for nutritional condition, there are other methods to validate results that could be used in future studies, such as the expression of genes related to growth, which can provide new tools for analyzing growth in fish [52]. The levels of gene expression related to myogenesis and ATP concentrations in rainbow trout are drastically reduced in response to starvation [50]. The analysis of microarrays also shows that starvation in rainbow trout also decreases the levels of gene expression related to lipid metabolism and immune response [53].

All the results obtained indicate that the marine basidiomycete strain *Rhodotorula* sp. CNYC4007 represents a potential feed and/or supplement for first feeding of zebrafish. Additionally, it can be incorporated in species like rotifers and *Artemia* sp. nauplii to be bioencapsulated to potentially improve the nutritional state of larvae. All trials carried out in our study considered a comparison between *Rhodotorula* sp. and commercial sources of fatty acid (formulated diets or emulsions) at different larval stages. In each case, zebrafish larvae fed with *Rhodotorula* sp. showed at least equal growth and nutritional condition than the larval treatment with the commercial alternative. These results are similar in magnitude to those reported in species with differences in fatty acid metabolism such as *Salmo salar* parr and *Sparus aurata* larvae, where fish oil replacement with alternative sources of DHA (*Schistochytrium* sp. and *Crypthecodinium cohnii*) was suggested [25,31]. These results are relevant, as in other species, nutritional studies on zebrafish have determined positive or negative influences of some food compounds [34] and correspond to the first step in the screening process for novel ingredients or additives with a potential use in aquaculture [35].

Previous studies on *Rhodotorula* yeast have proposed it as a source of carotenoids [54], bioactive substances [55] and lipids, but just for aliphatic18-carbon atom fatty acid chains [56]. The *Rhodotorula* sp. strain CNYC4007 is the only *Rhodotorula* species that has been reported as a DHA and EPA producer and to the best of our knowledge this is the first work in which a *Rhodotorula* yeast has been used as feed. Finally, this study seeks to contribute in the search for a sustainable source of PUFAs for feeding fish [57] to avoid the use of fish oil from over-exploited marine resources.

4. Materials and Methods

4.1. Obtaining and Producing the Strain Rhodotorula sp. CNYC4007

The strain used in this study has a 97% identity for the gene 18S ribosomal RNA (rRNA) of *Rhodotorula* sp. [27]. It was extracted from Caleta Maule, Biobío Region, Chile, and is stored in the National Collection of Yeast Cultures in the UK with the code CNYC4007. Two preparations were obtained from the strain and used in the experiments: lyophilized flour and microencapsulated, which were prepared with atomization drying using two stabilizers (maltodextrin and capsule), following the protocols established by Pino et al. [58]. Formulation and proximate composition of the strain *Rhodotorula* sp. and commercial diets C1 (Mikrovit Hi-Protein, Silesia, Poland) y C2-DHA (Algamac, Poland) used in experiments are detailed in Table 3.

Table 3. Proximate composition of the diets used as a control (C1), positive control (C2-DHA) and the strain *Rhodotorula* sp. CNYC4007.

Components (%)	C1	C2-DHA	CNYC4007 Microencapsulated	CNYC4007 Meal
Protein	49	17.6	n/a	n/a
Carbohydrate	n/a	15.9	92.5	36.0
Fat	8.5	56.2	0.3	7.6
Nitrogen compounds	n/a	n/a	1.4	6.1
Fiber	3	n/a	1.2	1.7
Moisture	6	2.1	3.7	83.1
Ash	n/a	8.2	0.5	7.6
Calories (kcal)	n/a	640	378.2	169.8

n/a = not available.

4.2. Obtaining and Maintaining Danio rerio Larvae

A breeding stock of *D. rerio* wild type strain was maintained at 27 ± 1 °C with a 12:12 h light:dark photoperiod and constant filtration. Males and females of the same age were kept together in a glass aquarium with a density not exceeding four to five fish/L. The fish were fed twice daily with *Artemia* sp. nauplii (Utah strain; Aquafauna Bio-Marine Inc., Hawthorne, CA, USA) and TetraMin tropical flakes. To obtain the larvae needed for the experiments, spawning was induced at three-day intervals by placing plastic receptacles in the culture systems. The fertilized eggs were removed from the receptacles and washed twice, first with a mixture of $1\times$ reconstituted saline solution (E3 medium: $NaCl_2$ 19 mM; KCl 9 mM; $CaCl_2$ 16 mM; $MgSO_4$ 17 mM; Merck & Co., Inc., Kenilworth, NJ, USA) and methylene blue 0.01% (*w/v*) and then with $1\times$ E3 medium. The healthy embryos were incubated in a temperature controlled camera (27 °C) until hatching occurred.

In preliminary observations on early development of zebra fish larvae we assessed the exact moment when the larvae had the adequate mouth size to begin the feeding with rotifers and *Artemia* sp. nauplii. At six days post-fertilization the larvae had completed yolk absorption and were used in first-feeding experiments (larvae at six days post-yolk absorption) or continued maturing until reaching the phase required for the subsequent experiments, when the larvae were 12 days old (larvae at 12 days post-yolk absorption). Plastic wells (700 mL) were used as rearing chambers, and installed in the aquariums. To ensure optimal water quality, the wells were equipped with mesh walls to allow for water circulation. The remains of feed and fecal matter were regularly removed from the bottom of the wells. Wells of different sizes were used for the feeding experiments according to larval size. The wells were placed in a 50-L tank at 26 °C, with constant oxygenation and a light/darkness ratio 14:10. The water was partially changed every two days to maintain optimal quality (total ammonia nitrogen (TAN) = 0.2 mg L^{-1}; conductivity = 201.9 μS cm^{-1}, pH = 7.5, dissolved oxygen = 7.32 mg L^{-1}).

4.3. Validation of the RNA/DNA Method for Danio rerio Larvae

An assay was conducted with first-feeding larvae (one day post-yolk absorption), when the larvae were 6 days old, to validate the RNA/DNA method to compare larvae fed with a standard commercial feed (C1; Mikrovit Hi-Protein Table 3) and larvae in starvation. Three larvae from each treatment (group) were randomly selected every day to determine RNA and DNA concentrations. The experiment lasted six days, when larvae were 11 days old, given that by this day there was 100% mortality of the larvae in starvation, because of which the statistical analysis was applied only up to the tenth day.

4.4. Estimating Larval Growth

Individual larvae were removed from the wells for analysis in all of the experiments. Cold anesthesia was used for observation and measurement. Larvae were placed individually in a petri dish over a gel pack bar frozen at $-20\,^{\circ}$C. Observations and measurements were made under a stereoscopic magnifying glass equipped with a digital camera (Canon EOS REBEL T3, Canon U.S.A., Inc., Huntington, NY, USA). Photographs of the larvae were analyzed using Image Pro-Plus 6.0 software, Media Cybernetics, Inc., Rockville, MD, USA. Magnifications of $10\times$ or $12\times$ was used to determine the total length depending on the size of the larvae. The distance from the mouth to the posterior point of the notochord was measured for pre-flexion larvae, while the distance from the mouth to the hypural plate (standard length) was used for more developed larvae. After measurement, the larvae were immediately placed in Eppendorf tubes and stored at $-80\,^{\circ}$C.

4.5. Experiments with First-Feeding Age Larvae

4.5.1. Feeding with Rhodotorula sp. CNYC4007 Lyophilized in Flour and in Microcapsules

Thirty first-feeding age larvae (one day post-yolk absorption), when the larvae were six days old, they were placed in 700-mL wells. The experiment involved three treatments: (1) larvae fed with standard commercial feed (control group; C1); (2) larvae fed with flour lyophilized of the strain *Rhodotorula* sp. CNYC4007; and (3) larvae fed with flour microencapsulated of *Rhodotorula* sp. CNYC4007 (Table 1). A preliminary experiment established a biomass of 30 mg (for 30 larvae) as an adequate daily ration of commercial feed and the lyophilized flour and 150 mg for microencapsulate. The experiment lasted nine days during which unconsumed feed was removed daily. Every second day three larvae per replicate were removed, measured (detail point 4) and stored at $-80\,^{\circ}$C individually for subsequent molecular analysis (detail point 7).

4.5.2. Larval Feeding with Bioencapsulation in Brachionus sp.

Rotifers used in this experiment were enriched following the protocol of Estudillo del Castillo et al. [16]. The assay involved three treatments with respective replicates: (1) larvae fed with non-enriched rotifers (control group; C1); (2) larvae fed with rotifers enriched with a commercial feed high in DHA (C2-DHA); and (3) larvae fed with rotifers enriched with flour lyophilized of *Rhodotorula* sp. CNYC4007. Each replicate involved 20 larvae in a 300-mL well with 0.2-mm mesh walls that allowed for the flow of water while preventing rotifers from escaping. On a daily basis, 1.000 enriched rotifers and 1.000 without enrichment were applied according to the treatment, following the individual ration. Six randomly selected larvae (three per replicate) were removed from each treatment, then observed, measured (see point 4) and stored individually at $-80\,^{\circ}$C for subsequent molecular analysis (see point 7).

4.6. Experiments with Larvae at 12 Days Post-Yolk Absorption

To carry out this experiment *Artemia* sp. (Utah strain; Aquafauna Bio-Marine Inc.) nauplii were enriched with *Rhodotorula* sp. CNYC4007 in the form of lyophilized flour. The time and conditions of the enrichment were defined by modifying the protocols from Silva [48]. Six hundred *Artemia* sp.

nauplii (8 h post-hatching) were placed in 15-mL tubes containing 5 mL of seawater. The tubes were incubated with continuous light and strong aeration at 26 °C. Three mg of C2-DHA or 3 mg of flour lyophilized with CNYC4007 were added, respectively, to the two tubes. The tubes containing nauplii with and without enrichment were kept under the same conditions. The enrichment time was 16 h, after which the nauplii were sieved (88 μm), washed and re-suspended in 2 mL of distilled water to proceed with feeding the larvae.

The experimental design involved three treatments: (1) Larvae fed with *Artemia* sp. nauplii without enrichment (control group; C1); (2) Larvae fed with *Artemia* sp. nauplii enriched with C2-DHA (positive control group); and (3) Larvae fed with *Artemia* sp. nauplii enriched with flour lyophilized with *Rhodotorula* sp. CNYC4007. Thirty larvae were placed in 700-mL containers (3 replicates per treatment). The experiment lasted nine days. Every second day three larvae per replicate were removed, observed, measured and then stored individually at −80 °C for subsequent molecular analysis. In all the trials (flour, rotifers and *Artemia* sp. nauplii) at the moment of photographing each larvae the presence of food in the gut was visually verified and then stored for posterior molecular analysis.

4.7. Obtaining the RNA/DNA Ratio Used as a Proxy for the Condition of Danio rerio Larvae

The protocol from Clemmensen [59], with modifications, was used to extract larval nucleic acid. The larvae were treated with Tris-EDTA (Ethylenediaminetetraacetic acid) buffer (Tris-HCl 0.05 M; NaCl 0.1 M; EDTA 0.01 M at pH 8.0) and SDS (sodium dodecyl sulfate) at 20% *p/v*, after which proteinase K (20 mg/mL) was added and homogenized with a glass swab until a uniform solution was observed. The solution was then centrifuged for one minute and placed in a thermo-regulated bath (50 °C) for 10 min, after which the samples were transferred to −20 °C for 20 min. The procedure of heating and cooling the samples was repeated three times, following which the samples were centrifuged. Chloroform, phenol and isoamyl alcohol (24:25:1) was added and homogenized in an extraction chamber. After centrifugation for 15 min, the supernatant was transferred to a sterile Eppendorf tube.

Nucleic acids were determined separately using the Quantifluor ONE dsDNA System (Promega Co., Fitchburg, WI, USA) for DNA and the Quantifluor RNA System (Promega Co., Fitchburg, WI, USA) for RNA. The samples were prepared according the suppliers protocols. Fluorescence was measured in a multi-detection microplate reader (BioTek, Winooski, VT, USA, model FL×800) under a blue optical channel with excitation of 492 nm and an emission of 540 nm.

4.8. Analysis of Fatty Acid Content in Artemia sp. nauplii

Polyunsaturated fatty acids were extracted from *Artemia* sp. nauplii by saponification reaction and quantified by HPLC following the methodology described by Li et al. [60]. To do this, 100 mg of *Artemia* sp. nauplii without enrichment, enriched with C2-DHA and flour lyophilized from the strain of *Rhodotorula* sp. CNYC4007 was used, each in triplicate. One mL of NaOH at 0.5 M in 96% of ethanol was added to the samples and homogenized with an Ultra Turrax for one minute. Samples were centrifuged at 7000 rpm for 5 min to eliminate solid residues after cell rupture. One mL of HCl at 0.6 N and 3 mL of ethyl acetate (LiChrosolv®, Rochester, NY, USA) was added to the supernatant and agitated with vortex for one minute and incubated for 30 min at room temperature. The treated samples were dried by a current of nitrogen ($N_{2(g)}$) to eliminate the organic solvent, lyophilized to eliminate the remains of water and stored at −20 °C until subsequent chromatographic analysis. Polyunsaturated fatty acids were quantified with a HPLC VWR™ HITACHI (VWR International Ltd., Lutterworth, UK) with an organizer (model L-2000), an ultraviolet (UV) detector (model L-2400), gradient editing pump (model L-2100/2130) and a 15-cm-by-4.6-mm LC-18 column (Supelco®, Sigma-Aldrich, Inc., Darmstadt, Germany). The mobile phase consisted of a flow of 1 mL per min^{-1} of gradient A (25% acetonitrile), to 50% of gradient B (100% acetonitrile), with a flow of 2 mL per min^{-1} for the first 15 min and then 1 mL per min^{-1} for another 15 min. DHA and EPA were identified through the construction of their respective calibration curves using HPLC grade standards (Sigma-Aldrich®, Darmstadt, Germany) with an injection volume of 10 µL [27].

4.9. Statistical Analysis

The Statistica 10 program was used to analyze the data obtained. In the first step the homogeneity of variance (Bartlett, London, UK) and normality of data (Kolmogorov-Smirnov test) was analyzed. A two-way ANOVA and multiple Tukey comparison test was applied for total larval length, DNA and RNA concentrations and the RNA/DNA ratio. The treatments were compared on the final day of the assay. The Kruskal-Wallis test was applied to data that did not present homogeneity of variance or normality. Mean was considered significant when probability was less than 0.05 ($p < 0.05$).

Acknowledgments: The authors are grateful for funding from COPAS Sur-Austral CONICYT PIA PFB31 of Universidad de Concepción. They also would like to thank the Marine Biotechnology Unit and the Project VIU120023 and VIU 15E0053 of Concurso Nacional de Valorización Universitaria, FONDEF of CONICYT.

Author Contributions: R. R. González-Saldía, N. Pino-Maureira, A. Llanos-Rivera y M. Barra conceived and designed the experiments; M. Barra performed the experiments; A. Llanos-Rivera, M. Barra y F. Cruzat analyzed the data; F. Cruzat also help to standardized RNA/DNA method; N. Pino-Maureira provided the biomass and microencapsulated of *Rhodotorula* sp. CNYC4007 for experiment; R. R. González-Saldía wrote the paper.

Conflicts of Interest: The authors declare no conflict of interest.

References

1. Food and Agriculture Organization (FAO). *El Estado Mundial de la Pesca y Acuicultura*; Departamento de Pesca y Acuicultura: Roma, Italy, 2004; pp. 1–115.
2. Food and Agriculture Organization (FAO). *El Estado Mundial de la Pesca y Acuicultura*; Departamento de Pesca y Acuicultura: Roma, Italy, 2012; pp. 1–251.
3. Tocher, D.R. Fatty acid requirement in ontogeny of marine and freshwater fish. *Aquac. Res.* **2010**, *41*, 717–732. [CrossRef]
4. Food and Agriculture Organization (FAO). *El Estado Mundial de la Pesca y Acuicultura*; Departamento de Pesca y Acuicultura: Roma, Italy, 2008; pp. 1–280.
5. Hardy, R.W.; Higgs, D.A.; Lalla, S.P.; Tacon, A.G.J. Alternative dietary protein and lipid sources for sustainable production of salmonids. *Fisken og Havet* **2001**, *8*, 44.
6. Izquierdo, M.S.; Turkmen, S.; Montero, D.; Zamorano, M.J.; Afonso, J.M.; Karalazos, V.; Fernández-Palacios, H. Nutritional programming through broodstock diets to improve utilization of very low fishmeal and fish oil diets in gilthead sea bream. *Aquaculture* **2015**, *449*, 18–26. [CrossRef]
7. Izquierdo, M.; Socorro, J.; Arantzamendi, L.; Hernandez-Cruz, C. Recent advances in lipid nutrition in fish larvae. *Fish Physiol. Biochem.* **2000**, *22*, 97–107. [CrossRef]
8. Hardy, R.W. Utilization on plant proteins in fish diets: Effects of global demand and supplies of fishmeal. *Aquac. Res.* **2010**, *41*, 770–776. [CrossRef]
9. Arslan, M.; Sirkecioglu, N.; Bayir, A.; Arslan, H.; Aras, M. The Influence of substitution of dietary fish oil with different vegetable oils on performance and fatty acid composition of brown trout, *Salmo trutta*. *Turk. J. Fish. Aquat. Sci.* **2012**, *12*, 575–583. [CrossRef]
10. Díaz, N. Efecto de la Relación EPA/DHA en Larvas de Puye (*Galaxias maculatus*, Jenyns. 1842), Cultivadas en Diferentes Salinidades. Tesis de Grado, Facultad de Acuicultura y Ciencias Veterinarias, Universidad Católica de Temuco, Temuco, Chile, 2004; p. 70.
11. Kaushik, S.; Georga, I.; Koumoundouros, G. Growth and body composition of zebrafish (*Danio rerio*) larvae fed a compound feed from first feeding onward: Toward implications on nutrients requirements. *Zebrafish* **2011**, *8*, 87–95. [CrossRef] [PubMed]
12. Li, K.; Ostensen, M.; Attramadal, K.; Winge, P.; Sparstad, T.; Bones, A.; Vadstein, O.; Kjørsvik, E.; Olsen, Y. Gene regulation of lipid and phospholipid metabolism in Atlantic cod (*Gadus morhua*) larvae. *Comp. Biochem. Physiol. Part B* **2015**, *190*, 16–26. [CrossRef] [PubMed]
13. Hamre, K.; Harboe, T. Artemia enriched with high n-3 HUFA may give a large improvement in performance of Atlantic halibut (*Hippoglossus hippoglossus* L.) larvae. *Aquaculture* **2008**, *277*, 239–243. [CrossRef]

14. Estudillo del Castillo, C.; Gapasin, R.; Leaño, E. Enrichment potential of HUFA-rich thraustochytrid *Schizochytrium mangrovei* for the rotifer *Brachionus plicatilis*. *Aquaculture* **2009**, *293*, 57–61. [CrossRef]
15. Satuino, C.; Hirayama, K. Fat-soluble vitamin requirements of the rotifer *Brachionus plicatilis*. In *The First Asian Fisheries Forum*; Maclean, L., Dizon, L.B., Hosillos, L.V., Eds.; Asian Fisheries Society: Manila, Philippines, 1986; pp. 619–622.
16. Hawkyard, M.; Sæle, Ø.; Nordgreen, A.; Langdon, C.; Hamre, K. Effect of iodine enrichment of *Artemia sp* on their nutritional value for larval zebrafish (*Danio rerio*). *Aquaculture* **2001**, *316*, 37–43. [CrossRef]
17. Ben Khemis, I.; de la Noue, J.; Audet, C. Feeding larvae of winter flounder *Pseudopleuronectes americanus* (Walbaum) with live prey or microencapsulated diet: Linear growth and protein, RNA and DNA content. *Aquac. Res.* **2000**, *31*, 377–386. [CrossRef]
18. Robinson, S.; Ware, D. Ontogenetic development of growth rates in larval Pacific herring, *Clupea harengus pallasi*, measured with RNA/DNA ratios in the Strait of Georgia, British Columbia. *Can. J. Fish. Aquat. Sci.* **1988**, *45*, 1422–1429. [CrossRef]
19. Chícharo, M.; Chícharo, L. RNA: DNA Ratio and Other Nucleic Acid Derived Indices in Marine Ecology. *Int. J. Mol. Sci.* **2008**, *9*, 1453–1471. [CrossRef] [PubMed]
20. Wing, N.; Campbell, P.; Dick, J.; Bell, J. Interactive effect of dietary palm oil concentration and water temperature on lipid digestibility in rainbow trout, *Oncorhynchus mykiss*. *Lipids* **2003**, *38*, 1031–1038.
21. Turchini, G.; Torstensen, B.; Ng, W. Fish oil replacement in finfish nutrition. *Rev. Aquac.* **2009**, *1*, 10–57. [CrossRef]
22. Valenzuela, A.; Sanhueza, J.; De la Barra, F. El aceite de pescado: Un desecho industrial transformado en un producto de alto valor comercial. *Grasas Aceites* **2011**, *XXII*, 84–98.
23. Morais, S.; Pratoomyot, J.; Taggart, J.; Bron, J.; Guy, D.; Gordon, J.; Tocher, D. Genotype-specific responses in Atlantic salmon (*Salmo salar*) subject to dietary fish oil replacement by vegetable oil: A liver transcriptomic analysis. *BMC. Genom.* **2011**, *12*, 255–259. [CrossRef] [PubMed]
24. Kiron, V. Fish immune system and its nutritional modulation for preventive health care. *Anim. Feed Sci.* **2012**, *173*, 111–133. [CrossRef]
25. Miller, M.; Nichols, P.; Carter, C. Replacement of fish oil with thraustochytrid *Schizochytrium sp* L oil in Atlantic salmon parr (*Salmo salar* L) diets. *Comp. Biochem. Physiol. Part A* **2007**, *148*, 382–392. [CrossRef] [PubMed]
26. Armenta, R.E.; Valentine, M.C. Single-Cell Oils as a Source of Omega-3 Fatty Acids: An Overview of Recent Advances. *J. Am. Oil Chem. Soc.* **2013**, *90*, 167–182. [CrossRef]
27. Pino, N.L.; Socias, C.; González, R.R. Marine fungoid producers of DHA, EPA and carotenoids from central and southern Chilean marine ecosystems. *Rev. Biol. Mar. Oceanogr.* **2015**, *50*, 507–520. [CrossRef]
28. Iida, I.; Nakahara, T.; Yocochi, T.; Kamisaka, Y.; Yagi, H.; Yamaoka, M.; Suzuki, O. Improvement of docosahexaenoic acid production in a culture of *Thraustochytrium aureum* by medium optimization. *J. Ferment. Bioeng.* **1996**, *81*, 76–78. [CrossRef]
29. Akanbi, T.O.; Barrow, C.J. Candida antarctica lipase A effectively concentrates DHA from fish and thraustochytrid oils. *Food Chem.* **2017**, *229*, 509–516. [CrossRef] [PubMed]
30. Fan, K.W.; Chen, F.; Jones, E.B.G.; Vrijmoed, L.L.P. Eicosapentaenoic and docosahexaenoic acids production by and okara-utilizing potential of thraustochytrids. *J. Ind. Microbiol. Biotechnol.* **2001**, *27*, 199–202. [CrossRef] [PubMed]
31. Ganuza, E.; Benitez-Santana, T.; Atalah, E.; Vega-Orellana, O.; Ganga, R.; Izquierdo, M. *Crypthecodinium cohnii* and *Schizochytrium sp* as potential substitutes to fisheries-derived oils from seabream (*Sparus aurata*) microdiets. *Aquaculture* **2008**, *277*, 109–116. [CrossRef]
32. Alestrom, P.; Holter, J.; Nourizadeh-Lillabadi, R. Zebrafish in functional genomics and aquatic biomedicine. *Trends Biotechnol.* **2006**, *24*, 15–21. [CrossRef] [PubMed]
33. Ulloa, P.; Iturra, P.; Neira, R.; Araneda, C. Zebrafish as a model organism for nutrition and growth: Towards comparative studies of nutritional genomics applied to aquacultured fishes. *Rev. Fish Biol. Fish.* **2011**, *21*, 649–666. [CrossRef]
34. Ribas, L.; Piferrer, F. The zebrafish (*Danio rerio*) as a model organism, with emphasis on applications for finfish aquaculture research. *Rev. Aquac.* **2014**, *6*, 209–240. [CrossRef]
35. Ulloa, P.E.; Medrano, J.F.; Feijoo, C.G. Zebrafish as animal model for aquaculture nutrition research. *Front. Genet.* **2014**, *5*, 313. [CrossRef] [PubMed]

36. Hedrera, M.I.; Galdames, J.A.; Jimenez-Reyes, M.F.; Reyes, A.E.; Avendaño-Herrera, R.; Romero, J.; Feijóo, C.G. Soybean Meal Induces Intestinal Inflammation in Zebrafish Larvae. *PLoS ONE* **2013**, *8*, e69983. [CrossRef] [PubMed]

37. Fulton, T. *Rate of Growth of Sea-Fishes*; Neill & Co.: Edinburgh, UK, 1902; Volume 1, p. 20.

38. Ikeda, T.; San, F.; Yamaguchi, A.; Matsuishi, T. RNA: DNA ratios of calanoid copepods from the epipelagic through abyssopelagic zones of the North Pacific Ocean. *Aquat. Biol.* **2007**, *1*, 99–108. [CrossRef]

39. Bulow, J.F. RNA-DNA ratios as indicators of growth rates in fish: A review. In *The Age and Growth of Fish*; Summerfelt, R.C., Hall, G.E., Eds.; The Iowa State University Press: Ames, IA, USA, 1987; pp. 45–64.

40. Caldarone, E.; Onge-Burns, J.; Buckley, L. Relationship of RNA/DNA ratio and temperature to growth in larvae of Atlantic cod *Gadus morhua*. *Mar. Ecol. Prog. Ser.* **2003**, *262*, 229–240. [CrossRef]

41. Buckley, L. RNA-DNA ratio: An index of larval fish growth in the sea. *Mar. Biol.* **1984**, *80*, 291–298. [CrossRef]

42. Olivar, M.; Díaz, M.; Chícharo, A. Tissue effect on RNA: DNA ratios of marine fish larvae. *Sci. Mar.* **2009**, *73*, 171–182. [CrossRef]

43. Chung, K.; Segnini, M. RNA-DNA ratio as physiological condition of rainbow trout fry fasted and fed. *Ital. J. Zool.* **1998**, *65*, 517–519. [CrossRef]

44. Narayan, S.; Archana, L.; Ranjan, R. Effects of dietary protein concentrations on growth and RNA: DNA ratio of rainbow trout (*Oncorhynchus mykiss*) cultured in Nuwakot district of Nepal. *Int. J. Fish. Aquat. Stud.* **2014**, *1*, 184–188.

45. Moreau, D.; Rosenberg, M. Microstructure and fat extractability in microcapsules based on whey proteins or mixtures of whey proteins and lactose. *Food Struct.* **1993**, *12*, 457–468.

46. Habi Mat Dian, N.; Sudin, N.; Affandi, Y. Characteristics of microencapsulated palm-based oil as affected by type of wall material. *J. Sci. Food Agric.* **1996**, *70*, 422–426. [CrossRef]

47. Navarro, J.; Henderson, J.; McEvoy, L.; Bell, M.; Amat, F. Lipid conversions during enrichment of Artemia. *Aquaculture* **1999**, *174*, 155–166. [CrossRef]

48. Silva, A. *Cultivo de Peces Marinos*; Departamento de Acuicultura, Facultad de Ciencias del Mar, Universidad Católica del Norte: Coquimbo, Chile, 2005.

49. Barclay, W.; Zeller, S. Nutritional enhancement of n-3 and n-6 fatty acids in rotifers and *Artemia* nauplii by feeding spray-dried *Schizochytrium* sp. *J. World Aquac. Soc.* **1996**, *27*, 314–322. [CrossRef]

50. Johansen, K.A.; Overturf, K. Alterations in expression of genes associated with muscle metabolism and growth during nutritional restriction and refeeding in rainbow trout. *Comp. Biochem. Physiol. Part B* **2006**, *144*, 119–127. [CrossRef] [PubMed]

51. Izquierdo, M.; Arakawa, T.; Takeuchi, R.; Haroun, R.; Watanabe, T. Effect of ω-3 HUFA levels in Artemia on growth of larval Japanese flounder (*Paralicthys olilaceus*). *Aquaculture* **1992**, *105*, 73–82. [CrossRef]

52. Masuda, Y.; Oku, H.; Okumura, T.; Nomura, K.; Kurokawa, T. Feeding restriction alters expression of some ATP related genes more sensitively than the RNA/DNA ratio in zebrafish, *Danio rerio*. *Comp. Biochem. Physiol. Part B* **2009**, *152*, 287–291. [CrossRef] [PubMed]

53. Salem, M.; Silverstain, J.; Rexroad, C.; Yao, J. Effect of starvation on global gene expression and proteolysis in rainbow trout (*Oncorhynchus mykiss*). *BMC. Genom.* **2007**, *8*, 328. [CrossRef] [PubMed]

54. Frengova, G.I.; Beshkova, D.M. Carotenoids from *Rhodotorula* and *Phaffia*: Yeasts of biotechnological importance. *J. Ind. Microbiol. Biotechnol.* **2009**, *36*, 163–180. [CrossRef] [PubMed]

55. MacDonald, M.C.; Arivalagan, P.; Barre, D.E.; MacInnis, J.A.; D'Cunha, G.B. *Rhodotorula glutinis* Phenylalanine/Tyrosine Ammonia Lyase Enzyme Catalyzed Synthesis of the Methyl Ester of para-Hydroxycinnamic Acid and its Potential Antibacterial Activity. *Front. Microbiol.* **2016**, *7*, 281. [CrossRef] [PubMed]

56. Taskin, M.; Ortucu, S.; Aydogan, M.N.; Arslan, N.P. Lipid production from sugar beet molasses under non-aseptic culture conditions using the oleaginous yeast *Rhodotorula glutinis* TR29. *Renew. Energy* **2016**, *99*, 198–204. [CrossRef]

57. Sprague, M.; Betancor, M.B.; Tocher, D.R. Microbial and genetically engineered oils as replacements for fish oil in aquaculture feeds. *Biotechnol. Lett.* **2017**, *39*, 1599–1609. [CrossRef] [PubMed]

58. Pino, N.; Gómez, C.; González, R.R. *Proceso de Obtención de Biomasa del Fungoide Marino CNYC 4007, Uso de la Biomasa y Suplemento Alimenticio que Contiene Esta Biomasa*; Solicitud CL201402916; Instituto de Propiedad Industrial: Santiago, Chile, 2014.

59. Clemmensen, C. RNA/DNA ratios of laboratory reared and wild herring larvae determined with a highly sensitive fluorescence method. *J. Fish Biol.* **1989**, *35*, 331–333. [CrossRef]

60. Li, Z.; Gu, T.; Kelder, B.; Kopchick, J. Analysis of fatty acids in mouse cells using Reversed-Phase High-Performance Liquid. *Chromatographia* **2001**, *54*, 463–467. [CrossRef]

MDPI AG

St. Alban-Anlage 66

4052 Basel, Switzerland

Tel. +41 61 683 77 34

Fax +41 61 302 89 18

http://www.mdpi.com

Marine Drugs Editorial Office

E-mail: marinedrugs@mdpi.com

http://www.mdpi.com/journal/marinedrugs

www.ingramcontent.com/pod-product-compliance
Lightning Source LLC
Chambersburg PA
CBHW041216220326
41597CB00033BA/5992